T5-CQI-887

SOLUTIONS MANUAL FOR

ROBERT A. ALBERTY

PHYSICAL CHEMISTRY

SIXTH EDITION

JOHN WILEY & SONS
New York Chichester Brisbane Toronto Singapore

Copyright © 1983 by John Wiley & Sons, Inc.

All rights reserved.

Reproduction or translation of any part of this work beyond that permitted by Section 107 or 108 of the 1976 United States Copyright Act without the permission of the copyright owner is unlawful. Requests for permission or further information should be addressed to the Permissions Department, John Wiley & Sons, Inc.

ISBN 0-471-87208-3
Printed in the United States of America

10 9 8 7 6 5 4 3 2 1

PREFACE

This manual gives solutions of the problems in the first set of the text R. A. Alberty, PHYSICAL CHEMISTRY, 6th ed., Wiley, New York, 1983, and it gives answers for the second set.

Working problems is an important part of learning physical chemistry. Not all knowledge of physical chemistry is quantitative, but much of it is. Since physical chemistry utilizes physics and mathematics to predict and interpret chemical phenomena, there are many opportunities to use quantitative methods.

The availability of hand-held electronic calculators has made it much easier to work physical chemistry problems and has made it possible to include more difficult problems.

The units of physical quantities are usually shown in solving problems in this manual. It is important to develop the habit of using units and cancelling them to obtain the units for the answer because this helps prevent errors.

In this book, as in your handwritten lecture notes, there is no distinction between italic (sloping) and roman (upright) type, but it is important to note that in the printed literature italic type is used for symbols for physical quantities and roman type is used for units.

This problem set has been built up over a long period of years (70 years if we go back to the first edition of OUTLINES OF THEORETICAL CHEMISTRY by Dr. Frederick H. Getman) and many individuals have contributed to it. Dr. Farrington Daniels developed many problems for physical chemistry instruction in the 45 years he was involved with OUTLINES OF THEORETICAL CHEMISTRY AND PHYSICAL CHEMISTRY.

Robert A. Alberty

Cambridge, Massachusetts
August 1982

CONTENTS

I. THERMODYNAMICS

II. QUANTUM CHEMISTRY

III. CHEMICAL DYNAMICS

IV. STRUCTURES

PART ONE

THERMODYNAMICS

CHAPTER 1: First Law of Thermodynamics

1.1 Assuming that the atmosphere is isothermal at 0 °C and that the average molar mass of air is 29 g mol^{-1}, calculate the atmospheric pressure at 20,000 ft above sea level.

SOLUTION

$$h = (2.0 \times 10^4 \text{ ft})(12 \text{ in ft}^{-1})(2.54 \text{ cm in}^{-1})(10^{-2} \text{ m cm}^{-1})$$
$$= 6096 \text{ m}$$

$$P = P\, e^{-gMh/RT}$$

$$P = (1.013 \text{ bar}) \exp\left[\frac{-(9.8 \text{ m s}^{-2})(29 \times 10^{-3} \text{ kg mol}^{-1})(6096 \text{ m})}{(8.314 \text{ J K}^{-1} \text{ mol}^{-1})(273 \text{ K})}\right]$$

$$= 0.472 \text{ bar}$$

1.2 Calculate the second virial coefficient of hydrogen at 0 °C from the fact that the molar volumes at 50.7, 101.3, 202.6, and 303.9 bar are 0.4634, 0.2386, 0.1271, and 0.09004 L mol^{-1}, respectively.

SOLUTION $\quad \dfrac{PV}{RT} = 1 + \dfrac{B}{V} + \dfrac{C}{V^2} + \cdot \cdot \cdot$

P/bar	50.7	101.3	202.6	303.9
V/L mol^{-1}	0.4634	0.2386	0.1271	0.09004
PV/RT	1.035	1.064	1.134	1.205
(1/V)/mol L^{-1}	2.158	4.191	7.868	11.106

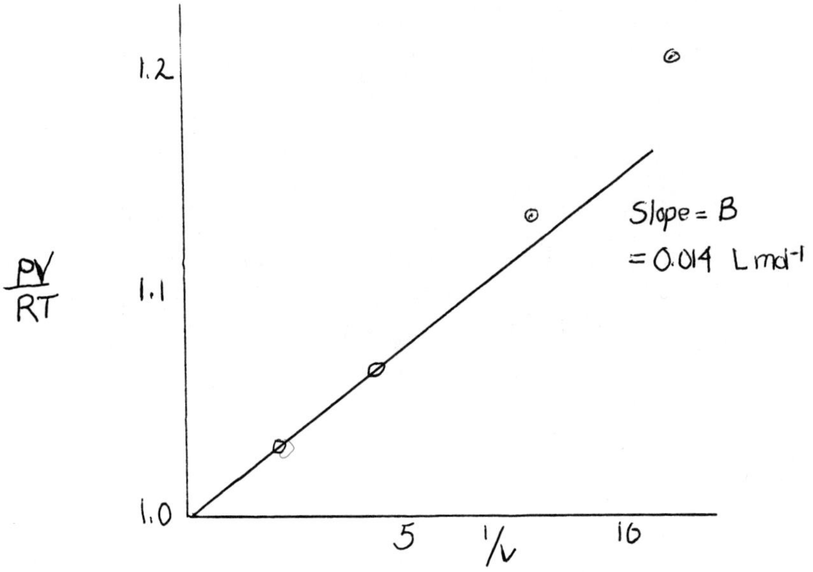

1.3 The second virial coefficient B of methyl isobutyl ketone is -1580 cm^3 mol^{-1} at 120 °C. Compare its compressibility factor at this temperature with that of a perfect gas at 1 bar.

<u>SOLUTION</u> $$Z = \frac{PV}{RT} = 1 + \frac{B}{V} = 1 + \frac{BP}{RT}$$

In the B/V term, V may be replaced by RT/P if $Z \approx 1$, since the approximation is made in a small correction term. At 1 atm

$$Z = 1 + \frac{(-1.58 \text{ L mol}^{-1})(1 \text{ bar})}{(0.08314 \text{ L bar K}^{-1} \text{ mol}^{-1})(393.15 \text{ K})}$$

$$= 0.952$$

The compressibility factor for a perfect gas is of course unity.

1.4 Using Fig. 1.4 calculate the compressibility factor Z for NH_3(g) at 400 K and 50 bar.

<u>SOLUTION</u> $B = -110$ cm^3 mol^{-1}

$$B' = \frac{B}{RT} = \frac{-(110\ \text{cm}^3\ \text{mol}^{-1})(10^{-3}\ \text{L cm}^{-3})}{(0.08314\ \text{L bar K}^{-1}\ \text{mol}^{-1})(400\ \text{K})}$$

$$= -3.31 \times 10^{-3}\ \text{bar}^{-1}$$

$$Z = 1 + B'P = 1 - (3.31 \times 10^{-3}\ \text{bar}^{-1})(50\ \text{bar})$$

$$= 0.835$$

1.5 Show that the virial equation written in terms of pressure has the following form for a van der Waals gas $Z = 1 + [b - (a/RT)](P/RT)$ if terms in P^2, P^3, etc., are neglected.

SOLUTION

$$P = \frac{RT}{V-b} - \frac{a}{V^2} \qquad \frac{PV}{RT} = \frac{V}{V-b} - \frac{a}{VRT}$$

Adding and subtracting $\frac{V-b}{V-b}$ on the right side of the equation

$$\frac{PV}{RT} = 1 + \frac{b}{V-b} - \frac{a}{VRT}$$

Since the second and third terms are small correction terms, b can be ignored in comparison with V; and V in the second and third term can be replaced by RT/P.

$$Z = 1 + \frac{bP}{RT} - \frac{aP}{(RT)^2} = 1 + \left(b - \frac{a}{RT}\right)\left(\frac{P}{RT}\right)$$

1.6 Show that for a gas of spherical molecules b in the van der Waals equation is four times the molecular volume times Avogadro's constant.

SOLUTION

The molecular volume for a spherical molecule is

$$\frac{4}{3}\pi\left(\frac{d}{2}\right)^3 = \frac{\pi}{6}d^3$$

where d is the diameter. Since the center of a second spherical molecule cannot come within a distance d of the center of the first spherical molecule, the excluded volume per pair of molecules is

$$\frac{4}{3}\pi d^3$$

The constant b in van der Waals equation is the excluded volume per molecule times Avogadro's constant

$$b = \frac{2}{3}\pi d^3 N_A = 4\left(\frac{\pi}{6} d^3\right)N_A$$

1.7 How much work is done when a person weighing 75 kg (165 lbs.) climbs the Washington monument, 555 ft high? How many kilojoules must be supplied to do this muscular work, assuming that 25% of the energy produced by the oxidation of food in the body can be converted into muscular mechanical work?

SOLUTION

$w = mgh$

work = (mass)(acceleration of gravity)(height)

$$= (75\text{kg})(9.806\text{m s}^{-2})(555\text{ft})\times(12\text{in ft}^{-1})\times(2.54\times10^{-2}\text{m in}^{-1})$$

$$= 1.244\times10^5\text{ J}$$

The energy needed is four times greater than the work done.

$$E = 4\,(1.244\times10^5\text{ J})$$

$$= 497.6\text{ kJ}$$

1.8 Derive the expression for the reversible isothermal work for a van der Waals gas.

SOLUTION

$$w = -\int_{V_1}^{V_2} PdV \qquad P = \frac{RT}{V-b} - \frac{a}{V^2}$$

$$w = -\int_{V_1}^{V_2} \frac{RT}{v-b}\, dV + \int_{V_1}^{V_2} \frac{a}{V^2}\, dV$$

$$= - RT\ell n \frac{V_2-b}{V_1-b} + a\left(\frac{1}{V_1} - \frac{1}{V_2}\right)$$

1.9 Are the following expressions exact differentials?

(a) $xy^2dx - x^2y\, dy$ (b) $\frac{dx}{y} - \frac{x}{y^2}\, dy$

SOLUTION

(a) Taking the cross derivatives

$$\frac{\partial xy^2}{\partial y} = 2xy \qquad \frac{\partial(-x^2y)}{\partial x} = -2xy$$

Since the cross derivatives are unequal, the expression is not an exact differential.

(b) $\frac{\partial(1/y)}{\partial y} = -\frac{1}{y^2}$ $\frac{\partial(-x/y^2)}{\partial x} = -\frac{1}{y^2}$

Since the cross derivatives are equal, the expression is an exact differential.

1.10 A mole of liquid water is vaporized at 100 °C and 1.013 bar. The heat of vaporization is 40.69 kJ mol^{-1}. What are the values of (a) w_{rev}, (b) q, (c) ΔU, and (d) ΔH?

SOLUTION

(a) Assuming that water vapor is a perfect gas and that the volume of liquid water is negligible,

$$w = -P\Delta V = -RT$$
$$= -(8.314 \times 10^{-3}\ \text{kJ K}^{-1}\ \text{mol}^{-1})(373.15\ \text{K})$$
$$= -3.10\ \text{kJ mol}^{-1}$$

(b) The heat of vaporization is 40.69 kJ mol^{-1}, and, since heat is absorbed, q has a positive sign.

$$q = 40.69\ \text{kJ mol}^{-1}$$

(c) $\Delta U = q + w$
$$= (40.69 - 3.10)\ \text{kJ mol}^{-1}$$
$$= 37.59\ \text{kJ mol}^{-1}$$

(d) $\Delta H = \Delta U + \Delta(PV) = \Delta U + P\Delta V$
$$= \Delta U + RT$$
$$= 37.59\ \text{kJ mol}^{-1} + (8.314 \times 10^{-3}\ \text{kJ K}^{-1}\text{mol}^{-1})(373.15\ \text{K})$$
$$= 40.69\ \text{kJ mol}^{-1}$$

1.11 Calculate H°(2 000 K) - H°(0 K) for H(g).

SOLUTION

Equation 1.69 may be integrated to obtain

$$H^\circ(T_2) - H^\circ(T_1) = \int_{T_1}^{T_2} C_P^\circ\ dT$$

$$H^\circ(2\ 000\ \text{K}) - H^\circ(0\ \text{K}) = \int_0^{2\ 000} \frac{5}{2} R\ dT = \frac{5}{2} R(2\ 000)$$
$$= 41.572\ \text{kJ mol}^{-1}$$

Table A.2 yields 6.197 + 35.376 = 41.573 kJ mol^{-1}

(Note that for O(g) a slightly higher value is obtained because there is some absorption of heat by excitation to higher electronic levels.)

1.12 Considering H_2O to be a rigid nonlinear molecule, what value of C_P for the gas would be expected classically? If vibration is taken into account, what value is expected? Compare these values of C_P with the actual values at 298 and 3000 K in Table A.1.

SOLUTION

A rigid molecule has translational and rotational energy. The translational contribution to C_V is $\frac{3}{2}$ R = 12.471 J K^{-1} mol^{-1}.

Since H_2O is a nonlinear molecule, it has three rotational degrees of freedom, and so the rotational contribution to C_V is $\frac{3}{2}$ R = 12.471 J K^{-1} mol^{-1}.

Thus C_P for the rigid molecule is

$$C_P = C_V + R = 33.258 \text{ J K}^{-1} \text{ mol}^{-1}$$

Since H_2O is a nonlinear molecule, the number of vibrational degrees of freedom is 3N - 6 = 3. Since each vibrational degree of freedom contributes R to the heat capacity, classical theory predicts

$$\begin{aligned} C_P &= 33.258 \text{ J K}^{-1} \text{ mol}^{-1} + 3R \\ &= 58.201 \text{ J K}^{-1} \text{ mol}^{-1} \end{aligned}$$

The experimental value of C_P at 298 K is 33.577 J K^{-1} mol^{-1}, which is only slightly higher than the value expected for a rigid molecule. The experimental value of C_P at 3000 K is 55.664 J K^{-1} mol^{-1}, which is only slightly less than the classical expectation for a vibrating water molecule.

1.13 The equation for the molar heat capacity of n-butane is $C_P = 19.41 + 0.233\ T$ where C_P is given in $J\ K^{-1}\ mol^{-1}$. Calculate the heat necessary to raise the temperature of 1 mole from 25 to 300 °C at constant pressure.

SOLUTION

$$q = \int_{T_1}^{T_2} C_P\ dT = 19.41(T_2-T_1) + \frac{1}{2}(0.233)(T_2^2-T_1^2)$$

$$= 19.41\ (573.15-298.15) + \frac{1}{2}(0.233)(573.15^2-298.15^2)$$

$$= 33.31\ kJ\ mol^{-1}$$

1.14 One mole of nitrogen at 25 °C and 1 atm is expanded reversibly and isothermally to a pressure of 0.132 bar. (a) What is the value of w? What is the value of w if the temperature is 100 °C?

SOLUTION

(a) $$w = -\int_{P_1}^{P_2} PdV = -RT\ln\frac{P_1}{P_2}$$

$$= -(8.314\ J\ K^{-1} mol^{-1})(298.15\ K)\ln\frac{1}{0.132}$$

$$= -5027\ J\ mol^{-1}$$

(b) $$w = (-5027\ J\ mol^{-1})(373.15\ K)/(298.15\ K)$$

$$= -6292\ J\ mol^{-1}$$

1.15 Calculate the temperature increase and final pressure of helium if a mole is compressed adiabatically and reversibly from 44.8 L at 0 °C to 22.4 L.

SOLUTION

$$\gamma = C_P/C_V = (\tfrac{5}{2} R)/(\tfrac{3}{2} R) = \tfrac{5}{3}$$

$$\frac{T_1}{T_2} = \left(\frac{V_2}{V_1}\right)^{\gamma-1}$$

$$T_2 = (273.15\ \text{K})(44.8\ \text{L}/22.4\ \text{L})^{2/3}$$
$$= 433.6\ \text{K} \quad \text{or} \quad 160.4\ °\text{C}$$

Thus the temperature increase is 160.4 °C. The final temperature is given by

$$P = \frac{RT}{V} = \frac{(0.08314\ \text{L bar K}^{-1}\ \text{mol}^{-1})\ (433.6\ \text{K})}{22.4\ \text{L mol}^{-1}}$$
$$= 1.609\ \text{bar}$$

1.16 A mole of argon is allowed to expand adiabatically from a pressure of 10 bar and 298.15 K to 1 bar. What is the final temperature and how much work can be done?

SOLUTION

$$\gamma = C_P/C_V = (\tfrac{5}{2} R)/(\tfrac{3}{2} R) = \tfrac{5}{3}$$

$$(\gamma-1)/\gamma = 2/5$$

$$\frac{T_1}{T_2} = \left(\frac{P_1}{P_2}\right)^{(\gamma-1)/\gamma}$$

$$T_2 = (298.15\ \text{K})(1/10)^{2/5} = 118.70\ \text{K}$$

$$w = \int_{T_1}^{T_2} C_V\, dT = \frac{3}{2} R(T_2 - T_1)$$
$$= \frac{3}{2}(8.314\ \ \text{K}^{-1}\text{mol}^{-1})(118.70\ \text{K} - 298.15\ \text{K})$$
$$= \quad 2238\ \text{J mol}^{-1}$$

Thus the maximum work that can be done on the surroundings is 2238 J mol^{-1}.

1.17 A tank contains 20 liters of compressed nitrogen at 10 bar and 25 °C. Calculate the maximum work (in joules) which can be obtained when gas is allowed to expand to 1 bar pressure (a) isothermally and (b) adiabatically.

SOLUTION

(a) For the isothermal expansion of 1 mol

$$w_{rev} = -RT \ln \frac{P_2}{P_1}$$

$$= (8.314 \text{ J K}^{-1} \text{ mol}^{-1})(298.15 \text{ K}) \ln 10$$

$$= -5708 \text{ J mol}^{-1}$$

There are

(10 bar)(20 L)/(0.08314 L bar K^{-1} mol^{-1})(298 K) = 8.07 mol.

Therefore the maximum work done on the surroundings is 46.1 kJ.

(b) For the adiabatic expansion we will assume that $\gamma = C_P/C_V$ has the value it has at room temperature. From Table A.2

$$\gamma = 29.125/(29.125 - 8.314) = 1.399$$

$$\frac{T_1}{T_2} = \left(\frac{P_1}{P_2}\right)^{(\gamma-1)/\gamma}$$

$$T_2 = (298.15 \text{ K})(1/10)^{0.285} = 154.7 \text{ K}$$

$$w = \int_{T_1}^{T_2} C_V \, dT = C_V(T_2 - T_1)$$

$$= (20.811 \text{ J K}^{-1} \text{ mol}^{-1})(154.7 - 298.15 \text{ K})$$

$$= -2.99 \text{ kJ mol}^{-1}$$

For 8.07 moles the maximum work done on the surroundings is 24.1 kJ.

1.18 In an adiabatic calorimeter, oxidation of 0.4362 gram of naphthalene caused a temperature rise of

1.707 °C. The heat capacity of the calorimeter and water was 10,290 JK^{-1}. If corrections for oxidation of the wire and residual nitrogen are neglected, what is the enthalpy of combustion of naphthalene per mole?

SOLUTION

$$\Delta H = \frac{(10{,}290\ JK^{-1})(1.707\ K)(128.19\ g\ mol^{-1})}{(0.4362\ g)(1000\ cal\ kcal^{-1})}$$

$$= -5163\ kJ\ mol^{-1}$$

1.19 The following reactions might be used to power rockets.

(1) $H_2(g) + \frac{1}{2}\,O_2(g) = H_2O(g)$

(2) $CH_3OH(\ell) + 1\frac{1}{2}\,O_2(g) = CO_2(g) + 2H_2O(g)$

(3) $H_2(g) + F_2(g) = 2HF(g)$

(a) Calculate the enthalpy changes at 25 °C for each of these reactions per kilogram of reactants (b) Since the thrust is greater when the molar mass of the exhaust gas is lower, divide the heat per kilogram by the molar mass of the product (or the average molar mass in the case of reaction 2) and arrange the above reactions in order of effectiveness on the basis of thrust.

SOLUTION

(a)(1) $\Delta H = -241.818\ kJ\ mol^{-1}$

$= (-241.818\ kJ\ mol^{-1})(1000\ g\ kg^{-1})/(18\ g\ mol^{-1})$

$= -13.4\ MJ\ kg^{-1}$

(2) $\Delta H = -393.509 + 2(-241.818) + 238.66$

$= -638.49\ kJ\ mol^{-1}$

$= (-638.49 \text{ kJ mol}^{-1})(1000 \text{ g kg}^{-1})/(80 \text{ g mol}^{-1})$

$= -7.98 \text{ MJ kg}^{-1}$

(3) $\Delta H = 2(-271.1) = -542.2 \text{ kJ mol}^{-1}$

$= (-542.2 \text{ kJ mol}^{-1})(1000 \text{ g kg}^{-1})/(40 \text{ g mol}^{-1})$

$= -13.6 \text{ MJ kg}^{-1}$

(b) (1) $-13.4/18 = -0.744$

(2) $$\frac{-7.98}{\frac{1}{3}(44 + 2 \times 18)} = -0.299$$

(3) $-13.6/20 = -0.680$

$\therefore$ (1) > (3) > (2)

1.20 Calculate the enthalpy of formation of PCl_5(cr), given the heats of the following reactions at 25 °C.

$$2P(cr) + 3Cl_2(g) = 2PCl_3(\ell) \quad \Delta H° = -6.35.13 \text{ kJ mol}^{-1}$$

$$PCl_3(\ell) + Cl_2(g) = PCl_5(cr) \quad \Delta H° = -137.28 \text{ kJ mol}^{-1}$$

SOLUTION

Multiplying the second reaction by 2 and adding the two reactions yields

$$2P(cr) + 5Cl_2(g) = 2PCl_5(cr) \quad \Delta H° = -909.69 \text{ kJ mol}^{-1}$$

$$\Delta H_f°[PCl_5(cr)] = (-909.69 \text{ kJ mol}^{-1})/2 = -454.85 \text{ kJ mol}^{-1}$$

1.21 Calculate $\Delta H°$ for the dissociation $O_2(g) = 2O(g)$ at 0, 298, and 3000 K. In Section 12.1 the enthalpy change for dissociation at 0 K will be found to be equal to the spectroscopic dissociation energy D_0.

SOLUTION

$\Delta H°$ (0 K) = 2(246.785) = 493.570 kJ mol^{-1}

$\Delta H°$ (298 K) = 2(249.170) = 498.340 kJ mol^{-1}

$\Delta H°$ (3000 K) = 2(256.722) = 513.444 kJ mol^{-1}

The spectroscopic dissociation energy of O_2 is given as 5.115 eV in Table 13.4. This can be converted to kJ mol^{-1} by multiplying by 96.485 kJ V^{-1} mol^{-1} to obtain 493.521 kJ mol^{-1}.

1.22 Compare the enthalpies of combustion of $CH_4(g)$ to $CO_2(g)$ and $H_2O(g)$ at 298 and 2000 K.

$$CH_4(g) + 2O_2(g) = CO_2(g) + 2H_2O(g)$$

SOLUTION

$$\Delta H°(298\ K) = -393.522 + 2(-241.827) - (-74.873)$$
$$= -802.303\ kJ\ mol^{-1}$$
$$\Delta H°(2000\ K) = -396.639 + 2(-251.668) - (-92.462)$$
$$= -807.513\ kJ\ mol^{-1}$$

1.23 Calculate $\Delta H°_{298}$ for

$$H_2(g) + F_2(g) = 2HF(g)$$
$$H_2(g) + Cl_2(g) = 2HCl(g)$$
$$H_2(g) + Br_2(g) = 2HBr(g)$$
$$H_2(g) + I_2(g) = 2HI(g)$$

SOLUTION

(a) 2(-271.1) = -542.2 kJ mol^{-1}

(b) 2(-92.31) = -184.62 kJ mol^{-1}

(c) 2(-36.40) - -30.91 = -103.71 kJ mol^{-1}

(d) 2(26.48) - 62.44 = -9.48 kJ mol^{-1}

1.24 Methane may be produced from coal in a process represented by the following steps, where coal is approximated by graphite:

$$2C(cr) + 2H_2O(g) = 2CO(g) + 2H_2(g)$$

$$CO(g) + H_2O(g) = CO_2(g) + H_2(g)$$

$$CO(g) + 3H_2(g) = CH_4(g) + H_2O(g)$$

The sum of these three reactions is

$$2C(cr) + 2H_2O(g) = CH_4(g) + CO_2(g)$$

What is $\Delta H°$ at 500 K for each of these reactions? Check that the sum of the $\Delta H°$'s of the first three reactions is equal to $\Delta H°$ for the fourth reaction. From the standpoint of heat balance would it be better to develop a process to carry out the overall reactions in three separate reactors, or in a single reactor?

SOLUTION

$$\Delta H°_{500} = 2(-110.02) - 2(-243.83) = 267.62 \text{ kJ mol}^{-1}$$

$$\Delta H°_{500} = -393.68 - (-110.02) - (-243.83) = -39.83 \text{ kJ mol}^{-1}$$

$$\Delta H°_{500} = -80.82 - 243.83 - (-110.02) = -214.63 \text{ kJ mol}^{-1}$$

$$\Delta H°_{500} = -80.82 - 393.68 - 2(-243.83) = 13.16 \text{ kJ mol}^{-1}$$

Since the first reaction is very endothermic, there is an advantage in carrying the subsequent reactions out in the same reactor so that they can provide heat.

1.25 Compare the enthalpy of combustion of $CH_4(g)$ to $CO_2(g)$ and $H_2O(\ell)$ at 298 K with the sum of the enthalpies of combustion of graphite and $2H_2(g)$, from which $CH_4(g)$ can in principle be produced.

SOLUTION

$$CH_4(g) + 2O_2(g) = CO_2(g) + 2H_2O(\ell)$$

$$\Delta H°(298\ K) = -393.51 + 2(-285.83) - (-74.81)$$
$$= -890.36\ kJ\ mol^{-1}$$

$$2H_2(g) + O_2(g) = 2H_2O(\ell) \quad \Delta H° = -571.66\ kJ\ mol^{-1}$$

$$C(graphite) + O_2(g) = CO_2(g) \quad \Delta H° = -393.51\ kJ\ mol^{-1}$$

The sum of the enthalpy changes for the last two reactions ($-965.17\ kJ\ mol^{-1}$) is more negative than the enthalpy change for the first reaction by the enthalpy of formation of $CH_4(g)$.

1.26 Calculate the heat of hydration of $Na_2SO_4(s)$ from the integral heats of solution of $Na_2SO_4(s)$ and $Na_2SO_4 \cdot 10\ H_2O(s)$ in infinite amounts of H_2O, which are $-2.34\ kJ\ mol^{-1}$ and $78.87\ kJ\ mol^{-1}$, respectively. Enthalpies of hydration cannot be measured directly because of the slowness of the phase transition.

SOLUTION

$$Na_2SO_4(s) = Na_2SO_4(ai) \quad \Delta H° = -2.34\ kJ\ mol^{-1}$$

$$Na_2SO_4(ai) = Na_2SO_4 \cdot 10\ H_2O \quad \Delta H° = -78.87\ kJ\ mol^{-1}$$

$$Na_2SO_4(s) + 10\ H_2O(\ell) = Na_2SO_4 \cdot 10\ H_2O(s)$$
$$\Delta H° = -81.21\ kJ\ mol^{-1}$$

1.27 Calculate the integral heat of solution of one mole of $HCl(g)$ in $200\ H_2O(\ell)$.

$$HCl(g) + 200\ H_2O(\ell) = HCl\ in\ 200\ H_2O$$

SOLUTION

$$\Delta H^\circ_{298} = \Delta H^\circ_{f,\mathrm{HCl\ in\ 200\ H_2O}} - \Delta H_{f,\mathrm{HCl(g)}}$$
$$= -166.272 - (-92.307)$$
$$= -73.965 \text{ kJ mol}^{-1}$$

1.28 Calculate the enthalpies of reaction at 25 °C for the following reactions in dilute aqueous solutions:

(a) $HC\ell(ai) + NaBr(ai) = HBr(ai) + NaC\ell(ai)$

(b) $CaC\ell_2(ai) + Na_2CO_3(ai) = CaCO_3(s) + 2\ NaC\ell(ai)$

SOLUTION

(a) $\Delta H^\circ = 0$ because all of the reactants and products are completely ionized.

$$H^+ + C\ell^- + Na^+ + Br^- = H^+ + Br^- + Na^+ + C\ell^-$$

(b) $Ca^{2+}(ai) + CO_3^{2-}(ai) = CaCO_3(s)$

$$\Delta H^\circ = -1206.92 - (-542.83) - (-677.14)$$
$$= 13.05 \text{ kJ mol}^{-1}$$

1.29 The heat capacities of a gas may be represented by $C_P = a + bT + cT^2$

For N_2, $a = 26.984$, $b = 5.910 \times 10^{-3}$, and $c = -3.377 \times 10^{-7}$, when C_P is expressed in J K^{-1} mol^{-1}. How much heat is required to heat a mole of N_2 from 300 K to 1000 K?

SOLUTION

$$q = \int_{300}^{1000} (26.984 + 5.910 \times 10^{-3}T - 3.377 \times 10^{-7}T^2)dT$$
$$= 26.984(1000 - 300) + \frac{1}{2}(5.910\text{x}10^{-3})(1000^2 - 300^2) - \frac{1}{3}(3.377\text{x}10^{-7})(1000^3 - 300^3)$$
$$= 21.468 \text{ kJ mol}^{-1}$$

1.30 What is the heat evolved in freezing water at -10 °C given that

$H_2O(\ell) = H_2O(cr)$ $\qquad \Delta H°(273\ K) = -6004\ J\ mol^{-1}$

and

$C_p(H_2O,\ell) = 75.3\ J\ K^{-1}\ mol^{-1}$ and $C_p(H_2O,s) = 36.8\ J\ K^{-1}\ mol^{-1}$

SOLUTION

$$\Delta H°(263\ K) = H°(273K) + [C_{P,H_2O(cr)} - C_{P,H_2O(\ell)}] \times (263\ K - 273\ K)$$

$$= -6004\ J\ mol^{-1} + (-38.5\ J\ K^{-1} mol^{-1})(-10)$$

$$= -5619\ J\ mol^{-1}$$

1.31 0.843 bar

1.32 1.78×10^{-44} bar

1.33 $\alpha = 1/T \qquad \kappa = 1/P$

1.35 $21.7\ cm^3\ mol^{-1}$

1.36 0.324 nm

1.37 (a) 20.08
(b) 1642
(c) 2775 bar

1.39 $0.51\ L\ mol^{-1}$

1.40 914 m

1.41 $3.887\ J\ mol^{-1}$

1.44 (a) 1992, (b) -23,300, (c) -23,300, (d) -21,300 $J\ mol^{-1}$

1.45 106.780 $kJ\ mol^{-1}$

1.46 At 298 K HI is a rigid diatomic molecule. At 200 K its vibration is almost completely excited. At 298 K I_2 is almost completely vibrationally excited. At 2000 K it absorbs more energy per degree because of electronic excitation.

1.47

	CO	CO_2	NH_3	CH_4
Cp(classical)	37.413	62.354	83.136	108.077
Cp(3000 K)	37.217	62.229	79.496	101.391

The values are in $JK^{-1}\ mol^{-1}$.

1.48 40,874 J mol^{-1}

1.49 (a) 13.075, (b) 10.598 kJ mol^{-1}

1.50 (a) -74.5 L bar mol^{-1} (b) 7448 J mol^{-1} (c) 0 (d) 0

1.51 (a) 567 K (b) 9.42 bar (c) 5527 J mol^{-1}

1.52 (a) 0.495, (b) 0.307 bar

1.53 (a) 41.84, (b) 25.56 kJ

1.54 2.91 kWh 0.172 4.4 kg

1.55

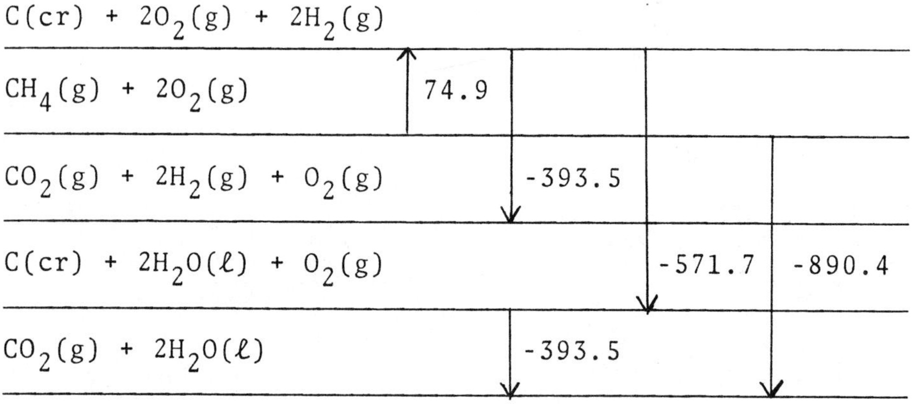

1.56 (a) -253.42, (b) -249.70 kJ mol^{-1}

1.57 -41.51 kJ mol^{-1}

1.58 12.01 kJ mol^{-1}

1.59 -585 ± 8 kJ mol^{-1} and -589 ± 2 kJ mol^{-1}

1.60 (a) -120.9, (b) -50.2, (c) -19.92, (d) -45.2 kJ mol^{-1}

1.61 225.756 kJ mol^{-1} is absorbed.

1.62 -214.627, -800.521, -1015.148 kJ mol^{-1}

1.63 432.074, 435.998, 459.578 kJ mol^{-1}

1.64 214.627, -97.92, 116.71 kJ mol^{-1}. Single reactor

1.65 (a) -94.1, (b) -359.0 kJ mol^{-1}

1.66 2432.6 J g^{-1}

CHAPTER 2. Second and Third Laws of Thermodynamics

2.1 Theoretically, how high could a gallon of gasoline lift an automobile weighing 2800 lb against the force of gravity, if it is assumed that the cylinder temperature is 2200 K and the exit temperature 1200 K? (Density of gasoline = 0.80 g cm^{-3}; 1 lb = 453.6 g; 1 ft = 30.48 cm; 1 liter = 0.2642 gal. Heat of combustion of gasoline = 46.9 kJ g^{-1}.)

SOLUTION

$$q = \frac{(46.9 \times 10^3 \text{ J g}^{-1})(1 \text{ gal.})(10^3 \text{cm}^3 \text{ L}^{-1})(0.80 \text{ g cm}^{-3})}{0.2642 \text{ gal. L}^{-1}}$$

$$= 14.2 \times 10^7 \text{ J}$$

$$w = q\,\frac{T_2 - T_1}{T_2} = \frac{(14.2 \times 10^7 \text{ J})(2200 \text{ K} - 1200 \text{ K})}{(2200 \text{ K})}$$

$$= 6.45 \times 10^7 \text{ J}$$

$$= mgh = (2800 \text{ lb})(0.4536 \text{ kg lb}^{-1})(9.8 \text{ m s}^{-2}) \times (0.3048 \text{ m ft}^{-1})h$$

$$h = 17{,}000 \text{ ft}$$

2.2 (a) What is the maximum work that can be obtained from 1000 J of heat supplied to a water boiler at 100 °C if the condenser is at 20 °C?

(b) If the boiler temperature is raised to 150 °C by the use of superheated steam under pressure, how much more work can be obtained?

SOLUTION

(a) $w = q\dfrac{T_2 - T_1}{T_2} = (1000 \text{ J})\dfrac{80 \text{ K}}{373.1 \text{ K}} = 214 \text{ J}$

(b) $w = (1000\ J)\ \dfrac{130\ K}{423.1\ K} = 307\ J$ or 93 J more than (a)

2.3 What is the entropy change for the freezing of one mole of water at 0 °C? The heat of fusion is 333.5 J g^{-1}.

SOLUTION

$$\Delta S = \frac{\Delta H}{T} = \frac{-(333.5\ J\ g^{-1})(18.015\ g\ mol^{-1})}{273.15\ K}$$
$$= -22.00\ J\ K^{-1}\ mol^{-1}$$

2.4 Calculate the increase in entropy of a mole of silver that is heated at constant pressure from 0 to 30 °C if the value of C_P in this temperature range is considered to be constant at 25.48 J K^{-1} mol^{-1}.

SOLUTION

$$\Delta S = C_P\ \ell n\frac{T_2}{T_1} = (25.48\ J\ K^{-1}\ mol^{-1})\ \ell n\frac{303}{273}$$
$$= 2.657\ J\ K^{-1}\ mol^{-1}$$

2.5 Calculate the change in entropy of a mole of aluminum which is heated from 600 °C to 700 °C. The melting point of aluminum is 660 °C, the heat of fusion is 393 J g^{-1}, and the heat capacities of the solid and liquid may be taken as 31.8 and 34.3 J K^{-1} mol^{-1}, respectively.

SOLUTION

$$\Delta S = \int_{T_1}^{T_f} \frac{C_{P,s}}{T}\,dT + \frac{\Delta H_f}{T_f} + \int_{T_f}^{T_2} \frac{C_{P,\ell}}{T}\,dT$$

$$= C_{P,s}\,\ell n\frac{T_f}{T_1} + \frac{\Delta H_f}{T_f} + C_{P,\ell}\,\ell n\frac{T_2}{T_f}$$

$$= (31.8\ J\ K^{-1}mol^{-1})\ell n\frac{933K}{873K} + \frac{(27gmol^{-1})(393Jg^{-1})}{933K}$$

$$+ (34.3JK^{-1}mol^{-1})\ell n\frac{973K}{933K}$$

$$= 14.92\ J\ K^{-1}\ mol^{-1}$$

2.6 A mole of steam is condensed at 100 °C and the water is cooled to 0 °C and frozen to ice. What is the entropy change of the water? Consider that the average specific heat of liquid water is 4.2 J K^{-1} g^{-1}. The heat of vaporization at the boiling point and the heat of fusion at the freezing point are 2258.1 and 333.5 J g^{-1}, respectively.

SOLUTION

$$\Delta S = -\frac{\Delta H_{vap}}{T_{vap}} + \int_{373K}^{273K}\frac{C_P}{T}\,dT - \frac{\Delta H_{fus}}{T_{fus}}$$

$$= -\frac{(2258.1\ J\ g^{-1})(18.016\ g\ mol^{-1})}{373.15\ K}$$

$$+(75.379JK^{-1}mol^{-1})\int_{373}^{273} d\ell nT - \frac{(333.5Jg^{-1})(18.016gmol^{-1})}{273.15\ K}$$

$$= -154.4\ J\ K^{-1}\ mol^{-1}$$

2.7 Calculate the increase in entropy of nitrogen when it is heated from 25 to 1000 °C (a) at constant pressure, and (b) at constant volume. Given: $C_P = 26.9835 + 5.9622 \times 10^{-3}T - 3.377 \times 10^{-7}T^2$ in J K^{-1} mol^{-1}.

SOLUTION

(a) $$\Delta S = \int_{298}^{1273} \frac{C_P}{T} dT = \int_{298}^{1273} \left(\frac{26.9835}{T} + 5.9622 \times 10^{-3} - 3.377 \times 10^{-7} T\right) dT$$

$$= 26.9835 \ln\frac{1273}{298} + 5.9622\times10^{-3}(1273 - 298) - \left(\frac{1}{2}\right)3.377\times10^{-7}(1273^2 - 298^2)$$

$$= 45.25 \text{ J K}^{-1} \text{ mol}^{-1}$$

(b) $$\Delta S = \int_{298}^{1273} \frac{C_P - 8.314}{T} dT = \int_{298}^{1273} \frac{C_P}{T} dT - 8.314 \ln\frac{1273}{298}$$

$$= 45.25 - 8.314 \ln\frac{1273}{298} = 33.18 \text{ J K}^{-1} \text{ mol}^{-1}$$

2.8 Calculate the entropy change in joules for a hundredfold isothermal expansion of a mole of perfect gas.

SOLUTION

$$dS = \frac{dq_{rev}}{T} = \frac{PdV}{T} = \frac{RdV}{V}$$

$$S = R \ln\frac{V_2}{V_1} = (8.314 \text{ J K}^{-1} \text{ mol}^{-1}) \ln 100$$

$$= 38.3 \text{ J K}^{-1} \text{ mol}^{-1}$$

2.9 In the reversible isothermal expansion of a perfect gas at 300 K from 1 to 10 liters, where the gas has an initial pressure of 20.27 bar, calculate (a) ΔS for the gas and (b) ΔS for all systems involved in the expansion.

SOLUTION

(a) $$n = \frac{PV}{RT} = \frac{(20.27 \text{ bar})(1 \text{ L})}{(0.08314 \text{ L bar K}^{-1} \text{ mol}^{-1})(300 \text{ K})}$$

$$= 0.812 \text{ mol}$$

$$\Delta S = nR\ln\frac{V_2}{V_1}$$

$$= (0.812 \text{ mol})(8.314 \text{ J K}^{-1} \text{ mol}^{-1}) \ln\frac{10}{1}$$

$$= 15.56 \text{ J K}^{-1}$$

(b) $\Delta S = 0$ since the process is carried out reversibly. The heat gained by the gas is equal to the heat lost by the heat reservoir, and both bodies are at the same temperature.

2.10 One mole of ammonia (considered to be a perfect gas) initially at 25 °C and 1 bar pressure is heated at constant pressure until the volume has trebled. Calculate (a) q, (b) w, (c) ΔH, (d) ΔU, and (e) ΔS. Given:
$C_P = 25.895 + 32.999 \times 10^{-3}T - 30.46 \times 10^{-7}T^2$ in J K^{-1} mol^{-1}.

SOLUTION

(a) $$q = \int_{T_1}^{T_2} C_P \, dT$$

$$= \int_{298}^{894} [25.895 + 32.999 \times 10^{-3}T - 30.46 \times 10^{-7} T^2]dT$$

$$= (25.895)(596) + \frac{32.999\text{x}10^{-3}}{2}(894^2 - 298^2) - \frac{30.46\text{x}10^{-3}}{3}(894^3 - 298^3)$$

$$= 26.4 \text{ kJ mol}^{-1}$$

(b) $$w = -P\Delta V = -R(T_2 - T_1) = -(8.314 \text{ J K}^{-1} \text{ mol}^{-1})(596 \text{ K})$$
$$= -4.96 \text{ kJ mol}^{-1}$$

(c) $\Delta H = q_P = 26.4 \text{ kJ mol}^{-1}$

(d) $\Delta U = q + w = 26.4 - 5.0 = 21.4 \text{ kJ mol}^{-1}$

(e) $S = \int_{298}^{894} \frac{C_P}{T}\, dT$

$= \int_{298}^{894} [\frac{25.895}{T} + 32.999\text{x}10^{-3} - 30.46\text{x}10^{-7}T]dT$

$= 2.303(25.895)\log\frac{894}{298} + 32.999\text{x}10^{-3}(894 - 298)$

$- \frac{30.46\text{x}10^{-7}}{2}(894^2 - 298^2)$

$= 46.99\ \text{J K}^{-1}\ \text{mol}^{-1}$

2.11 A mole of perfect gas is expanded isothermally and reversibly from 30 L to 100 L at 300 K. (a) What are the values of ΔU, ΔS, w and q? (b) If the expansion is carried out irreversibly by allowing the gas to expand into an evacuated container, what are the values of ΔU, ΔS, w, and q?

SOLUTION

(a) $\Delta U = 0$

$\Delta S = R\ell n\frac{V_2}{V_1}$

$= (8.314\ \text{J K}^{-1}\ \text{mol}^{-1})\ell n\frac{100\ \text{L}}{30\ \text{L}}$

$= 10.010\ \text{J K}^{-1}\ \text{mol}^{-1}$

$w = -RT\ell n\frac{V_2}{V_1}$

$= -(8.314\ \text{J K}^{-1}\ \text{mol}^{-1})(300\ \text{K})\ell n\frac{100\ \text{L}}{30\ \text{L}}$

$= -3.003\ \text{kJ mol}^{-1}$

$q = 3.003\ \text{kJ mol}^{-1}$

(b) $\Delta U = 0$

$\Delta S = 10.010\ \text{J K}^{-1}\ \text{mol}^{-1}$

$w = 0$

$q = 0$

2.12 The temperature of a mole of perfect gas is increased from 300 K to 500 K. What is the change in entropy (a) if the volume is held constant and (b) if the pressure is held constant?

SOLUTION

(a) $\Delta S = C_V \ln\frac{T_2}{T_1}$

$= \frac{3}{2}(8.314\ \text{J K}^{-1}\ \text{mol}^{-1})\ln\frac{500\ \text{K}}{300\ \text{K}}$

$= 6.371\ \text{J K}^{-1}\ \text{mol}^{-1}$

(b) $\Delta S = C_P \ln\frac{T_2}{T_1}$

$= \frac{5}{2}(8.314\ \text{J K}^{-1}\ \text{mol}^{-1})\ln\frac{500\ \text{K}}{300\ \text{K}}$

$= 10.618\ \text{J K}^{-1}\ \text{mol}^{-1}$

2.13 The pressure on a perfect gas is increased from 1 bar to 10 bar at constant temperature. What is the change in entropy?

SOLUTION

$\Delta S = -R\ln\frac{P_2}{P_1}$

$= -8.314\ \text{J K}^{-1}\ \text{mol}^{-1}\ \ln\frac{10\ \text{bar}}{1\ \text{bar}}$

$= -19.144\ \text{J K}^{-1}\ \text{mol}^{-1}$

2.14 What is the change in entropy of a mole of liquid at 25 °C when the pressure is raised to 1000 bar? The coefficient of thermal expansion α is $1.237 \times 10^{-3}\ \text{K}^{-1}$, the density is 0.879 g cm^{-3}, and the molar mass is 78.11 g mol^{-1}.

SOLUTION

$\Delta S = -V\alpha\Delta P$

$$= -\left(\frac{78.11 \text{ gmol}^{-1}}{0.879 \text{ gcm}^{-3}}\right)(10^{-2}\text{mcm}^{-1})^3(1.237\text{x}10^{-3}\text{K}^{-1})(1000 \text{ bar})$$

$$= -10.99 \text{ J K}^{-1} \text{ mol}^{-1}$$

2.15 Ten moles of H_2 and two moles of D_2 are mixed at 25 °C. What is the value of $\Delta S°$?

SOLUTION

$$\Delta S = -R(n_1 \ln X_1 + n_2 \ln X_2)$$

$$= -(8.314 \text{ J K}^{-1} \text{ mol}^{-1})(10\ln\frac{10}{12} + 2\ln\frac{2}{12})$$

$$= 44.95 \text{ J K}^{-1} \text{ mol}^{-1}$$

2.16 Calculate the molar entropy of liquid chlorine at its melting point, 172.12 K, from the following data obtained by W. F. Giauque and T. M. Powell.

T/K	15	20	25	30	35	40	50	60
$C_p/JK^{-1}mol^{-1}$	3.72	7.74	12.09	16.69	20.79	23.97	29.25	33.47

T/K	70	90	110	130	150	170	172.12
$C_p/JK^{-1}mol^{-1}$	36.32	40.63	43.81	47.24	51.04	55.10	M.P.

The heat of fusion is 6406 J mol^{-1}. Below 15 K it may be assumed that C_p is proportional to T^3.

SOLUTION

For T < 15 K, $C_p = CT^3$, $C = \dfrac{3.72 \text{ J K}^{-1} \text{ mol}^{-1}}{(15 \text{ K})^3}$

$$S_{15K} = \int_0^{15K} CT^3 \frac{dT}{T} = [CT^3/3]_0^{15}$$

$$= \frac{3.72}{15^3} \times \frac{15^3}{3} = 1.24 \text{ J K}^{-1} \text{ mol}^{-1}$$

The contributions from 15 K to the melting point

are calculated using $C_P\Delta T/T$, where C_P is the average for the range and T is the average for the range.

K	T_{avg}	$C_{P,avg}$	ΔT
15-20	17.5	5.73	5
20-25	22.5	9.92	5
25-30	27.5	14.39	5
30-35	32.5	18.74	5
35-40	37.5	22.38	5
40-50	45	26.61	10
50-60	55	31.36	10
60-70	65	34.89	10
70-90	80	38.47	20
90-110	100	42.22	20
110-130	120	45.52	20
130-150	140	49.14	20
150-170	160	53.07	20
170-172.12	171.06	55.10	2.12

$$S_{liquid,Tm} = 1.24 + \sum \frac{C_P\Delta T}{T} + \frac{\Delta H_{fus}}{Tm}$$

$$= 1.24 + 69.29 + \frac{6406}{172.12}$$

$$= 107.75 \text{ J K}^{-1} \text{ mol}^{-1}$$

2.17 Calculate the molar entropy of carbon disulfide at 25 °C from the following heat-capacity data and the heat of fusion, 4389 J mol^{-1}, at the melting point (161.11 K).

T/K	15.05	20.15	29.76	42.22	57.52	75.54	89.37
C_p/JK^{-1}mol^{-1}	6.90	12.01	20.75	29.16	35.56	40.04	43.14

T/K	99.00	108.93	119.91	131.54	156.83	161-298
C_p/JK^{-1}mol^{-1}	45.94	48.49	50.50	52.63	56.62	75.48

SOLUTION

$$S^\circ(298.15K) = \frac{C_P(15.05K)}{3} + \int_{15.05K}^{161.11K} \frac{C_P}{T}dT + \frac{\Delta H_{fus}}{161.11K} + \int_{161.11}^{298.15} \frac{C_P}{T}dT$$

The first integral may be approximated by multiplying the average value of C_P/T for each temperature interval by the width of the interval. Thus the first contribution is

$$\frac{1}{2}\left(\frac{6.90}{15.05} + \frac{12.01}{20.15}\right)(20.15 - 15.05) = 2.69 \text{ J K}^{-1}\text{mol}^{-1}$$

The first integral has the value 74.69 J $K^{-1}mol^{-1}$

Thus

$$S^\circ(298.15K) = \frac{6.90}{3} + 74.69 + \frac{4389}{161.11} + 75.45\ln\frac{298.15}{161.11}$$
$$= 150.67 \text{ J K}^{-1} \text{ mol}^{-1}$$

2.18 Using molar entropies from Table A.1, calculate ΔS° for the following reactions at 25 °C.

(a) $H_2(g) + \frac{1}{2}O_2(g) = H_2O(\ell)$

(b) $H_2(g) + Cl_2(g) = 2HCl(g)$

(c) Methane(g) + $\frac{1}{2}O_2(g)$ = methanol(ℓ)

SOLUTION

(a) $\Delta S^\circ = 69.91 - 130.68 - \frac{1}{2}(205.13)$
$= -163.34 \text{ J K}^{-1} \text{ mol}^{-1}$

(b) $\Delta S^\circ = 2(186.908) - 130.684 - 223.066$
$= 20.066 \text{ J K}^{-1} \text{ mol}^{-1}$

(c) $\Delta S^\circ = 126.8 - 186.264 - \frac{1}{2}(205.138)$
$= -162.0 \text{ J K}^{-1} \text{ mol}^{-1}$

2.19 What is $\Delta S°$ for $H_2(g) = 2H(g)$ at 298, 1000, and 3000 K?

SOLUTION

$\Delta S°(298K) = 2(114.604) - 130.574$
$= 98.634\ J\ K^{-1}\ mol^{-1}$

$\Delta S°(1000K) = 2(139.758) - 166.113$
$= 113.403\ J\ K^{-1}\ mol^{-1}$

$\Delta S°(3000K) = 2(162.594) - 202.778$
$= 122.410\ J\ K^{-1}\ mol^{-1}$

2.20 What is $\Delta S°$ (298K) for

$$H_2O(\ell) = H^+(ao) + OH^-(ao)$$

Why is this change negative and not positive?

SOLUTION

$\Delta S = -10.75 - 69.92 = -80.67\ J\ K^{-1}\ mol^{-1}$

The ions polarize neighboring water molecules and attract them. For this reason the product state is more ordered than the reactant state.

2.21 From electromotive force measurements it has been found that $\Delta S°$ for the reaction

$$\tfrac{1}{2}H_2(g) + AgCl(cr) = HCl(aq) + Ag(cr)$$

is $-62.4\ J\ K^{-1}\ mol^{-1}$ at 298.15 K. What is the value of $S°[Cl^-(aq)]$?

SOLUTION

Since HCl is completely dissociated in aqueous solution and $S°[H^+(ao)] = 0$,

$$\Delta S° = S°[Cl^-(ao)] + S°[Ag(cr)] - \tfrac{1}{2}S°[H_2(g)] - S°[AgCl(cr)]$$

$$-62.4 = S°[Cl^-(ao)] + 42.55 - \tfrac{1}{2}(130.684) - 96.2$$

$$S°[Cl^-(ao)] = 56.6\ J\ K^{-1}\ mol^{-1}$$

2.22 The gas is not returned to its initial state.

2.23 (a) 16.1% (b) 34.8% (c) 37.1%
(d) 37.3% (e) 47.1% (f) 62.0%

2.24 0.079 J 2.25 9.96 J K^{-1} mol^{-1}

2.26 Since this quantity is always positive, the change is spontaneous.

2.27 (a) 8.50, (b) 76.61, (c) 5.00 J K^{-1} mol^{-1}

2.28 125.39 J K^{-1} mol^{-1}

2.29 $\Delta S = R\ell n[(V_2-b)/(V_1-b)]$

2.30 $\left(\frac{\partial S}{\partial V}\right)_T = \left(\frac{\partial P}{\partial T}\right)_V = \frac{R}{V-b}$ 2.31 19.262 J K^{-1} mol^{-1}

2.32 $\Delta U = -2.252$ kJ mol^{-1} $\Delta S = 0$
$q = 0$ $w = -2.252$ kJ mol^{-1}

2.33 -0.0253 J K^{-1} mol^{-1} 2.34 9.13 J K^{-1} mol^{-1}

2.35 For the system $\Delta S = 0.021$ J K^{-1} mol^{-1}. Therefore the change is spontaneous.

2.36 115.5 J K^{-1} mol^{-1} 2.37 288.7 J K^{-1} mol^{-1}

2.39 (a) 80.479, (b) -32.93, (c) -38.9 J K^{-1} mol^{-1}

2.40 -5.06, 45.426 J K^{-1} mol^{-1}

2.41 316.6 J K^{-1} mol^{-1}

CHAPTER 3. Gibbs Energy and Chemical Potential

3.1 Calculate $\Delta G°$ for $H_2O(g,25°C) = H_2O(\ell,25°C)$
The vapor pressure of water at 25 °C is 3168 Pa.

SOLUTION

$$\Delta G° = -RT\ell n\frac{a_{H_2O(\ell)}}{a_{H_2O(g)}}$$

$$= -(8.314\ J\ K^{-1}\ mol^{-1})(298.15\ K)\ \ell n\frac{101,325\ Pa}{3,168\ Pa}$$

$$= -8.59\ kJ\ mol^{-1}$$

This process is carried out in two reversible steps.

3.2 Calculate the change in Gibbs energy for the process $H_2O(\ell,-10\ °C) = H_2O(cr,-10\ °C)$
The vapor pressure of water at -10 °C is 286.5 Pa, and the vapor pressure of ice at -10 °C is 260.0 Pa. The process may be carried out by the following reversible steps:

1. A mole of water is transferred at -10 °C from liquid to saturated vapor (P = 286.5 Pa). $\Delta G = 0$, since the two phases are in equilibrium.
2. The water vapor is allowed to expand from 286.5 to 260.0 Pa at -10 °C.
3. A mole of water is transferred at -10 °C from vapor at P = 260.0 Pa to ice at -10 °C.

SOLUTION

Since the first and third steps are transfers at equilibrium, there are no Gibbs energy changes associated with them.

$$\Delta G_2 = -RT\ell n\frac{P_1}{P_2}$$

$$= (8.314\ J\ K^{-1}\ mol^{-1})(263.15\ K)\ell n\frac{286.5\ Pa}{260.0\ Pa}$$

$$= -212.5\ J\ mol^{-1}$$

3.3 The change in Gibbs energy for the conversion of aragonite to calcite at 25 °C is -1046 J mol^{-1}. The density of aragonite is 2.93 g cm^{-3} at 25 °C and the density of calcite is 2.71 g cm^{-3}. At what pressure at 25 °C would these two forms of $CaCO_3$ be in equilibrium?

SOLUTION

aragonite = calcite $\Delta G° = -1046 \text{ J mol}^{-1}$

$$\Delta V = \frac{100.09 \text{ g mol}^{-1}}{2.71 \text{ g cm}^{-3}} - \frac{100.09 \text{ g mol}^{-1}}{2.93 \text{ g cm}^{-3}}$$

$$= 2.77 \text{ cm}^3 \text{ mol}^{-1} = 2.77 \times 10^{-6} \text{ m}^3 \text{ mol}^{-1}$$

$$\left(\frac{\partial \Delta G}{\partial P}\right)_T = \Delta V$$

$$\int_1^2 d\Delta G = \int_1^P \Delta V dP$$

$$\Delta G_2 - \Delta G_1 = \Delta V(P - 1)$$

$$0 + 1046 \text{ J mol}^{-1} = (2.77 \times 10^{-6} \text{m}^3 \text{mol}^{-1})(P - 1)$$

$$P = 3780 \text{ bar}$$

3.4 A mole of perfect gas is compressed isothermally from 1 to 5 bar at 100 °C. (a) What is the Gibbs energy change? (b) What would have been the Gibbs energy change if the compression had been carried out at 0 °C?

SOLUTION

(a) $\Delta G° = RT \ell n \frac{P_2}{P_1}$

$= (8.314 \text{ J K}^{-1} \text{ mol}^{-1})(373.15 \text{ K}) \ell n 5$

$= 4993 \text{ J mol}^{-1}$

(b) $\Delta G° = (8.314 \text{ J K}^{-1} \text{ mol}^{-1})(273.15 \text{ K}) \ell n 5$

$= 3655 \text{ J mol}^{-1}$

3.5 The value of $(\partial U/\partial V)_T$ is equal to zero for a perfect gas, but what is it for a van der Waals gas?

SOLUTION

$$P = \frac{RT}{V-b} - \frac{a}{V^2}$$

$$\left(\frac{\partial P}{\partial T}\right)_V = \left(\frac{R}{V-b}\right)$$

$$\left(\frac{\partial U}{\partial V}\right)_T = T\left(\frac{\partial P}{\partial T}\right)_V - P$$

$$= \frac{RT}{V-b} - \frac{RT}{V-b} + \frac{a}{V^2} = \frac{a}{V^2}$$

The internal energy of a van der Waals gas increases as it expands because the attractive interactions between molecules decrease.

3.6 What is the difference between the molar heat capacity of iron at constant pressure and constant volume at 25 °C? Given: $\alpha = 35.1 \times 10^{-6}\ K^{-1}$, $\kappa = 0.52 \times 10^{-6}\ bar^{-1}$, and the density is 7.86 g cm^{-3}.

SOLUTION

$$V = \frac{55.847\ g\ mol^{-1}}{7.86\ g\ cm^{-3}} = \frac{7.11\ cm^3 mol^{-1}}{(10^2 cm\ m^{-1})^3} = 7.11x10^{-6}\ m^3 mol^{-1}$$

$$C_P - C_V = \alpha^2 TV/\kappa$$

$$= \frac{(35.1x10^{-6}K^{-1})^2(298K)(7.11x10^{-6}m^3mol^{-1})(10^5\ Pa\ bar^{-1})}{(0.52 \times 10^{-6}\ bar^{-1})}$$

$$= 0.51\ J\ K^{-1}\ mol^{-1}$$

3.7 When pressure is applied to a liquid, its volume decreases. Assuming that the isothermal compressibility $\kappa = -\frac{1}{V}\left(\frac{\partial V}{\partial P}\right)_T$

is independent of pressure, derive an expression for the volume as a function of pressure.

SOLUTION

$$\int_{V_1}^{V_2} \frac{dV}{V} = -\kappa \int_{P_1}^{P_2} dP$$

$$\ell n \frac{V_2}{V_1} = -\kappa (P_2 - P_1)$$

$$V_2 = V_1 \, e^{-\kappa(P_2 - P_1)}$$

3.8 When a liquid is compressed its Gibbs energy is increased. To a first approximation the increase in Gibbs energy can be calculated using $(\partial G/\partial P)_T = V$, assuming a constant molar volume. What is the change in Gibbs energy for liquid water when it is compressed to 1000 bar?

SOLUTION

$$\int_{G_1}^{G_2} dG = \int_{P_1}^{P_2} VdP$$

$$\begin{aligned} \Delta G &= V\Delta P \\ &= (18 \times 10^{-6} \text{ m}^3 \text{ mol}^{-1})(999 \text{ bar})(10^5 \text{ Pa bar}^{-1}) \\ &= 1.798 \text{ kJ mol}^{-1} \end{aligned}$$

3.9 Show that when a liquid is compressed, the change in Gibbs energy is given by $\Delta G = V\Delta P - \frac{1}{2} \kappa V(\Delta P)^2$ if $\kappa\Delta P \ll 1$. The isothermal compressibility is represented by κ.

SOLUTION

$$\int_{G_1}^{G_2} dG = \int_{P_1}^{P_2} VdP = \int_{P_1}^{P_2} V\, e^{-\kappa(P-P_1)}\, dP$$

$$\Delta G = -\frac{V}{\kappa}\left[e^{-\kappa(P_2-P_1)} - 1\right]$$

If $\kappa\Delta P \ll 1$, $e^{-\kappa\Delta P} = 1 - \kappa\Delta P + \frac{1}{2}(\kappa\Delta P)^2 - \ldots$

$$\Delta G = \frac{V}{\kappa}\left[\kappa\Delta P - \frac{1}{2}\kappa^2(\Delta P)^2 + \cdots\right]$$

$$= V\Delta P - \frac{1}{2}\kappa V(\Delta P)^2$$

3.10 The standard entropy of $H_2(g)$ at 298 K is given in Table A.1. What is the entropy at 10 bar and 100 bar, assuming perfect gas behavior?

SOLUTION

$$S = S° - R\, \ell n \frac{P}{P°}$$

At 10 atm $S = 130.684 - (8.31441)\ \ell n\ 10$
$= 111.539\ \mathrm{J\ K^{-1}\ mol^{-1}}$

At 100 atm $S = 130.684 - (8.31441)\ \ell n\ 100$
$= 92.395\ \mathrm{J\ K^{-1}\ mol^{-1}}$

3.11 At 298 K, $S° = 205.138\ \mathrm{J\ K^{-1}\ mol^{-1}}$ for $O_2(g)$. What is the entropy of $O_2(g)$ at 100 bar, assuming that it is a perfect gas?

SOLUTION

$$S = S° - R\, \ell n \frac{P}{P°} = 205.138 - 8314\ \ell n\ 100$$

$$= 166.849\ \mathrm{J\ K^{-1}\ mol^{-1}}$$

3.12 A mole of helium is compressed isothermally and reversibly at 100 °C from a pressure of 2 to 10 bar. Calculate (a) q, (b) w, (c) ΔG, (d) ΔA, (e) ΔH, (f) ΔU, and (g) ΔS.

SOLUTION

(a) $\Delta U = q + w = 0$

$$q = -w = -RT\ln\frac{P_2}{P_1}$$

$$= -(8.314\ \text{JK}^{-1}\text{mol}^{-1})(373.15\text{K})\ln\frac{10\ \text{bar}}{2\ \text{bar}}$$

$$= -4993\ \text{J mol}^{-1}$$

(b) $w = -q = 4993\ \text{J mol}^{-1}$

(c) $\Delta G = \int_{P_1}^{P_2} V dP = RT\ln\frac{P_2}{P_1} = 4993\ \text{J mol}^{-1}$

(d) $\Delta A = w_{max} = 4993\ \text{J mol}^{-1}$

(e) $\Delta H - \Delta U + \Delta(PV) = 0$ (f) $\Delta U = 0$

(g) $\Delta S = \Delta H - \frac{\Delta G}{T} = 0 - \frac{4933\ \text{J mol}^{-1}}{373.15\ \text{K}}$

$$= -13.38\ \text{J K}^{-1}\text{mol}^{-1}$$

3.13 From tables giving ΔG_f°, ΔH_f°, and C_P for $H_2O(\ell)$ and $H_2O(g)$ at 298 K, calculate (a) the vapor pressure of $H_2O(\ell)$ at 25 °C, and (b) the boiling point at 1 atm.

SOLUTION

(a) $H_2O(\ell) = H_2O(g)$

$$\Delta G^\circ = -228.572 - (-237.129) = 8.557\ \text{kJ mol}^{-1}$$

$$\Delta G^\circ = -RT\ln(P/P^\circ)$$

$$P/P^\circ = \exp\frac{-8557\ \text{J mol}^{-1}}{(8.314\ \text{J K}^{-1}\ \text{mol}^{-1})(298.15\ \text{K})}$$

$$= 3.168 \times 10^{-2}$$

$$P = (3.168 \times 10^{-2})(10^5\ \text{Pa}) = 3.17 \times 10^3\ \text{Pa}$$

(b) $\Delta H°(298.15\ K) = -241.818 - (-285.830) = 44.012\ kJ\ mol^{-1}$

In the absence of data on the dependence of C_P on T we will calculate $\Delta H°(T)$ from

$$\Delta H°(T) = 44{,}012 + \int_{273.15}^{T} (33.577 - 75.291)dT$$

$$= 44.012 - 41.714\ (T - 273.15)$$

$$\frac{\partial(\Delta G°/T)}{\partial T} = - \frac{\Delta H°(T)}{T^2}$$

$$\frac{\Delta G°}{T} = -\int [\frac{-44{,}012}{T^2} - \frac{41.714}{T} + \frac{(41.714)(273.15)}{T^2\ \ T^2}] dT$$

$$= \frac{44.012}{T} + 41.714 \ell n T + \frac{(41.714)(298.15)}{T} + I$$

where I is an integration constant which can be evaluated from $\Delta G°$ (298.15 K).

$$\frac{8590}{298.15} = \frac{44.012}{298.15} + 41.714 \ell n 298.15 + \frac{(41.714)(298.15)}{(298.15)} + I$$

$I = -398.191$

At the boiling point $\Delta G° = 0$ and so we need to solve the following equation by successive approximations.

$$0 = \frac{44{,}012}{T} + 41.714 \ell n T + \frac{(41.714)(298.15)}{T} - 398.191$$

Trying T = 373 K   RHS = 0.163

Trying T = 374 K   RHS = -0.130

Therefore the standard boiling point calculated in this way is close to 373.5 K.

3.14 One mole of a perfect gas is allowed to expand reversibly and isothermally (25°C) from a pressure of 1 bar to a pressure of 0.1 bar: (a) What is the change in Gibbs energy? (b) What would be the change in Gibbs energy if the process occurred irreversibly?

SOLUTION

(a) Equation 3.39 $\left(\frac{\partial G}{\partial P}\right)_T = V = \frac{RT}{P}$

$$\Delta G = RT\ln\frac{P_2}{P_1}$$

$$= (8.314\ \text{J K}^{-1}\ \text{mol}^{-1})(298.15\ \text{K})\ln 0.1$$

$$= -5708\ \text{J mol}^{-1}$$

(b) $G = -5708\ \text{J mol}^{-1}$ because G is a state function and depends only on the initial state and the final state.

3.15 (a) Calculate the work done against the atmosphere when 1 mol of toluene is vaporized at its boiling point, 111 °C. The heat of vaporization at this temperature is 361.9 J g^{-1}. For the vaporization of 1 mol, calculate (b) q, (c) ΔH, (d) ΔU, (e) ΔG, and (f) ΔS.

SOLUTION

(a) Assuming that toluene vapor is a perfect gas and that the volume of the liquid is negligible, the work done against the atmosphere is

$$w = P\Delta V = RT$$

$$= (8.314\ \text{J K}^{-1}\ \text{mol}^{-1})(384\ \text{K})$$

$$= 3193\ \text{J mol}^{-1}$$

NOTE: The work done on the toluene is $-3193\ \text{J mol}^{-1}$.

(b) $q_p = \Delta H_{vap} = (361.9\ \text{J g}^{-1})(92.13\ \text{g mol}^{-1})$
$= 33{,}342\ \text{J mol}^{-1}$

(c) $\Delta H = 33{,}342\ \text{J mol}^{-1}$

(d) $\Delta U = q + w = 33{,}342 - 3193 = 30{,}149\ \text{J mol}^{-1}$

(e) $\Delta G = 0$ because the evaporation is reversible at the boiling point and 1 atm.

(f) $\Delta S = \frac{q_{rev}}{T} = \frac{33{,}342\ \text{J mol}^{-1}}{384\ \text{K}} = 86.8\ \text{J K}^{-1}\ \text{mol}^{-1}$

3.16 One mole of a perfect gas at 300 K has an initial pressure of 15 bar and is allowed to expand isothermally to a pressure of 1 bar. Calculate (a) the maximum work that can be obtained from the expansion, (b) ΔU, (c) ΔH, (d) ΔG, and (e) ΔS.

SOLUTION

(a) $w = RT\ell n\frac{P_2}{P_1} = (8.314\ \mathrm{J\ K^{-1}\ mol^{-1}})(300\mathrm{K})\ell n\frac{1}{15}$

$= -6.754\ \mathrm{kJ\ mol^{-1}}$

(b) $\Delta U = 0$ because $\partial U/\partial T = 0$ for a perfect gas

(c) $\Delta H = \Delta V + \Delta(PV) = 0$

(d) $\Delta G = RT\ell n\frac{P_2}{P_1} = -6.754\ \mathrm{kJ\ mol^{-1}}$

(e) $\Delta S = R\ell n\frac{P_1}{P_2} = (8.314\ \mathrm{J\ K^{-1}\ mol^{-1}})(\ell n15)$

$= 22.51\ \mathrm{J\ K^{-1}\ mol^{-1}}$

3.17 Show that if the compressibility factor is given by $Z = 1 + B'P$ the fugacity is given by $f = Pe^{Z-1}$. If Z is not very different from unity, $e^{Z-1} = 1 + (Z - 1) + \cdots \cong Z$ so that $f = PZ$. Using this approximation, what is the fugacity of $H_2(g)$ at 50 bar and 298 K using its van der Waals constants?

SOLUTION

$$\ell n\left(\frac{f}{P^\circ}\right) = \ell n\left(\frac{P}{P^\circ}\right) + \int_\circ^P \frac{Z-1}{P}\,dP$$

$$= \ell n\left(\frac{P}{P^\circ}\right) + \int_\circ^P B'dP$$

$$= \ell n\left(\frac{P}{P^\circ}\right) + B'P$$

$$f = P\,e^{Z-1}$$

$$Z = 1 + \left(b - \frac{a}{RT}\right)\left(\frac{P}{RT}\right)$$

$$= 1 + \left[0.02661 - \frac{0.2476}{(0.08314)(298)}\right]\frac{50}{(0.08314)(298)}$$

$$= 1.034$$

$$f = (50\ \text{bar})\ e^{0.034} = 51.7\ \text{bar}$$

3.18 For a real gas the chemical potential is given by $G = G° + RT\ell n\left(\frac{f}{P°}\right)$ For a van der Waals gas it may be shown that to a first approximation

$$\ell n\left(\frac{f}{P°}\right) = \ell n\left(\frac{P}{P°}\right) + \left[b - \left(\frac{a}{RT}\right)\right]\frac{P}{RT} \qquad \text{so that}$$

$$\mu = \mu° + RT\ell n\left(\frac{P}{P°}\right) + \left[b - \left(\frac{a}{RT}\right)\right] P$$

What are the corresponding relations for S, A, U, H and V?

SOLUTION

$$S = -\left(\frac{\partial\mu}{\partial T}\right)_P = S° - R\ell n\left(\frac{P}{P°}\right) - \frac{aP}{RT^2}$$

$$A = \mu - P\left(\frac{\partial\mu}{\partial P}\right)_T = \mu° + RT\ell n\left(\frac{P}{P°}\right) - RT$$

$$U = A - T\left(\frac{\partial\mu}{\partial T}\right)_P = H° - RT - \frac{aP}{RT} = U° - \frac{aP}{RT}$$

$$H = U + PV = U° + \frac{aP}{RT} + RT + bP - \frac{aP}{RT}$$

$$= H° + \left(b - \frac{2a}{RT}\right)P$$

$$V = V = \frac{RT}{P} + b - \frac{a}{RT}$$

3.19 Calculate the partial molar volume of zinc chloride in 1-molal $ZnCl_2$ solution using the following data.

% by weight of $ZnCl_2$	2	6	10	14	18	20
Density/g cm^{-3}	1.0167	1.0532	1.0891	1.1275	1.1665	1.1866

SOLUTION

Taking the first solution as an example, 1 g of solution contains 0.02 g of $ZnCl_2$ (M = 136.28 g mol^{-1}) and 0.98 g of H_2O. The weight of solution containing 1000 g of water is

$$\frac{1000}{0.98} = 1020 \text{ g}$$

$$\text{molality } ZnCl_2 = \frac{(0.02)(1020 \text{ g})}{136.28 \text{ g mol}^{-1}}$$

$$= 0.1497 \text{ m}$$

Volume of solution containing 1000 g of H_2O

$$\frac{1020 \text{ g}}{1.0167 \text{ g cm}^{-3}} = 1003.2 \text{ cm}^3$$

Wt%	Molality	Volume containing 1000 g of H_2O
2	0.1497 m	1003.2
6	0.4683	1010.1
10	0.8152	1020.2
14	1.194	1031.3
18	1.610	1045.5
20	1.834	1053.4

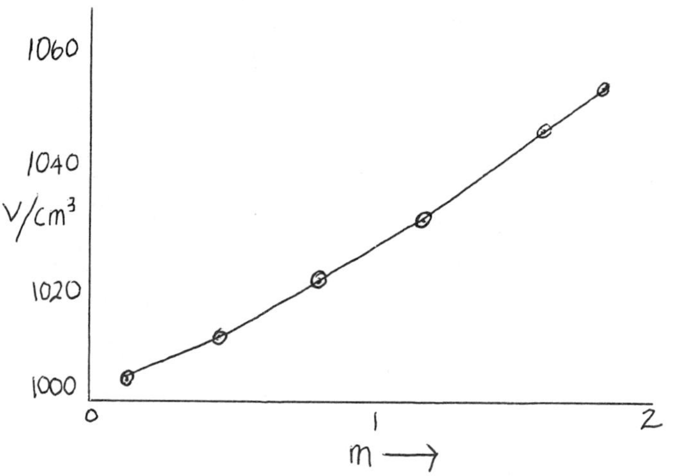

The slope of this plot at m=1 molar is 29.3 cm^3 mol^{-1} and so this is the partial molar volume of $ZnCl_2$.

3.20 For a solution of ethanol and water at 20 °C which has a mole fraction of ethanol of 0.2, the partial molar volume of water is 17.9 cm^3 mol^{-1} and the partial molar volume of ethanol is 55.0 cm^3 mol^{-1}. What volumes of pure ethanol and water are required to make a liter of this solution? At 20 °C the density of ethanol is 0.789 g cm^{-3} and the density of water is 0.998 g cm^{-3}.

SOLUTION

Component 1 is water, and component 2 is ethanol.

$$V = n_1\bar{V}_1 + n_2\bar{V}_2 = 4n_2\bar{V}_1 + n_2\bar{V}_2 = n_2(4\bar{V}_1 + \bar{V}_2)$$

$$1000\ cm^3 = n_2[(4)(17.9\ cm^3 mol^{-1}) + 55.0\ cm^3 mol^{-1}]$$

$$n_2 = 7.90\ mol = \frac{g_2}{46.07\ g\ mol^{-1}}$$

$$g_2 = 363.9\ g$$

$$\text{Volume of pure ethanol} = \frac{363.9\ \text{g}}{0.789\ \text{g cm}^{-3}} = 461\ \text{cm}^3$$

$$n_1 = 4n_2 = 31.60 = \frac{g_1}{18.016} \qquad g_1 = 569.3\ \text{g}$$

$$\text{Volume of pure water} = \frac{569.3\ \text{g}}{0.998\ \text{g cm}^{-3}} = 570\ \text{cm}^3$$

3.21 Calculate ΔG and ΔS for the formation of a quantity of air containing 1 mol of gas by mixing nitrogen and oxygen at 298.15 K. Air may be taken to be 80% nitrogen and 20% oxygen by volume.

SOLUTION

$$\Delta G_{mix} = RT(X_1 \ell n X_1 + X_2 \ell n X_2)$$

$$= (8.314\ \text{JK}^{-1}\text{mol}^{-1})(298.15\text{K})(0.8 \ell n 0.8 + 0.2 \ell n 0.2)$$

$$= -1239\ \text{J mol}^{-1}$$

$$\Delta S_{mix} = -R(X_1 \ell n X_1 + X_2 \ell n X_2)$$

$$= 4.159\ \text{J K}^{-1}\ \text{mol}^{-1}$$

3.22 Using experimental data from Fig. 3.6 calculate $(\partial S/\partial \ell)_T$ and $(\partial U/\partial \ell)_T$ using only the data at 10 °C and 70 °C, assuming these quantities are independent of temperature. These thermodynamic quantities do depend on the elongation so calculate them for 200% and 400% elongation. (For f and T simply use averages for this temperature range.)

SOLUTION At 200% elongation

$$\left(\frac{\partial S}{\partial \ell}\right)_T = -\left(\frac{\partial f}{\partial T}\right)_\ell = -\frac{(10.8 - 9.0)(10^5\ \text{N m}^{-2})}{(70 - 10)\text{K}}$$

$$= -3.0 \times 10^3 \text{ N K}^{-1} \text{ m}^{-2}$$

$$\left(\frac{\partial U}{\partial \ell}\right)_T = f - T\left(\frac{\partial f}{\partial T}\right)_\ell = 9.5\times10^5\text{Nm}^{-2} - (313\text{K})(3.0\times10^3\text{NK}^{-1}\text{ m}^{-2})$$

$$= 0.1 \text{ N m}^{-2}$$

At 400% elongation

$$\left(\frac{\partial S}{\partial \ell}\right)_T = -\frac{(26 - 21.3) \times 10^5 \text{ N m}^{-2}}{60 \text{ K}}$$

$$= -7.8 \times 10^3 \text{ N K}^{-1} \text{ m}^{-2}$$

$$\left(\frac{\partial U}{\partial \ell}\right)_T = 23\times10^5 \text{ N m}^{-2} - (313 \text{ K})(7.8\times10^3 \text{ N K}^{-1} \text{ m}^{-2})$$

$$= -1.4 \text{ N m}^{-2}$$

3.23 -113 J mol^{-1}

3.24 89.5 J mol^{-1}

3.25 $dU = TdS - PdV = \left(\frac{\partial U}{\partial S}\right)_V dS + \left(\frac{\partial U}{\partial V}\right)_S dV$

$$\therefore \left(\frac{\partial U}{\partial S}\right)_V = T$$

$$dH = TdS + VdP = \left(\frac{\partial H}{\partial S}\right)_P dS + \left(\frac{\partial H}{\partial P}\right)_S dP$$

$$\therefore \left(\frac{\partial H}{\partial S}\right)_V = T$$

$$\left(\frac{\partial U}{\partial S}\right)_V = \left(\frac{\partial H}{\partial S}\right)_P$$

$$\therefore \left(\frac{\partial H}{\partial P}\right)_S = V$$

$$dG = -SdT + VdP = \left(\frac{\partial G}{\partial T}\right)_P dT + \left(\frac{\partial G}{\partial P}\right)_T dP$$

$$\therefore \left(\frac{\partial G}{\partial P}\right)_T = G$$

$$\left(\frac{\partial H}{\partial P}\right)_S = \left(\frac{\partial G}{\partial P}\right)_T$$

3.26 $C_P - C_V = \dfrac{R^2T\,V^3}{RTV^3 - 2a(V-b)}$ where V is the molar volume

3.27 32.3 kJ mol^{-1}

3.28 $\left(\dfrac{\partial S}{\partial V}\right)_T = \dfrac{R}{V-b}$

$\Delta S = R\ell n\dfrac{V_2 - b}{V_1 - b}$

3.29 1.756 kJ mol^{-1} 3.30 168.858 J K^{-1} mol^{-1}

3.31 (a) 6820 J mol^{-1} (b) 6070 J mol^{-1}
(c) 0 J mol^{-1} (d) 75.7 J K^{-1} mol^{-1}

3.32 (a) -5230 J mol^{-1} (b) 5230 J mol^{-1}
(c) 0 J mol^{-1} (d) -5230 J mol^{-1}
(e) 19.142 J K^{-1} mol^{-1} (f) 0 J mol^{-1}
(g) 0 J mol^{-1} (h) 0 J mol^{-1}
(i) -5230 J mol^{-1} (j) 19.142 J K^{-1} mol^{-1}
(k) 0 J K^{-1} mol^{-1} (l) 19.142 J K^{-1} mol^{-1}

3.33 3100, -40,670, -40,670, -37,570, 0,
3100 J mol^{-1}
-108.8 J K^{-1} mol^{-1}

3.34 0, 19.16 J K^{-1} mol^{-1} 3.36 51.7 bar

3.37 0.79 cm^3 g^{-1} 3.38 (a) 0.204 cm^3 g^{-1}
(b) 19.4 cm^3 mol^{-1}

3.39 26.01 cm^3

3.40 $\Delta S = -R(X_1\ell nX_1 + X_2\ell nX_2 + X_3\ell nX_3)$

$\Delta G = RT(X_1\ell nX_1 + X_2\ell nX_2 + X_3\ell nX_3)$

3.41 -4732 J mol^{-1} 15.86 J K^{-1} mol^{-1}

3.42 6.0×10^{25}, 5.8×10^{25} m^{-3}

CHAPTER 4. Phase Equilibria

4.1 What is the maximum number of phases that can be in equilibrium at constant temperature and pressure in one-, two-, and three-component systems?

SOLUTION

$v = c - p + 2$

$0 = 1 - p + 2, \ p = 3$

$0 = 2 - p + 2, \ p = 4$

$0 = 3 - p + 2, \ p = 5$

4.2 The critical temperature of carbon tetrachloride is 283.1 °C. The densities in grams per cubic centimeter of the liquid ρ_ℓ and vapor ρ_v at different temperatures are as follows:

t/°C	100	150	200	250	270	280
ρ_ℓ	1.4343	1.3215	1.1888	0.9980	0.8666	0.7634
ρ_v	0.0103	0.0304	0.0742	0.1754	0.2710	0.3597

What is the critical molar volume of CCl_4? It is found that the mean of the densities of the liquid and vapor does not vary rapidly with temperature and can be represented by

$$\frac{\rho_\ell + \rho_v}{2} = AT + B$$ where A and B are constants.

The extrapolated value of the average density at the critical temperature is the critical density. The molar volume V_c at the critical point is equal to the molar mass divided by the critical density.

SOLUTION

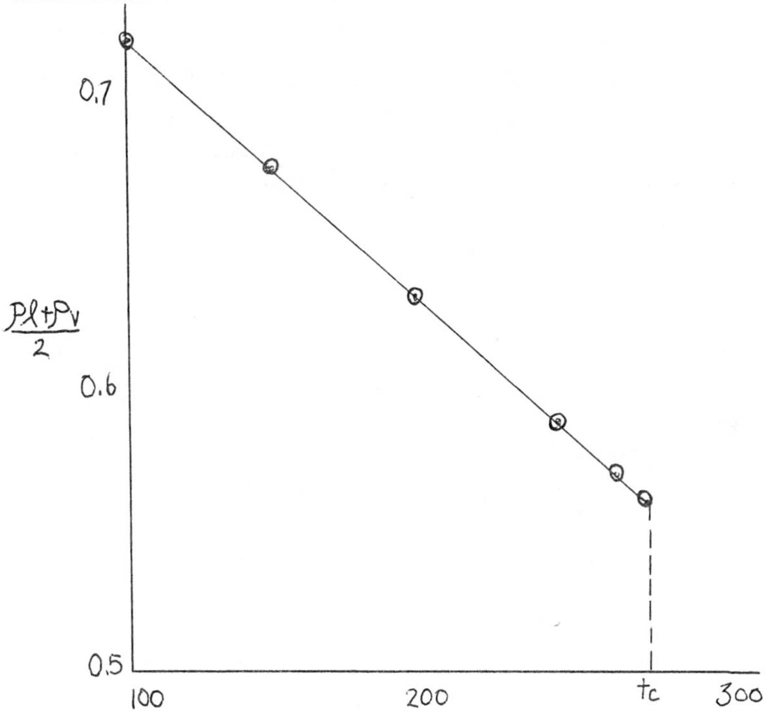

Extrapolating $\frac{\rho_\ell + \rho_v}{2}$ to t_c we obtain

$$\rho_c = 0.557 \text{ g cm}^{-3}$$

$$V_c = \frac{153.84 \text{ g mol}^{-1}}{0.557 \text{ g cm}^{-3}} = 276 \text{ cm}^3 \text{ mol}^{-1}$$

4.3 Ice has the unusual property of a melting point that is lowered by increasing pressure. Thus, one can skate on ice provided that the pressure exerted by one's skates is great enough to liquefy the ice under them. Would a 75-kg skater whose skates contact the ice with an area of 0.1 cm^2 to be able to skate at -3 °C?

SOLUTION

From Example 4.2, $\Delta T = -(7.5 \times 10^{-3} \text{ K bar}^{-1})\Delta P$

$$\Delta T = \frac{-(7.5 \times 10^{-3} \text{ K bar}^{-1})(75 \text{ kg})(9.80 \text{ m s}^{-2})}{(0.1 \text{ cm}^2)(10^{-2} \text{ m cm}^{-1})^2(10^5 \text{ Pa bar}^{-1})}$$

= -5.51 °C so the answer is yes.

4.4 n-Propyl alcohol has the following vapor pressures

t/°C	40	60	80	100
P_{sat}/kPa	6.69	19.6	50.1	112.3

Plot these data so as to obtain a nearly straight line, and calculate (a) the heat of vaporization, and (b) the boiling point at 1 bar.

SOLUTION

log P is plotted versus 1/T

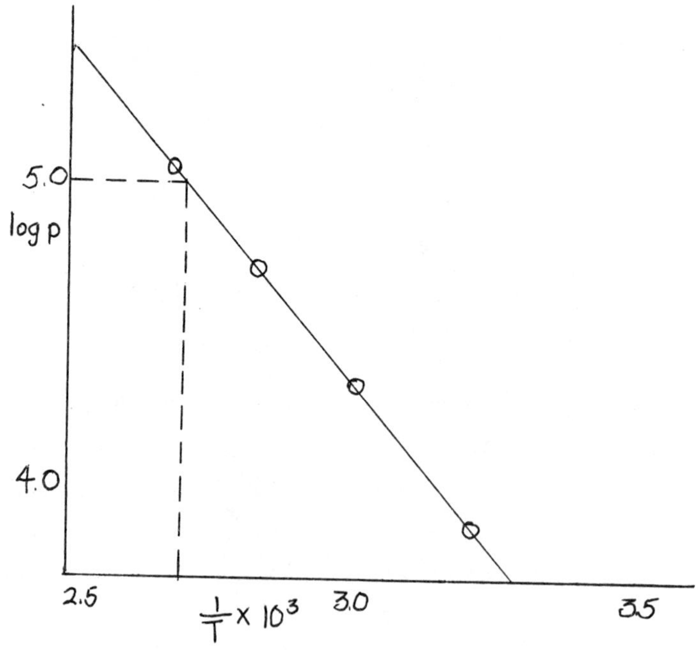

(a) slope $= -2.34\times10^3$ K $= \dfrac{\Delta H_{vap}}{2.303\ (8.314\ \text{J K}^{-1}\text{mol}^{-1})}$

$\Delta H_{vap} = 44.8\ \text{kJ mol}^{-1}$

(b) $\ln\dfrac{112.3}{100} = \dfrac{44{,}800}{8.3144}\left(\dfrac{1}{T} - \dfrac{1}{373.15}\right)$

$$T = \frac{1}{\frac{8.3144}{44{,}800}\ln\frac{112.3}{100} + \frac{1}{373.15}}$$

$= 370.2\ \text{K} = 97.0\ °\text{C}$

4.5 For uranium hexafluoride the vapor pressures (in Pa) for the solid and liquid are given by

$\ln P_S = 29.411 - 5893.5/T$

$\ln P_L = 22.254 - 3479.9/T$

Calculate the temperature and pressure of the triple point.

SOLUTION

$$29.411 - \frac{5893.5}{T} = 22.254 - \frac{3479.9}{T}$$

$T = 337.2\ \text{K} = 64.0\ °\text{C}$

$P = e^{29.411 - 5893.5/337.2} = 152.2\ \text{kPa}$

4.6 If $\Delta C_P = C_{P,vap} - C_{P,liq}$ is independent of temperature, then $\Delta H_{vap} = \Delta H_{o,vap} + T\Delta C_P$ where $\Delta H_{o,vap}$ is the hypothetical enthalpy of vaporization at absolute zero. Since ΔC_P is negative, ΔH_{vap} decreases as the temperature increases. Show that if the vapor is a perfect gas, the vapor pressure is given as a function of temperature by

$$\ln P = \frac{-\Delta H_{o,vap}}{RT} + \frac{\Delta C_P}{R} \ln T + \text{constant}$$

SOLUTION

$$\frac{d\ln P}{dT} = \frac{\Delta H_{o,vap} + T\Delta C_P}{RT^2} = \frac{\Delta H_{o,vap}}{RT^2} + \frac{\Delta C_P}{RT}$$

Integrating

$$\ln P = -\frac{\Delta H_{o,vap}}{RT} + \frac{\Delta C_P}{R} \ln T + \text{constant}$$

4.7 The heats of vaporization and of fusion of water are 2490 J g^{-1} and 333.5 J g^{-1} at 0 °C. The vapor pressure of water at 0 °C is 611 Pa. Calculate the sublimation pressure of ice at -15 °C, assuming that the enthalpy changes are independent of temperature.

SOLUTION

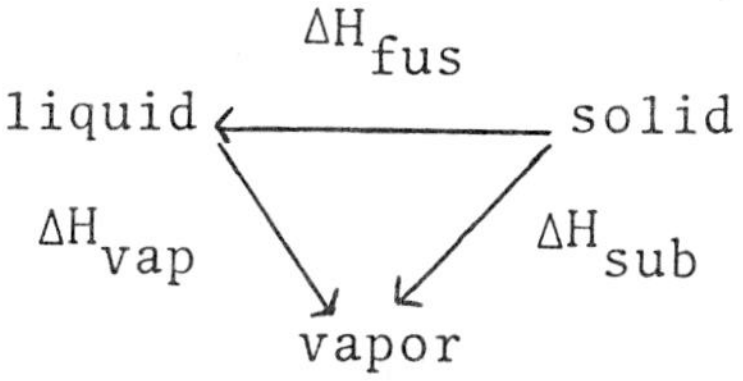

$$\Delta H_{sub} = \Delta H_{fus} + \Delta H_{vap}$$
$$= 333.5 + 2490$$
$$= 2824 \text{ J g}^{-1}$$

$$\ln \frac{P_2}{P_1} = \frac{\Delta H_{sub}(T_2 - T_1)}{RT_1T_2}$$

$$P_2 = P_1 \exp[\Delta H_{sub}(T_2 - T_1)/RT_1T_2]$$

$$= (611 \text{ Pa}) \exp\left[\frac{(2824 \times 18 \text{ J mol}^{-1})(-15 \text{ K})}{(8.314\text{K}^{-1}\text{mol}^{-1})(273.15\text{K})(258.15\text{K})}\right]$$

$$= 166 \text{ Pa}$$

4.8 The vapor pressure of toluene is 8.00 kPa at 40.3 °C and 2.67 kPa at 18.4 °C. Calculate

(a) the heat of vaporization, and (b) the vapor pressure at 25 °C.

SOLUTION

(a) $$\ln\frac{P_2}{P_1} = \frac{\Delta H_{vap}(T_2 - T_1)}{RT_1T_2}$$

$$\Delta H_{vap} = \frac{(8.314 JK^{-1}mol^{-1})(291.6K)(313.5K)}{21.9\ K}\ln\frac{8.00}{2.67}$$

$$= 38.1\ kJ\ mol^{-1}$$

(b) $$\ln\frac{P_2}{2.67kPa} = \frac{(38{,}100\ J\ mol^{-1})(6.6\ K)}{(8.314JK^{-1}mol^{-1})(298.15K)(291.6K)}$$

$$P_2 = 3.78\ kPa$$

4.9 The sublimation pressures of solid Cl_2 are 352 Pa at -112 °C and 35 Pa at -126.5 °C. The vapor pressures of liquid Cl_2 are 1590 Pa at -110 °C and 7830 Pa at -80 °C. Calculate (a) ΔH_{sub}, (b) ΔH_{vap}, (c) ΔH_{fus}, and (d) the triple point.

SOLUTION

(a) $$\Delta H_{sub} = \frac{RT_1T_2}{T_2-T_1}\ln\frac{P_2}{P_1}$$

$$= \frac{(8.314JK^{-1}mol^{-1})(161.15K)(146.65K)}{14.5\ K}\ln\frac{352}{35}$$

$$= 31.4\ kJ\ mol^{-1}$$

(b) $$\Delta H_{vap} = \frac{(8.314JK^{-1}mol^{-1})(173.15K)(193.15K)}{20\ K}\ln\frac{7830}{1590}$$

$$= 22.1\ kJ\ mol^{-1}$$

(c) $$\Delta H_{fus} = \Delta H_{sub} - \Delta H_{vap}$$

$$= 31.4 - 22.1 = 9.3\ kJ\ mol^{-1}$$

(d) For the solid

$$\ln P = \ln 352 + \frac{31{,}400}{8.314}\left(\frac{1}{161.15} - \frac{1}{T}\right) = 29.300 - \frac{3777}{T}$$

For the liquid

$$\ell n P = \ell n 1590 + \frac{22,100}{8.314}\left(\frac{1}{173.15} - \frac{1}{T}\right) = 22.723 - \frac{2658}{T}$$

At the triple point

$$29.300 - \frac{3777}{T} = 22.723 - \frac{2658}{T} \qquad T = \frac{1119}{6.577} = 170.1 \text{ K}$$

$$T = 107.1 \text{ K} - 273.15 \text{ K} = -103.0 \text{ °C}$$

4.10 The boiling point of n-hexane at 1 bar is 68.6 °C. Estimate (a) its molar heat of vaporization, and (b) its vapor pressure at 60 °C.

SOLUTION

(a) $\Delta H_{vap} = (88 \text{ J K}^{-1} \text{ mol}^{-1})(69 + 273)\text{K} = 30.1 \text{ kJ mol}^{-1}$

$$\text{(b) } \ell n\frac{P_2}{P_1} = \frac{\Delta H_{vap}}{R}\left(\frac{1}{T_1} - \frac{1}{T_2}\right)$$

$$\ell n\frac{1 \text{ bar}}{P} = \frac{30,100 \text{Jmol}^{-1}}{8.314 \text{JK}^{-1}\text{mol}^{-1}}\left(\frac{1}{333\text{K}} - \frac{1}{341.8\text{K}}\right)$$

$$P = 0.755 \text{ bar}$$

4.11 According to Trouton's rule the entropy of vaporization of a liquid at its boiling point is 88 J K^{-1} mol^{-1}. What is the change in boiling point expected for a liquid with a boiling point of (a) 100 °C and (b) 200 °C at 101,325 Pa in going to a reference state of 1 bar?

SOLUTION

(a) $\Delta H_{vap} = (88 \text{ J K}^{-1} \text{ mol}^{-1})(373 \text{ K}) = 32,800 \text{ J mol}^{-1}$

$$\ell n\frac{P_2}{P_1} = \frac{\Delta H_{vap}(T_2 - T_1)}{RT_1T_2}$$

$$\ln\frac{10^5}{101{,}325} = \frac{32{,}800(T_2 - T_1)}{(8.314)(373)^2}$$

$$T_2 - T_1 = -0.46\ °C$$

(b) $\Delta H_{vap} = (88 JK^{-1}mol^{-1})(473K) = 41{,}600\ J\ mol^{-1}$

$$T_2-T_1 = \frac{(8.314)(473)^2}{41{,}600}\ln\frac{10^5}{101{,}325} = -0.59\ °C$$

4.12 Ethanol and methanol form very nearly ideal solutions. The vapor pressure of ethanol is 5.93 kPa, and that of methanol is 11.83 kPa, at 20 °C. (a) Calculate the mole fraction of methanol and ethanol in a solution obtained by mixing 100 g of each. (b) Calculate the partial pressures and the total vapor pressure of the solution. (c) Calculate the mole fraction of methanol in the vapor.

SOLUTION

(a) $x_{C_2H_5OH} = \frac{100/146}{100/146 + 100/32} = 0.410$

$x_{CH_3OH} = \frac{100/32}{100/146 + 100/32} = 0.590$

(b) $P_{C_2H_5OH} = x_{C_2H_5OH}\ P^{\circ}_{C_2H_5OH}$

$= (0.410)(5930\ Pa) = 2430\ Pa$

$P_{CH_3OH} = x_{CH_3OH}\ P^{\circ}_{CH_3OH}$

$= (0.590)(11{,}830\ Pa) = 6980\ Pa$

$P_{total} = 2430\ Pa + 6980\ Pa = 9410\ Pa$

(c) $x_{CH_3OH,vapor} = \frac{6980\ Pa}{9410\ Pa} = 0.742$

4.13 The vapor pressure of a solution containing 13 g of a nonvolatile solute in 100 g of water at 28 °C is 3.6492 kPa. Calculate the molar mass of the solute, assuming that the solution is ideal. The vapor pressure of water at this temperature is 3.7417 kPa.

SOLUTION

$$x_2 = \frac{n_2}{n_1+n_2} = \frac{P_1{}^\circ - P_1}{P_1{}^\circ} = \frac{3741.7 - 3649.2}{3741.7} = 0.0247$$

$$\frac{n_2}{n_1+n_2} = 0.0247 = \frac{13/M}{(100/18) + (13/M)}$$

$$M = 92.4 \text{ g mol}^{-1}$$

4.14 One mole of benzene (component 1) is mixed with two moles of toluene (component 2). At 60 °C the vapor pressures of benzene and toluene are 51.3 and 18.5 kPa, respectively. (a) As the pressure is reduced at what pressure will boiling begin? (b) What will be the composition of the first bubble of vapor?

SOLUTION

$$P = P_2{}^{sat} + (P_1{}^{sat} - P_2{}^{sat})x_1$$

$$= 18.5 \text{ kPa} + (51.3 \text{ kPa} - 18.5 \text{ kPa})(0.333)$$

$$= 29.4 \text{ kPa}$$

$$y_1 = \frac{x_1 P_1{}^{sat}}{P_2{}^{sat} + (P_1{}^{sat} - P_2{}^{sat})x_1}$$

$$= \frac{(0.333)(51.3 \text{ kPa})}{18.5 \text{ kPa} + (32.8 \text{ kPa})(0.333)}$$

$$= 0.581 \text{ kPa}$$

4.15 What is the difference in chemical potential of benzene (component 1) in toluene (component 2) at $x_1 = 0.5$ and $x_1 = 0.1$ at 25 °C, assuming ideal solutions?

SOLUTION

$$\mu_1 = \mu_1^* + RT\ln 0.5$$
$$\mu_1 = \mu_1^* + RT\ln 0.1$$
$$\Delta\mu_1 = (8.314\ \text{J K}^{-1}\ \text{mol}^{-1})(298\ \text{K})\ln\frac{0.5}{0.1}$$
$$= 3988\ \text{J mol}^{-1}$$

4.16 Use the Gibbs-Duhem equation to show that if one component of a binary liquid solution follows Raoult's law, the other component will, too.

SOLUTION

$$x_1 d\mu_1 + x_2 d\mu_2 = 0$$

If $\mu_1 = \mu_1^\circ + RT\ln x_1$

$$d\mu_1 = \frac{RT}{x_1}\,dx_1$$

Using equation 1

$$d\mu_2 = -\frac{x_1}{x_2}\,d\mu_1 = \frac{RT}{x_2}\,dx_1$$

Since $x_1 + x_2 = 1$, $dx_2 = -dx_1$

$$d\mu_2 = \frac{RT}{x_2}\,dx_2 = RT\,d\ln x_2$$

$$\mu_2 = \text{const} + RT\ln x_2$$

If $x_2 = 1$, $\text{const} = \mu_2^\circ$

$$\mu_2 = \mu_2^\circ + RT\ln x_2$$

4.17 What are the entropy change and Gibbs energy change on mixing to produce a benzene-toluene solution with 1/3 mole fraction benzene at 25 °C?

SOLUTION

$$\Delta S_{mix} = -R(\tfrac{1}{3}\ell n\tfrac{1}{3} + \tfrac{2}{3}\ell n\tfrac{2}{3}) = 5.293 \text{ J K}^{-1} \text{ mol}^{-1}$$

$$\begin{aligned}\Delta G_{mix} &= -T\Delta S_{mix} \\ &= (298.15 \text{ K})(5.293 \text{ J K}^{-1} \text{ mol}^{-1}) \\ &= -1577 \text{ J mol}^{-1}\end{aligned}$$

4.18 At 100 °C benzene has a vapor pressure of 180.9 kPa, and toluene has a vapor pressure of 74.4 kPa. Assuming that these substances form ideal binary solutions with each other, calculate the composition of the solution that will boil at 1 bar at 100 °C and the vapor composition.

SOLUTION

$$x_B(180.9 \text{ kPa}) + (1 - x_B)(74.4 \text{ kPa}) = 100 \text{ kPa}$$

$$x_B = 0.240$$

$$y_B = \frac{(0.240)(180.9 \text{ kPa})}{(0.240)(180.9 \text{ kPa}) + (0.760)(74.4 \text{ kPa})}$$

$$= 0.434$$

4.19 The vapor pressures of benzene and toluene have the following values in the temperature range between their boiling points at 1 bar:

$t/°C$	79.4	88	94	100	110.0
$P^{sat}_{C_6H_6}$/bar	1.000	1.285	1.526	1.801	
$P^{sat}_{C_7H_8}$/bar		0.508	0.616	0.742	1.000

(a) Calculate the compositions of the vapor and liquid phases at each temperature and plot the boiling point diagram. (b) If a solution containing 0.5 mole fraction benzene and 0.5 mole fraction toluene is heated, at what temperature will the first bubble of vapor appear and what will be its composition?

SOLUTION

(a) At 88 °C $x_B(1.285) + (1 - x_B)\ 0.508 = 1$

$x_B = 0.633$

$y_B = (0.633)(1.285)/1 = 0.814$

At 94 °C $x_B(1.526) + (1 - x_B)\ 0.616 = 1$

$x_B = 0.422$

$y_B = (0.422)(1.526)/1 = 0.644$

100 °C $x_B(1.801) + (1 - x_B)\ 0.742 = 1$

$x_B = 0.244$

$y_B = (0.244)(1.801)/1 = 0.439$

(b) Please see page 58.

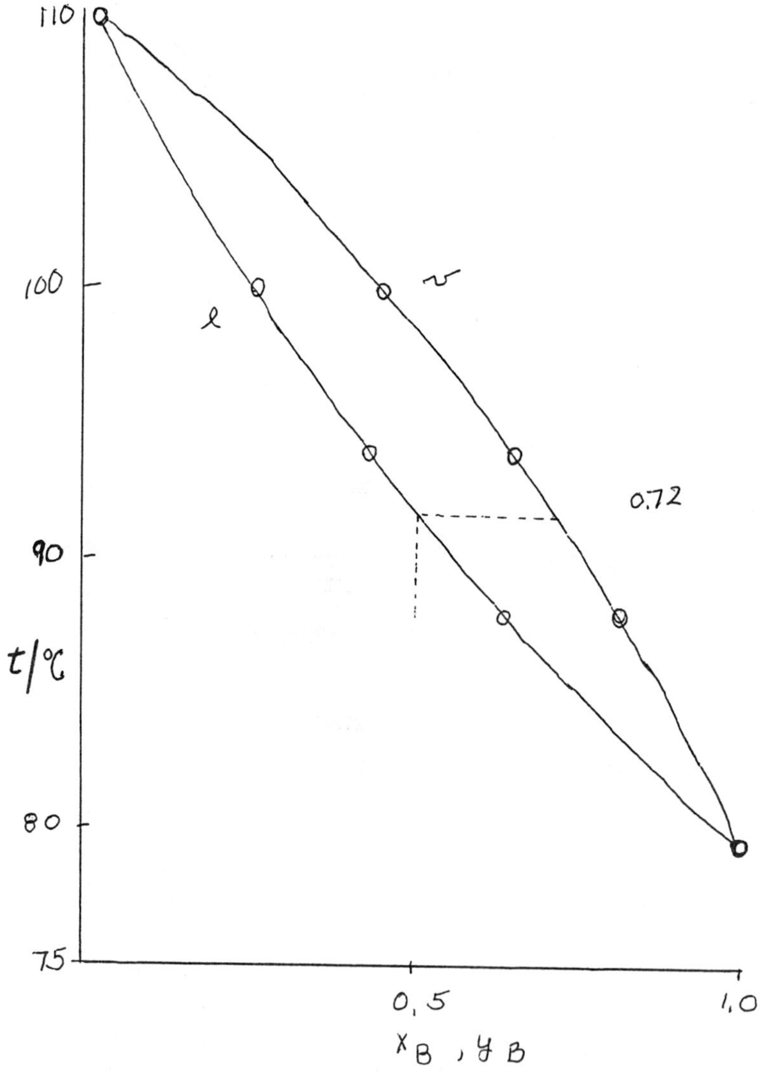

Bubble point = 92 °C $\qquad$ y_B = 0.72

4.20 The following table gives mole % acetic acid in aqueous solutions and in the equilibrium vapor at the boiling point of the solution at 1.013 bar.

B.P., °C	118.1	113.8	107.5	104.4	102.1	100.0
Mole % of acetic acid						
In liquid	100	90.0	70.0	50.0	30.0	0
In vapor	100	83.3	57.5	37.4	18.5	0

Calculate the minimum number of theoretical plates for the column required to produce an initial distillate of 28 mole % acetic acid from a solution of 80 mole % acetic acid.

SOLUTION

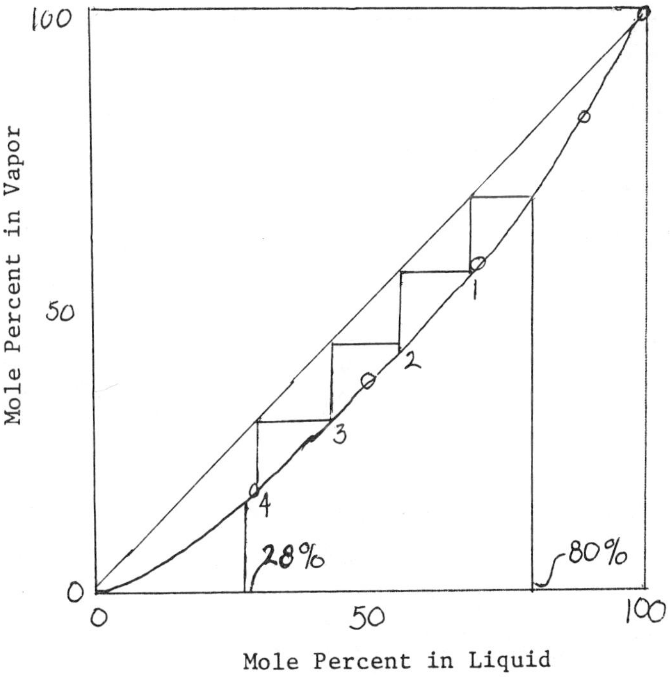

Since there are four steps, three theoretical plates are required in the column. The distilling pot counts as one plate.

4.21 If two liquids (1 and 2) are completely immiscible, the mixture will boil when the sum of the two partial pressures exceeds the applied pressure: $P = P_1^{sat} + P_2^{sat}$. In the vapor phase the ratio of the mole fractions of the two components is equal to the ratio of their vapor pressures.

$$\frac{P_1^{sat}}{P_2^{sat}} = \frac{x_1}{x_2} = \frac{g_1 M_2}{g_2 M_1}$$ where g_1 and g_2 are

the masses of components 1 and 2 in the vapor phase, and M_1 and M_2 are their molar masses. The boiling point of the immiscible liquid system naphthalene-water is 98 °C under a pressure of 97.7 kPa. The vapor pressure of water at 98 °C is 94.3 kPa. Calculate the weight percent of naphthalene in the distillate.

SOLUTION

$$\frac{g_1}{g_2} = \frac{P_1{}^\circ M_1}{P_2{}^\circ M_2} = \frac{(97.7 - 94.3)(128)}{(94.3)(18)} = 0.261$$

$$\text{Weight percent napthalene} = \frac{0.261}{1.261} = 0.207 \text{ or } 20.7\%$$

4.22 The protein human plasma albumin has a molar mass of 69,000 g mol^{-1}. Calculate the osmotic pressure of a solution of this protein containing 2 g per 100 cm^3 at 25 °C in (a) pascals, and (b) millimeters of water. The experiment is carried out using a salt solution for solvent and a membrane permeable to salt.

SOLUTION $$\pi = \frac{cRT}{M}$$

$$= \frac{(20\times10^{-3}\text{kgL}^{-1})(10^{3}\text{Lm}^{-3})(8.314\text{JK}^{-1}\text{mol}^{-1})(298.15\text{K})}{69 \text{ kg mol}^{-1}}$$

$$= 719 \text{ Pa}$$

$$h = \frac{\pi}{dg} = \frac{719 \text{ Pa}}{(1\text{gcm}^{-3})(10^{-3}\text{kg g}^{-1})(10^{2}\text{cm m}^{-1})(9.8\text{m s}^{-2})}$$

$$= 7.34 \times 10^{-2} \text{ m}$$

$$= 73.4 \text{ mm of water}$$

4.23 The following osmotic pressures were measured for solutions of a sample of polyisobutylene in benzene at 25 °C.

$c/10^{-2}$ g cm^{-3}	0.500	1.00	1.50	2.00
π/g cm^{-2}	0.505	1.03	1.58	2.15

Calculate the number average molar mass from the value of π/c extrapolated to zero concentration of the polymer.

SOLUTION $$\frac{\pi}{c} = \frac{RT}{M} + Bc$$

The ratio π/c is plotted versus c and extrapolated to $c = 0$.

c/g/100 cm^3	0.50	1.0	1.50	2.00
$\frac{\pi(100)}{c\ 1028}$/$\frac{\text{atm cm}^3}{\text{g}}$	0.0982	0.1002	0.1025	0.1046

$$\left(\frac{\pi}{c}\right)_{c=0} = \frac{(9.60\times10^{-2}\text{atmcm}^3\text{g}^{-1})(101325\text{Pa atm}^{-1}) \times}{(10^{3}\text{gkg}^{-1})(10^{-2}\text{mcm}^{-1})^{3}}$$

$$= 9.73 \text{ Pa(mol m}^{-3})^{-1}$$

$$\left(\frac{\pi}{c}\right)_{c=0} = \frac{RT}{M}$$

$$M = \frac{RT}{\left(\frac{\pi}{c}\right)_{c=0}}$$

$$M = \frac{(8.314\ \text{J K}^{-1}\ \text{mol}^{-1})(298\ \text{K})}{9.73\ \text{Pa}\ (\text{mol m}^{-3})^{-1}}$$

$$= 255\ \text{kg mol}^{-1}$$

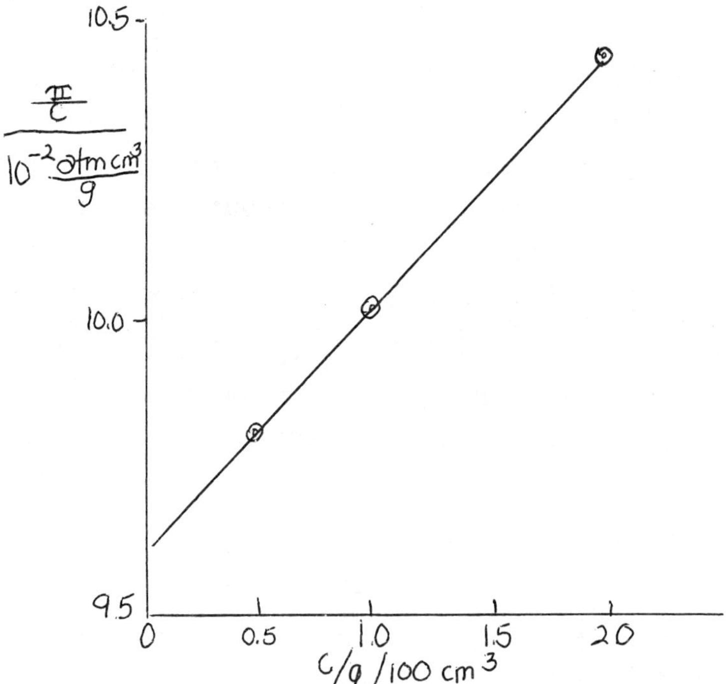

4.24 A sample of polymer contains 0.50 mole fraction with molar mass 100,000 g mol^{-1} and 0.50 mole fraction with molar mass 200,000 g mol^{-1}. Calculate (a) M_n, and (b) M_m.

SOLUTION

(a) $M_n = \frac{\Sigma n_i M_i}{\Sigma n_i} = \Sigma x_i M_i$ where x_i is mole fraction of the ith component

$$= 0.5(10^5) + 0.5(2x10^5) = 150,000 \text{ g mol}^{-1}$$

(b) $$M_m = \frac{\Sigma n_i M_i^2}{\Sigma n_i M_i}$$

Dividing numerator and denominator by Σn_i

$$M_m = \frac{\Sigma x_i M_i^2}{\Sigma x_i M_i} = \frac{0.5(10^{10}) + 0.5(4x10^{10})}{0.5(10^5) + 0.5(2x10^5)}$$

$$= 167,000 \text{ g mol}^{-1}$$

4.25 Calculate the osmotic pressure of a 1 mol L^{-1} sucrose solution in water from the fact that at 30 °C the vapor pressure of the solution is 4.1606 kPa. The vapor pressure of water at 30 °C is 4.2429 kPa. The density of pure water at this temperature (0.99564 g cm^{-3}) may be used to estimate $\overline{V}_1$ for a dilute solution. To do this problem, Raoult's law is introduced into equation 4.52.

SOLUTION $\overline{V}_1 \pi = RT\, \ell n\, x_1$

Substituting Raoult's law $P_1 = x_1 P_1°$

$$\pi = -\frac{RT}{\overline{V}_1} \ell n \frac{P_1}{P_1°}$$

$$\overline{V}_1 = \frac{18.02 \text{ g mol}^{-1}}{0.99564 \text{ g cm}^{-3}} = 1810 \text{ cm}^3 \text{ mol}^{-1}$$

$$= 0.01810 \text{ L mol}^{-1}$$

$$\pi = \frac{(0.08314\text{L bar K}^{-1}\text{mol}^{-1})(303.15\text{K})}{0.0180 \text{ L mol}^{-1}} \ell n \frac{4.2429\text{kPa}}{4.1606\text{kPa}}$$

$$= 27.3 \text{ bar}$$

4.26 For a solution of n-propanol and water, the following partial pressures in kPa are measured at

25 °C. Draw a complete pressure-composition diagram, including the total pressure. What is the composition of the vapor in equilibrium with a solution containing 0.5 mole fraction of n-propanol?

$x_{n\text{-propanol}}$	P_{H_2O}	$P_{n\text{-propanol}}$	$x_{n\text{-propanol}}$	P_{H_2O}	$P_{n\text{-propanol}}$
0	3.168	0.00	0.600	2.65	2.07
0.020	3.13	0.67	0.800	1.79	2.37
0.050	3.09	1.44	0.900	1.08	2.59
0.100	3.03	1.76	0.950	0.56	2.77
0.200	2.91	1.81	1.000	0.00	2.901
0.400	2.89	1.89			

SOLUTION

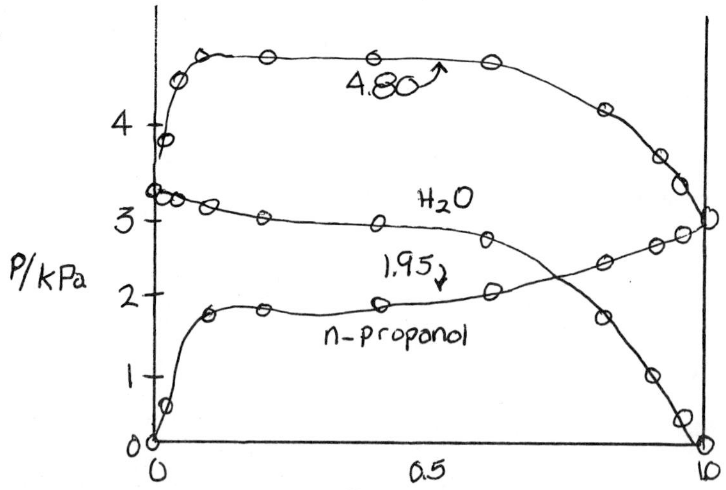

$x_{n\text{-propanol}}$

$$x_{vapor} = \frac{1.95 \text{ kPa}}{4.80 \text{ kPa}} = 0.406$$

4.27 Using the Henry law constants in Table 4.3, calculate the percentage (by volume) of oxygen and nitrogen in air dissolved in water at 25 °C. The air in equilibrium with the water at 1 bar pressure may be considered to be 20% oxygen and 80% nitrogen by volume.

SOLUTION

$$x_{N_2} = \frac{P_{N_2}}{K_{N_2}} = \frac{0.80 \times 10^5 \text{ Pa}}{8.68 \times 10^9 \text{ Pa}} = 9.22 \times 10^{-6}$$

$$x_{O_2} = \frac{P_{O_2}}{K_{O_2}} = \frac{0.20 \times 10^5 \text{ Pa}}{4.40 \times 10^9 \text{ Pa}} = 4.55 \times 10^{-6}$$

The percentage of oxygen in this dissolved air is given by

$$\frac{X_{O_2}}{X_{O_2} + X_{N_2}} \times 100 = \frac{4.55 \times 10^{-4}}{(4.55 + 9.22) \times 10^{-6}} = 33.0\%$$

The percentage of nitrogen is 100 - 33.0 = 67.0%

4.28 The following data on ethanol-chloroform solutions at 35 °C were obtained by G. Scatchard and C. L. Raymond [J. Am. Chem. Soc., 60, 1278 (1938)]:

$x_{EtOH,liq}$	0	0.2	0.4	0.6	0.8	1.0
$x_{EtOH,vap}$	0.0000	0.1382	0.1864	0.2554	0.4246	1.0000
Total pressure, kPa	39.345	40.559	38.690	34.387	25.357	13.703

Calculate the activity coefficients of ethanol in these solutions according to Convention I.

SOLUTION

$$\gamma_{EtOH} = \frac{x_{vap}\, P_{vap}}{x_{liq}\, P^\circ_{liq}}$$

$x_{EtOH} = 0.2$

$$= \frac{(0.1382)(40.559)}{(0.2)(13.703)} = 2.045$$

$$x_{EtOH} = 0.4 \quad \gamma_{EtOH} = \frac{(0.1864)(38.690)}{(0.4)(13.703)} = 1.316$$

$$x_{EtOH} = 0.6 \quad \gamma_{EtOH} = \frac{(0.2554)(34.287)}{(0.6)(13.703)} = 1.065$$

$$x_{EtOH} = 0.8 \quad \gamma_{EtOH} = \frac{(0.4246)(25.357)}{(0.8)(13.703)} = 0.982$$

$$x_{EtOH} = 1.0 \quad \gamma_{EtOH} = \frac{(1.000)(13.703)}{(1.000)(13.703)} = 1.000$$

4.29 Using the data in Problem 4.26, calculate the activity coefficients of water and n-propanol at 0.20, 0.40, 0.60, and 0.80 mole fraction n-propanol, using Convention II and considering n-propanol to be the solvent.

<u>SOLUTION</u>

$$x_1 = 0.2 \quad \gamma_1 = \frac{1.81}{(0.2)(2.901)} = 3.12$$

$$x_1 = 0.4 \quad \gamma_1 = \frac{1.89}{(0.4)(2.901)} = 1.63$$

$$x_1 = 0.6 \quad \gamma_1 = \frac{2.07}{(0.6)(2.901)} = 1.19$$

$$x_1 = 0.8 \quad \gamma_1 = \frac{2.37}{(0.8)(2.901)} = 1.02$$

To obtain the Henry law constant for H_2O (Component 2) plot P_2/x_2 versus x_2.

At $x_2 = 0.05$, $\frac{P_2}{x_2} = \frac{0.56}{0.05} = 11.2$

At $x_2 = 0.10$, $\frac{P_2}{x_2} = \frac{1.08}{0.1} = 10.8$

Thus as $x_2 \longrightarrow 0, P_2 = 11.6\ x_2$

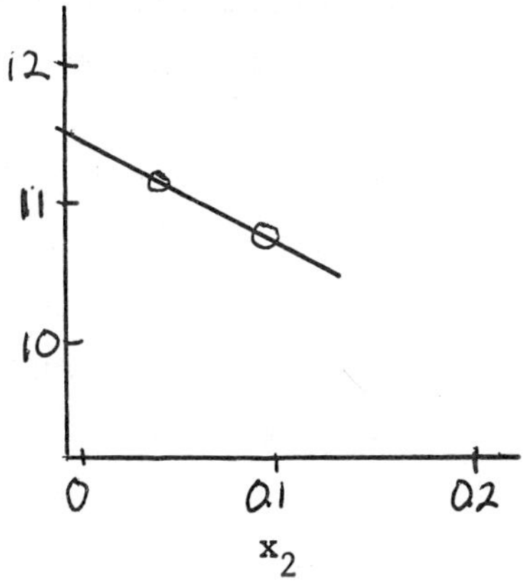

$\left.\begin{matrix} x_2=0.2 \\ x_1=0.8 \end{matrix}\right\} \gamma_2 = \frac{1.79}{0.2(11.6)} = 0.772$

$\left.\begin{matrix} x_2=0.4 \\ x_1=0.6 \end{matrix}\right\} \gamma_2 = \frac{2.65}{0.4(11.6)} = 0.571$

$\left.\begin{matrix} x_2=0.6 \\ x_1=0.4 \end{matrix}\right\} \gamma_2 = \frac{2.89}{0.6(11.6)} = 0.415$

$\left.\begin{matrix} x_2=0.8 \\ x_1=0.2 \end{matrix}\right\} \gamma_2 = \frac{2.91}{0.8(11.6)} = 0.314$

4.30 Show that the equations for the bubble-point line and dew-point line for nonideal solutions are given by

$$x_1 = \frac{P - \gamma_2 P_2^{sat}}{\gamma_1 P_1^{sat} - \gamma_2 P_2^{sat}}$$

$$y_1 = \frac{P\gamma_1 P_1^{sat} - \gamma_1\gamma_2 P_1^{sat} P_2^{sat}}{P\gamma_1 P_1^{sat} - P\gamma_2 P_2^{sat}}$$

<u>SOLUTION</u>

$$P = P_1 + P_2 = x_1\gamma_1 P_1^{sat} + (1 - x_1)\gamma_2 P_2^{sat}$$

$$= \gamma_2 P_2^{sat} + x_1(\gamma_1 P_1^{sat} - \gamma_2 P_2^{sat})$$

$$x_1 = \frac{P - \gamma_2 P_2^{sat}}{\gamma_1 P_1^{sat} - \gamma_2 P_2^{sat}}$$

$$y_1 = \frac{P_1}{P_1 + P_2} = \frac{x_1\gamma_1 P_1^{sat}}{\gamma_2 P_2^{sat} + x_1(\gamma_1 P_1^{sat} - \gamma_2 P_2^{sat})}$$

Substituting the expression for x_1 yields the desired expression for y_1.

4.31 Calculate the solubility of p-dibromobenzene in benzene at 20 °C and 40 °C assuming ideal solutions are formed. The enthalpy of fusion of p-dibromobenzene is 13.22 kJ mol^{-1} at its melting point (86.9 °C).

SOLUTION

At 20 °C $$\ell n x_2 = \frac{\Delta H_{2f}(T - T_{2f})}{RTT_{2f}}$$

$$= \frac{(13{,}220 \text{ J mol}^{-1})(-66.9 \text{ K})}{(8.314 \text{ J K}^{-1} \text{ mol}^{-1})(360.1 \text{ K})(293.2 \text{ K})}$$

$$x_2 = 0.365$$

At 40 °C $$\ell n x_2 = \frac{(13{,}220 \text{ J mol}^{-1})(-46.9 \text{ K})}{(8.314 \text{ J K}^{-1} \text{ mol}^{-1})(360.1 \text{ K})(313.2 \text{ K})}$$

$$x_2 = 0.516$$

4.32 Calculate the solubility of naphthalene at 25 °C in any solvent in which it forms an ideal solution. The melting point of naphthalene is 80 °C, and the heat of fusion is 19.29 kJ mol^{-1}. The actual measured solubility of naphthalene in benzene is $x_1 = 0.296$.

SOLUTION

$$-\ell n x_1 = \frac{\Delta H_{fus,1}(T_{0,1} - T)}{RTT_{0,1}}$$

$$\ln x_1 = \frac{-(19{,}290\ \text{J mol}^{-1})(353\ \text{K} - 298\ \text{K})}{(8.314\ \text{J K}^{-1}\ \text{mol}^{-1})(353\ \text{K})(298\ \text{K})}$$

$$x_1 = 0.297$$

4.33 If 68.4 g of sucrose ($M = 342\ \text{g mol}^{-1}$) is dissolved in 1000 g of water: (a) What is the vapor pressure at 20 °C? (b) What is the freezing point? The vapor pressure of water at 20 °C is 2.3149 kPa.

SOLUTION

(a) $$x_2 = \frac{\frac{68.4}{342}}{\frac{68.4}{342} + \frac{1000}{18}} = 3.59 \times 10^{-3}$$

$$\frac{P_1^\circ - P_1}{P_1^\circ} = x_2$$

$$\frac{2.3149 - P_1}{2.3149} = 3.59 \times 10^{-3}$$

$$P_1 = 2.3149(1 - 3.59 \times 10^{-3})$$

$$= 2.3066\ \text{kPa}$$

(b) $$\Delta T_f = K_f m = 1.86\ (0.2) = -0.372$$

$$T_f = -0.372\ ^\circ\text{C}$$

4.34 The phase diagram for magnesium-copper at constant pressure shows that two compounds are formed: $MgCu_2$ which melts at 800 °C, and Mg_2Cu which melts at 580 °C. Copper melts at 1085 °C, and Mg at 648 °C. The three eutectics are at 9.4% by weight Mg(680 °C), 34% by weight Mg (560 °C), and 65% by weight Mg(380 °C). Construct the phase diagram. State the variance for each area and eutectic point.

SOLUTION

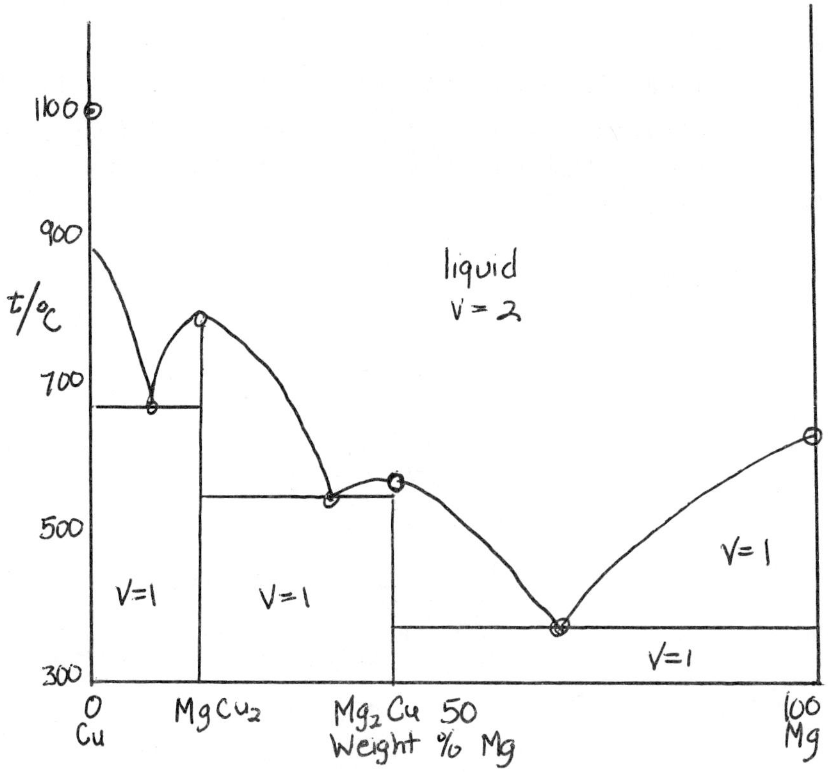

$v = 2 - p + 1$ In the liquid region $v = 2$, in the two-phase regions $v = 1$, and at the eutectic points $v = 0$.

4.35 For the ternary system benzene-isobutanol-water at 25 °C and 1 bar the following compositions have been obtained for the two phases in equilibrium.

Water-Rich Phase		Benzene-Rich Phase	
Isobutanol, wt.%	Water, wt.%	Isobutanol, wt.%	Benzene, wt.%
2.33	97.39	3.61	96.20
4.30	95.44	19.87	79.07
5.23	94.59	39.57	57.09
6.04	93.83	59.48	33.98
7.32	92.64	76.51	11.39

Plot these data on a triangular graph, indicating the tie lines. (a) Estimate the compositions of the phases that will be produced from a mixture of 20% isobutanol, 55% water, and 25% benzene. (b) What will be the composition of the principal phase when the first drop of the second phase separates when water is added to a solution of 80% isobutyl alcohol in benzene?

SOLUTION

(a) H_2O layer:

5.23% isobutanol
94.5% H_2O

Benzene layer:

39.57% isobutanol
57.09% benzene

(b) 10% H_2O

72% isobutanol
18% benzene

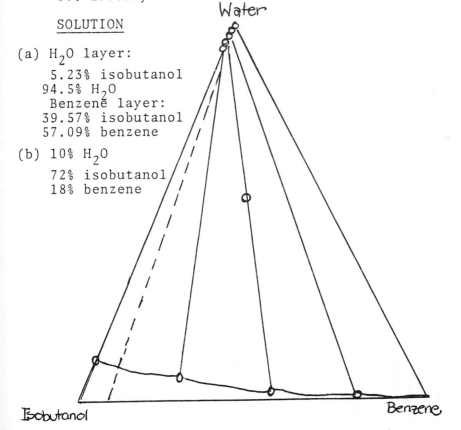

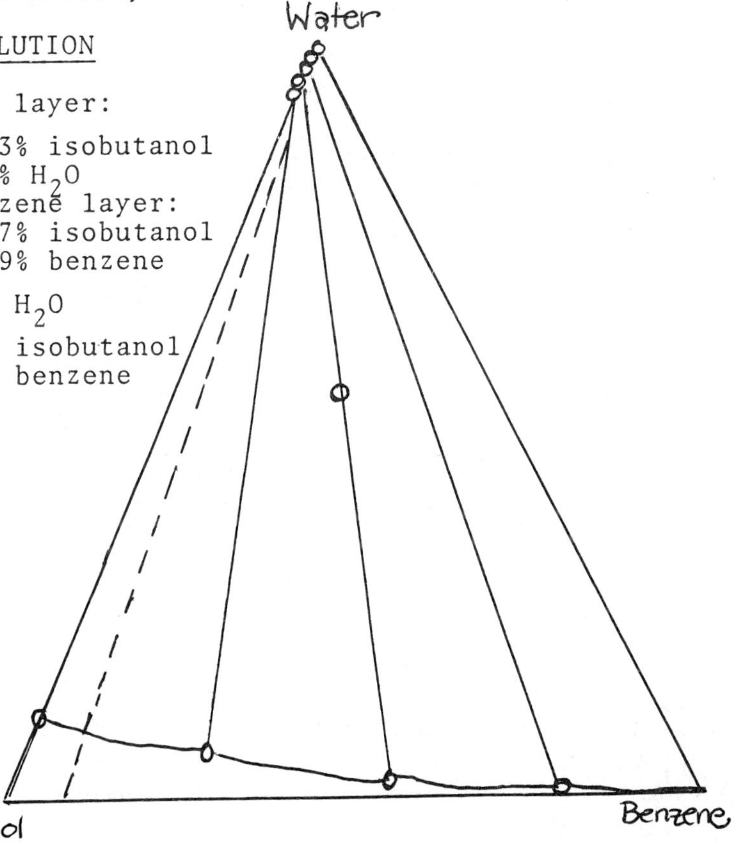

4.36 The following data are available from the system nickel sulfate-sulfuric acid-water at 25 °C. Sketch the phase diagram on triangular coordinate paper, and draw appropriate tie lines.

Liquid Phase		Solid Phase
$NiSO_4$,wt.%	H_2SO_4,wt.%	
28.13	0	$NiSO_4 \cdot 7H_2O$
27.34	1.79	$NiSO_4 \cdot 7H_2O$
27.16	3.86	$NiSO_4 \cdot 7H_2O$
26.15	4.92	$NiSO_4 \cdot 6H_2O$
15.64	19.34	$NiSO_4 \cdot 6H_2O$
10.56	44.68	$NiSO_4 \cdot 6H_2O$
9.65	48.46	$NiSO_4 \cdot H_2O$
2.67	63.73	$NiSO_4 \cdot H_2O$
0.12	91.38	$NiSO_4 \cdot H_2O$
0.11	93.74	$NiSO_4$
0.08	96.80	$NiSO_4$

SOLUTION

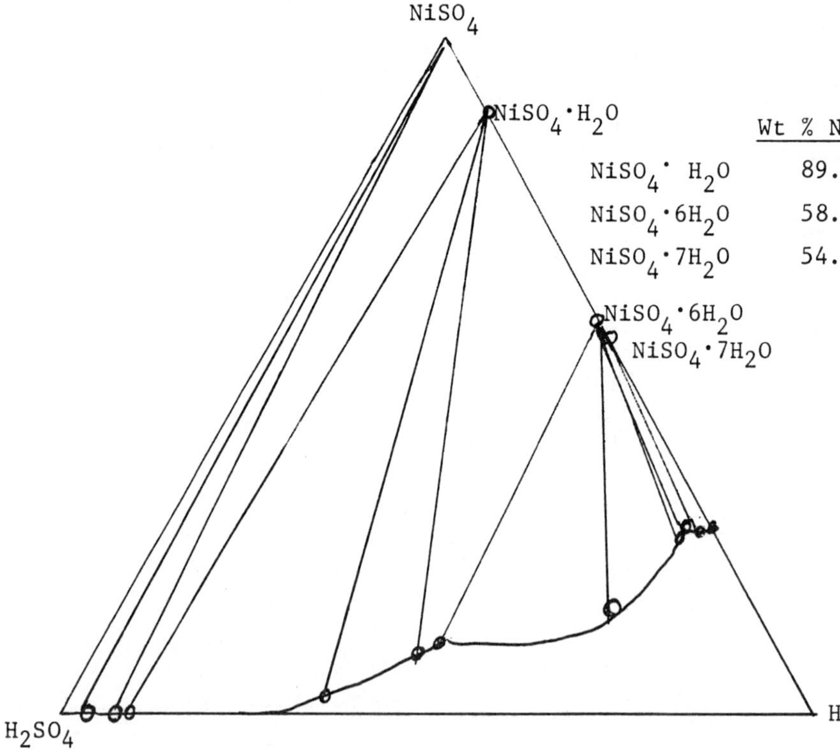

	Wt % $NiSO_4$
$NiSO_4 \cdot H_2O$	89.6
$NiSO_4 \cdot 6H_2O$	58.8
$NiSO_4 \cdot 7H_2O$	54.9

4.37 4

4.38 $0.274\ g\ cm^{-3}$ 131 °C

4.39 0.28

4.40 (a) -38.81 °C
(b) -16.6 °C

4.41 1.67 kPa
0.167 kPa

4.42 96 °C

4.43 (a) 2.82 kJ
(b) $50.1\ Pa\ K^{-1}$
(c) 360 Pa 387 Pa

4.44 (a) 1623 kPa
(b) 1428 kPa

4.45 0.076 Pa

4.46 $27.57\ kJ\ mol^{-1}$

4.47 (a) 5.8 °C 3.28 kPa (b) $9.90\ kJ\ mol^{-1}$

4.49 (a) $x_{CHCl_3} = 0.635$ $x_{CCl_4} = 0.365$
(b) 20.92 kPa

4.50 (a) $x_{BrCH_2CH_2Br} = 0.802$
(b) $x_{BrCH_2CH_2Br} = 0.425$

4.51 (a) 3.1615 kPa (b) 3.1615 kPa
(c) 3.1557 kPa (d) 3.1615 kPa

4.52 (a) $x_B = 0.72$ $x_T = 0.28$ 3.162 kPa
(b) $x_B = 0.537$

4.53 0.194 24.67 kPa

4.54 $\Delta G_{mix} = RT(x_1 \ell n x_1 + x_2 \ell n x_2) + w x_1 x_2$

$\Delta S_{mix} = -R(x_1 \ell n x_1 + x_2 \ell n x_2)$

$\Delta H_{mix} = w x_1 x_2$ $\Delta V_{mix} = 0$

4.55 (a) At -31.2 °C $x_{propane} = 0.560$
At -16.3 °C $x_{propane} = 0.196$
(b) At -31.2 °C $x_{propane,vap} = 0.885$
At -16.3 °C $x_{propane,vap} = 0.578$

4.56 (a) $x_{EtOH} = 0.69$ (b) $x_{EtOH} = 1.00$
(c) $x_{EtOH} = 0.69$ (d) $x_{EtOH} = 0.88$

4.57 (a) $x_{n\text{-propanol, vap}} = 0.37$
(b) $x_{n\text{-propanol, vap}} = 0.59$

4.58 122,000 g mol^{-1}

4.59 1810 Pa

4.60 (a) 36,800 g mol^{-1}
(b) 48,500 g mol^{-1}

4.61 $x_{C_6H_6} > 0.55$

4.62 562 g

4.63 2.36×10^{-3} K

4.64

$x_{n\text{-propanol}}$	0.2	0.4	0.6	0.8
γ_{H_2O}	1.14	1.52	2.09	2.82
$\gamma_{n\text{-propanol}}$	0.248	0.129	0.094	0.081

4.65

x_{CHCl_3}	0.2	0.4	0.6	0.8
γ_{CHCl_3}	0.58	0.70	0.84	0.96
$\gamma_{acetone}$	0.98	0.89	0.74	0.61

4.66 $\gamma_{acetone} = 1.67$ $f_{cs_2} = 1.38$

4.68 and 4.69: Please see page 75

4.70 See solution for 4.69

4.71 0.108

4.72 $x_{Cd} = 0.844$ or 74.4 g/100g of solution

4.73 251 °C

4.74 12.7 kJ mol^{-1}

4.75 and 4.76: Please see page 76

4.68

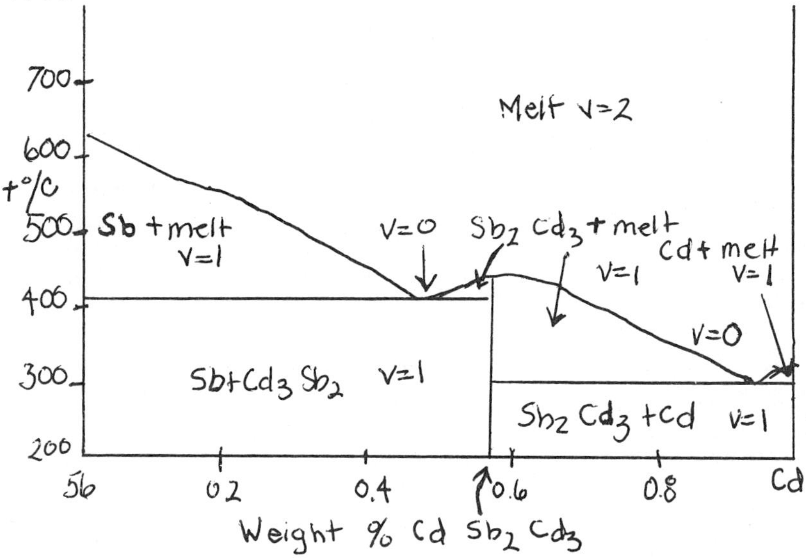
Melt V=2
Sb + melt
V=1
V=0
Sb2 Cd3 + melt
V=1
Cd + melt
V=1
V=0
Sb+Cd3 Sb2
V=1
Sb2 Cd3 + Cd
V=1
700
600
t°/C
500
400
300
200
Sb
0.2
0.4
0.6
0.8
Cd
Weight % Cd Sb2 Cd3

4.69

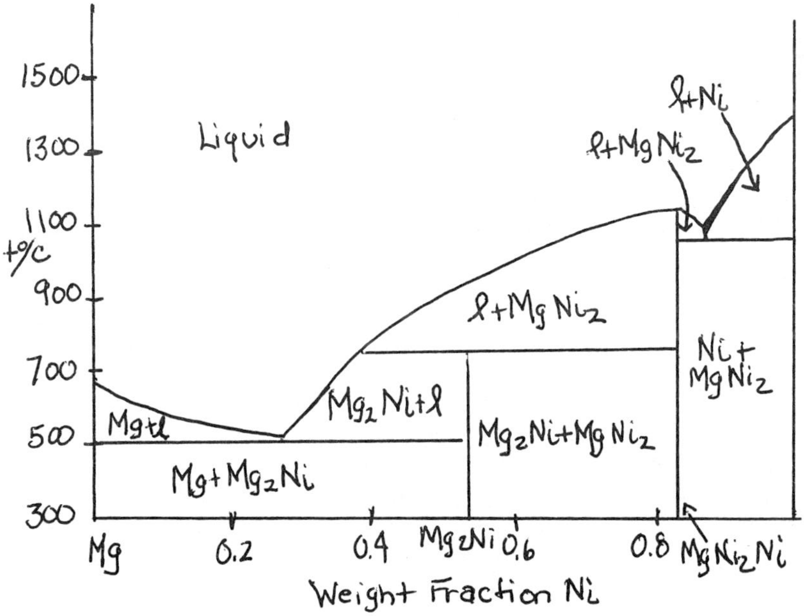
Liquid
l+Ni
l+Mg Ni2
l+Mg Ni2
Ni+
Mg Ni2
Mg2 Ni+l
Mg+l
Mg2Ni+Mg Ni2
Mg+Mg2Ni
1500
1300
1100
t°/C
900
700
500
300
Mg
0.2
0.4
Mg2Ni
0.6
0.8
MgNi2
Ni
Weight Fraction Ni

4.75

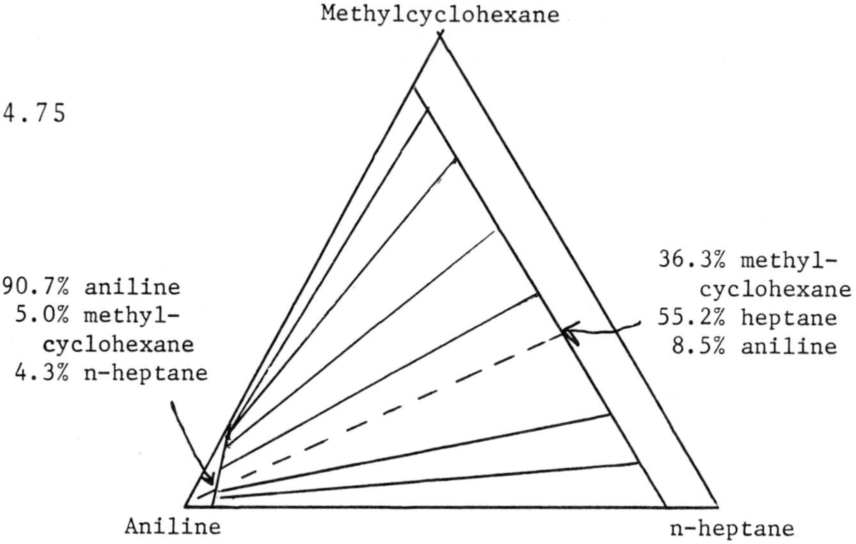
Methylcyclohexane
90.7% aniline
5.0% methyl-
cyclohexane
4.3% n-heptane
36.3% methyl-
cyclohexane
55.2% heptane
8.5% aniline
Aniline
n-heptane

4.76

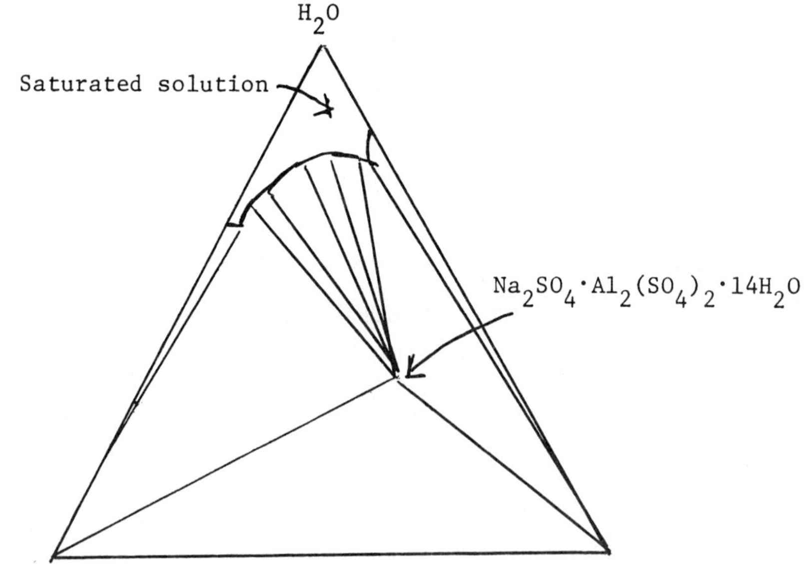
H_2O
Saturated solution
$Na_2SO_4 \cdot Al_2(SO_4)_2 \cdot 14H_2O$
Na_2SO_4
$Al_2(SO_4)_3$

CHAPTER 5. Chemical Equilibrium

5.1 For the reaction $N_2(g) + 3H_2(g) = 2NH_3(g)$ $K_P = 1.60 \times 10^{-4}$ at 400 °C. Calculate (a) $\Delta G°$ and (b) ΔG when the pressures of N_2 and H_2 are maintained at 10 and 30 bar, respectively, and NH_3 is removed at a partial pressure of 3 bar. (c) Is the reaction spontaneous under the latter conditions?

SOLUTION

(a) $$\Delta G° = -RT\ln K_P$$
$$= -(8.314\ JK^{-1}mol^{-1})(673K)\ \ln 1.60 \times 10^{-4}$$
$$= 48.9\ kJ\ mol^{-1}$$

(b) $$\Delta G = \Delta G° + RT\ \ln \frac{P^2_{NH_3}}{P_{N_2}\ P^3_{H_2}}$$
$$= 48.9 + (8.314 \times 10^{-3})(673)\ \ln \frac{3^3}{10(30)^3}$$
$$= 10.2\ kJ\ mol^{-1}$$

(c) No

5.2 A 1:3 mixture of nitrogen and hydrogen was passed over a catalyst at 450 °C. It was found that 2.04% by volume of ammonia was formed when the total pressure was maintained at 10.13 bar [A. T. Larson and R. L. Dodge, J. Am. Chem. Soc., 45, 2918 (1923)]. Calculate the value of K_P for $\frac{3}{2}H_2(g) + \frac{1}{2}N_2(g) = NH_3(g)$ at this temperature.

SOLUTION

At equilibrium $P_{H_2} + P_{N_2} + P_{NH_3} = 10.13$ bar

$P_{N_2} = (10.13 \text{ bar})(0.0204) = 0.207$ bar

$P_{H_2} + P_{N_2} = 10.13 \text{ bar} - 0.207 \text{ bar} = 9.923$ bar

$P_{H_2} = 3P_{N_2}$ because this initial ratio is not changed by reaction

$$P_{N_2} = \frac{9.923 \text{ bar}}{4} = 2.481 \text{ bar}$$

$$P_{H_2} = \frac{3}{4}(9.923 \text{ bar}) = 7.442 \text{ bar}$$

$$K_P = \frac{\left(P_{NH_3}/P°\right)}{\left(P_{H_2}/P°\right)^{3/2}\left(P_{N_2}/P°\right)^{1/2}}$$

$$= \frac{0.207}{7.442^{3/2}\ 2.481^{1/2}} = 6.47 \times 10^{-3}$$

5.3 Water vapor is passed over coal (assumed to be pure graphite in this problem) at 1000 K. Assuming that the only reaction occurring is the water gas reaction,

$$C(\text{graphite}) + H_2O(g) = CO(g) + H_2(g) \qquad K_P = 2.52$$

calculate the equilibrium pressures of H_2O, CO, and H_2 at a total pressure of 1 bar. (Actually the water gas shift reaction $CO(g) + H_2O(g) = CO_2(g) + H_2(g)$ occurs in addition, but it is considerably more complicated to take this subsequent reaction into account.)

SOLUTION

$$K_P = \frac{(P_{CO}/P°)(P_{H_2}/P°)}{(P_{H_2O}/P°)} = \frac{x^2}{y} = \frac{x^2}{1-2x} = 2.52$$

$$2x + y = 1$$
$$x^2 = 2.52 - 5.04x$$
$$x^2 + 5.04x - 2.52 = 0$$
$$x = \frac{-5.04 \pm \sqrt{5.04^2 + 4(2.52)}}{2} = 0.458 = \frac{P_{CO}}{P°} = \frac{P_{H_2}}{P°}$$
$$1 - 2x = 0.084 = \frac{P_{H_2O}}{P°}$$

$P_{H_2O} = 0.084$ bar $\quad P_{CO} = 0.458$ bar $\quad P_{H_2} = 0.458$ bar

5.4 How many moles of phosphorus pentachloride must be added to a liter vessel at 250 °C to obtain a concentration of 0.1 mol of chlorine per liter?

Given:
$$K_P = \frac{(P_{PCl_3}/P°)(P_{Cl_2}/P°)}{(P_{PCl_5}/P°)} = 1.80$$

SOLUTION

$$K_c = K_P\left(\frac{P°}{c°RT}\right)^{\Sigma\nu_i}$$
$$= 1.80\left[\frac{10^5\ \text{Pa}}{(10^3\text{mol m}^{-3})(8.314\text{JK}^{-1}\text{mol}^{-1})(523.1\text{K})}\right]$$
$$= 4.14 \times 10^{-2} = \frac{(0.1)^2}{x - 0.1}$$
$$x = 0.342\ \text{mol L}^{-1}$$

5.5 Calculate the total pressure that must be applied to a mixture of three parts of hydrogen and one part nitrogen to give a mixture containing 10% ammonia at equilibrium at 400 °C. At 400 °C, $K_P = 1.60 \times 10^{-4}$ for the reaction
$N_2(g) + 3H_2(g) = 2NH_3(g)$.

SOLUTION

$$K_P = \frac{(P_{NH_3}/P°)^2}{(P_{N_2}/P°)(P_{H_2}/P°)} = 1.60 \times 10^{-4} \quad \underline{\text{or}}$$

$$K_P = \frac{P_{NH_3}^2}{P_{N_2}\ P_{H_2}^3} = 1.60 \times 10^{-4}\ \text{bar}^{-2}$$

$$= \frac{(0.1P)^2}{(\frac{1}{4}0.9P)(\frac{3}{4}0.9P)^3} = \frac{(0.01)4^4}{3^3(0.9)^4\ P^2}$$

$P = 30.1$ bar

5.6 In the synthesis of methanol by $CO(g) + 2H_2(g) = CH_3OH(g)$ at 500 K, calculate the total pressure required for a 90% conversion to methanol if CO and H_2 are initially in a 1:2 ratio. Given: $K_P = 6.09 \times 10^{-3}$.

SOLUTION

	$CO(g)$ +	$2H_2(g)$ =	$CH_3OH(g)$	
initial moles	1	2	0	
equil. moles	0.1	0.2	0.9	total 1.2

$$K_P = \frac{(P_{CH_3OH}/P°)}{(P_{CO}/P°)(P_{H_2}/P°)^2} = 6.09 \times 10^{-3}$$

$$= \frac{\frac{0.9}{1.2}\frac{P}{P°}}{\frac{0.1}{1.2}\frac{P}{P°}\left(\frac{0.2}{1.2}\frac{P}{P°}\right)^2}$$

$$\frac{P}{P°} = \sqrt{\frac{(0.9)(1.2)^2}{(0.1)(0.04)(6.09 \times 10^{-3})}} = 231$$

$P = 231$ bar = total pressure for 90% conversion to CH_3OH

5.7 An evacuated tube containing 5.96×10^{-3} mol L^{-1} of solid iodine is heated to 973 K. The experimentally determined pressure is 0.496 [M. L. Perlman and G. K. Rollefson, J. Chem. Phys., 9, 362 (1941)]. Assuming perfect-gas behavior, calculate K_P for $I_2(g) = 2I(g)$.

SOLUTION

$$P = \frac{n}{V}RT$$

$$0.496 \text{ bar} = (1+\xi)(5.96\times10^{-3}\text{molL}^{-1})(.08314\text{Lbar K}^{-1}\text{mol}^{-1})(973\text{K})$$

$$\xi = \frac{0.496 \text{ bar}}{(5.96\times10^{-3}\text{molL}^{-1})(.08314\text{Lbar}^{-1}\text{ K}^{-1}\text{mol}^{-1})(973\text{K})} - 1$$

$$= 0.0287$$

$$K_P = \frac{4\xi^2(P/P°)}{1-\xi^2} = \frac{4(0.0287)^2(0.496)}{1 - 0.0287^2}$$

$$= 1.64 \times 10^{-3}$$

5.8 The value of K_P for $N_2O_4(g) = 2NO_2(g)$ is 0.143 at 25 °C. What is the value of K_c?

SOLUTION

$$K_c = K_P\left(\frac{P°}{c°RT}\right)^{\Sigma\nu_i}$$

$$= 0.143\left[\frac{10^5 \text{ Nm}^{-2}}{(10^3\text{molm}^{-3})(8.314\text{JK}^{-1}\text{mol}^{-1})(298.15\text{K})}\right]$$

$$= 5.77 \times 10^{-3}$$

5.9 At 1273 K and at a total pressure of 30.4 bar the equilibrium in the reaction $CO_2(g) + C(s) = 2CO(g)$ is such that 17 mole % of the gas is CO_2. (a) What percentage would be CO_2 if the total pressure were 20.3 bar? (b) What would be the effect on the equilibrium of adding N_2 to the reaction mixture in a closed vessel until the partial pressure of N_2 is 10 bar? (c) At what pressure of the reactants will 25% of the gas be CO_2?

SOLUTION

(a) $P_{CO_2} = (30.4\ \text{bar})(0.17) = 5.2\ \text{bar}$

$P_{CO} = (30.4\ \text{bar})(0.83) = 25.2\ \text{bar}$

$$K_P = \frac{(25.2)^2}{5.2} = 122$$

Let x = mole fraction CO_2

$$K_P = \frac{[20.3(1 - x)]^2}{20.3x} = 122 \qquad x = 0.127$$

Percentage CO_2 at equilibrium = 12.7%

(b) No effect for perfect gases because the partial pressures of the reactants are not affected.

(c) $$K_P = \frac{[0.75(P/P°)]^2}{0.25(P/P°)} = 122$$

$P = 54\ \text{bar}$

5.10 At 2000 °C water is 2% dissociated into oxygen and hydrogen at a total pressure of 1 bar.
(a) Calculate K_P for

$$H_2O(g) = H_2(g) + \tfrac{1}{2}O_2(g)$$

(b) Will the degree of dissociation increase or decrease if the pressure is reduced? (c) Will the degree of dissociation increase or decrease if argon gas is added, holding the total pressure equal to 1 bar? (d) Will the degree of dissociation change if the pressure is raised by addition of argon at constant volume to the closed system containing partially dissociated water vapor? (e) Will the degree of dissociation increase or decrease if oxygen gas is added while holding the total pressure constant at 1 bar?

SOLUTION

(a) $H_2O(g) = H_2(g) + \frac{1}{2} O_2(g)$

	$H_2O(g)$	$H_2(g)$	$\frac{1}{2} O_2(g)$	
Initially	1	0	0	
Equilibrium	$1-\xi$	ξ	$\xi/2$	Total = $1 + \xi/2$

$$P_{H2O} = \frac{1 - \xi}{1 + \frac{\xi}{2}} P \qquad P_{H_2} = \frac{\xi}{1 + \frac{\xi}{2}} P \qquad P_{O_2} = \frac{\xi/2}{1 + \frac{\xi}{2}} P$$

$$K_P = \frac{\left[\frac{\xi/2}{1+\xi/2}\frac{P}{P°}\right]^{1/2}\left[\frac{\xi}{1+\xi/2}\frac{P}{P°}\right]}{\frac{1 - \xi}{1+\xi/2}\frac{P}{P°}} = \frac{\xi^{3/2}(P/P°)^{1/2}}{\sqrt{2}\,(1+\frac{\xi}{2})^{1/2}(1-\xi)}$$

$$= \frac{0.02^{3/2}\ 1^{1/2}}{\sqrt{2}\ (1.01)^{1/2}\ (0.98)} = 2.03 \times 10^{-3}$$

(b) If the total pressure is reduced, the degree of dissociation will increase because the reaction will produce more molecules to fill the volume.

(c) If argon is added at constant pressure, the degree of dissociation will increase because the partial pressure due to the reactants will decrease.

(d) If argon is added at constant volume, the degree of dissociation will not be changed because the partial pressure due to the reactants will not change.

(e) If oxygen is added at constant total pressure, the degree of dissociation of H_2O will decrease because the reaction will be pushed to the left.

5.11 The equilibrium constant for the reaction $SO_2(g) + \frac{1}{2} O_2(g) = SO_3(g)$ at 1000 K is given by

$$K_P = \frac{(P_{SO_3}/P°)}{(P_{SO_2}/P°)(P_{O_2}/P°)^{1/2}} = 1.85$$

What is the ratio P_{SO_3}/P_{SO_2} (a) when the partial pressure of oxygen at equilibrium is 0.3 bar, (b) when the partial pressure of oxygen at equilibrium is 0.6 bar? (c) What is the effect on the equilibrium if the total pressure of the mixture of gases is increased by forcing in nitrogen at constant volume?

<u>SOLUTION</u>

(a) $$\frac{P_{SO_3}}{P_{SO_2}} = K_P\left(\frac{P_{O_2}}{P°}\right)^{1/2} = 1.85(0.3)^{1/2} = 1.01$$

(b) $$\frac{P_{SO_3}}{P_{SO_2}} = 1.85(0.6)^{1/2} = 1.43$$

(c) No effect if the gases behave as perfect gases.

5.12 At 55 °C and 1 bar the average molar mass of partially dissociated N_2O_4 is 61.2 g mol^{-1}. Calculate (a) ξ and (b) K_P for the reaction $N_2O_4(g) = 2NO_2(g)$. (c) Calculate ξ at 55 °C if the total pressure is reduced to 0.1 bar.

<u>SOLUTION</u>

(a) $$\xi = \frac{M_1 - M_2}{M_2 \, \Sigma\nu_i} = \frac{92.0 - 61.2}{61.2} = 0.503$$

(b) $$K_P = \frac{4\,\xi^2(P/P°)}{1 - \xi^2} = \frac{4\,(0.503)^2}{1 - 0.503^2} = 1.36$$

(c) $$\frac{\xi^2}{1 - \xi^2} = \frac{K_P}{4(P/P°)} = \frac{1.36}{4(0.1)}$$

$$\xi = 0.879$$

5.13 When N_2O_4 is allowed to dissociate to form NO_2 at 25 °C at a total pressure of 1 bar, it is 18.5% dissociated at equilibrium, and so K_P = 0.143. (a) If N_2 is added to the system at constant volume, will the equilibrium shift? (b) If the system is allowed to expand as N_2 is added at a constant total pressure of 1 bar, what will be the equilibrium degree of dissociation when the N_2 partial pressure is 0.6 bar?

SOLUTION

(a) If the gas mixture is ideal, the equilibrium will not shift.

(b) $$K_P = \frac{4\xi^2}{1 - \xi^2}(P_{NO_2} + P_{N_2O_4}) = 0.143$$

$$= \frac{4\xi^2}{1 - \xi^2}(0.4)$$

$$\xi = 0.286$$

5.14 Under what total pressure at equilibrium must PCl_5 be placed at 250 °C to obtain a 30% conversion into PCl_3 and Cl_2? For the reaction $PCl_5(g) = PCl_3(g) + Cl_2(g)$, K_P = 1.80 at 250 °C.

SOLUTION

$$K_P = \frac{\xi^2(P/P°)}{1 - \xi^2} = 1.80$$

$$\frac{P}{P°} = \frac{K_P(1 - \xi^2)}{\xi^2} = \frac{1.80(1 - 0.3^2)}{0.3^2} \qquad P = 18.2 \text{ bar}$$

5.15 A liter reaction vessel containing 0.233 mol of N_2 and 0.341 mol of PCl_5 is heated to 250 °C. The total pressure at equilibrium is 29.33 bar. Assuming that all the gases are ideal, calculate K_P for the only reaction that occurs.

$$PCl_5(g) = PCl_3(g) + Cl_2(g)$$

SOLUTION

$$n = \frac{PV}{RT} = \frac{(29.33\ \text{bar})(1\ \text{L})}{(0.08314\ \text{L bar K}^{-1}\ \text{mol}^{-1})(523.15\ \text{K})}$$

$$= 0.674\ \text{mol}$$

Moles of reactants = 0.674 - 0.233 = 0.441

= (0.341 - x) + 2x

x = moles of PCl_3 = moles of Cl_2 = 0.100

$$K_P = \frac{(P_{PCl_3}/P°)(P_{Cl_2}/P°)}{P_{PCl_5}/P°}$$

$$= \frac{\left(\frac{0.100}{0.674}\right)^2 (29.33)^2}{\left(\frac{0.241}{0.679}\right)(29.33)} = 1.81$$

5.16 At 250 °C PCl_5 is 80% dissociated at a pressure of 1.013 bar, and so K_P = 1.80. What is the percentage dissociation at equilibrium after sufficient nitrogen has been added at constant pressure to produce a nitrogen partial pressure of 0.9 bar? The total pressure is maintained at 1 bar.

SOLUTION

$$K_P = \frac{0.8^2\ (1.013)}{1 - 0.8^2} = 1.80$$

$$K_P = \frac{\xi^2(0.1)}{1 - \xi^2} = 1.80$$

ξ = 0.973 or 97.3% dissociated

5.17 Hydrogen is produced on a large scale from methane. Calculate the equilibrium constant K_P for the production of H_2 from CH_4 at 1000 K using the reaction $CH_4(g) + H_2O(g) = CO(g) + 3H_2(g)$

SOLUTION

$$\Delta G° = -200.297 - 19.460 + 192.576 = -27.181 \text{ kJ mol}^{-1}$$
$$= -(8.31441\text{x}10^{-3} \text{ kJK}^{-1}\text{mol}^{-1})(1000\text{K})\ell n K_P$$
$$K_P = 26.3$$

5.18 From the ΔG_f° of $Br_2(g)$ at 25 °C, calculate the vapor pressure of $Br_2(\ell)$. The pure liquid at 1 bar and 25 °C is taken as the standard state.

SOLUTION

$$Br_2(\ell) = Br_2(g) \qquad K_P = P_{Br_2}/P°$$
$$\Delta G° = -RT\ell n(P_{Br_2}/P°)$$
$$\frac{P_{Br_2}}{P°} = \exp\left[-\frac{3110 \text{ J mol}^{-1}}{(8.314\text{JK} \quad \text{mol} \quad)(298.15\text{K})}\right] = 0.285$$
$$P_{Br_2} = 0.285 \text{ bar}$$

5.19 In order to produce more hydrogen from "synthesis gas" ($CO + H_2$) the water gas shift reaction is used. $CO(g) + H_2O(g) = CO_2(g) + H_2(g)$ Calculate K_P at 1000 K and the equilibrium extent of reaction starting with an equimolar mixture of CO and H_2O.

SOLUTION

$$\Delta G° = -395{,}924 - (-200{,}297) - (-192{,}576) = -3051 \text{ J mol}^{-1}$$
$$= (8.314\text{JK}^{-1}\text{mol}^{-1})(1000\text{K})\ell n K_P$$

$$K_P = 1.44 = \frac{P_{H_2}\,P_{CO_2}}{P_{CO}\,P_{H_2O}} = \frac{x^2}{(1 - x)^2}$$

$x = 0.545$, fractional conversion of reactants to products

(Note that this reaction is exothermic so that there will be a larger extent of reaction at lower temperatures. In practice this reaction is usually carried out at about 700 K.)

5.20 Acetic acid is produced on a large scale by the carbonylation of methanol at about 500 K and 25 bar using a rhodium catalyst. What is K_y under these conditions? (ΔG_f° for acetic acid gas at 500 K is -335.28 kJ mol^{-1}.)

SOLUTION

$CO(g) + CH_3OH(g) = CH_3CO_2H(g)$

$\Delta G^\circ = -335.28 + 155.68 + 133.52 = -46.00$ kJ mol^{-1}

$$K = 6.39 \times 10^4 = \frac{y_{AA}P}{y_{CO}y_{CH_3OH}P^2}$$

$$K_y = \frac{y_{AA}}{y_{CO}\,y_{CH_3OH}} = (6.39 \times 10^4)(25) = 1.60 \times 10^6$$

5.21 What are the percentage dissociations of $H_2(g)$, $O_2(g)$, and $I_2(g)$ at 2000 K and a total pressure of 1 bar?

SOLUTION

$H_2(g) = 2H(g)$

$$\begin{aligned}\Delta G^\circ &= 2(106{,}653 \text{ J mol}^{-1})\\ &= -RT\ell n K_P\\ &= -(8.31441 \text{ J K}^{-1} \text{ mol}^{-1})(2000 \text{ K})\ell n K_P\end{aligned}$$

$$K_P = 2.69 \times 10^{-6} = \frac{4\xi^2}{1 - \xi^2}$$

$$\xi = \left(\frac{K_P}{K_P + 4}\right)^{1/2} = \left(\frac{2.69 \times 10^{-6}}{4}\right)^{1/2} = 0.000819$$

or 0.0819%

$O_2(g) = 2O(g)$

$\Delta G° = 2(121{,}448 \text{ J mol}^{-1})$

$K_P = 4.53 \times 10^{-7}$ $\xi = 0.0337\%$

$I_2(g) = 2I(g)$

$\Delta G° = 2(-29{,}637 \text{ J mol}^{-1})$

$K_P = 35.32$ $\xi = 94.8\%$

5.22 Calculate the degree of dissociation of $H_2O(g)$ into $H_2(g)$ and $O_2(g)$ at 2000 K and 1 bar. (Since the degree of dissociation is small, the calculation may be simplified by assuming that $P_{H_2O} = 1$ bar.)

SOLUTION

$H_2O(g) = H_2(g) + \frac{1}{2} O_2(g)$

$\Delta G° = 134{,}456 \text{ J mol}^{-1}$

$= -(8.31441 \text{ J K}^{-1} \text{ mol}^{-1})(2000 \text{ K})\ell n K_P$

$K_P = 3.079 \times 10^{-4}$

$$= \frac{(P_{H_2}/P°)(P_{O_2}/P°)^{1/2}}{(P_{H_2O}/P°)}$$

$$= \left(\frac{P_{H_2}}{P°}\right)\left(\frac{P_{H_2}}{2P°}\right)^{1/2} = \frac{1}{\sqrt{2}}\left(\frac{P_{H_2}}{P°}\right)^{3/2}$$

$$\left(\frac{P_{H_2}}{P°}\right) = (\sqrt{2}\, K_P)^{2/3} = 0.0058$$

5.23 At 500 K CH_3OH, CH_4 and other hydrocarbons can be formed from CO and H_2. Until recently the main source of the CO mixture for the synthesis of CH_3OH was methane.

$$CH_4(g) + H_2O(g) = CO(g) + 3H_2(g)$$

When coal is used as the source, the "synthesis gas" has a different composition.

$$C(graphite) + H_2O(g) = CO(g) + H_2(g)$$

Suppose we have a catalyst that catalyzes only the formation of CH_3OH. (a) What pressure is required to convert 25% of the CO to CH_3OH at 500 K if the "synthesis gas" comes from CH_4? (b) If the synthesis gas comes from coal?

SOLUTION

(a)	CO	+	$2H_2$	=	CH_3OH	
Initial	1		3		0	
Equil.	$1-\xi$		$3-2\xi$		ξ	Total = $4-2\xi$

$$K_P = \frac{y_{CH_3OH}}{y_{CO}\, y_{H_2}^2\, P^2} = \frac{\xi(4-2\xi)^2}{(1-\xi)(3-2\xi)^2\, P^2}$$

$$G° = -133.52 - (-155.68) = 22.16 \text{ kJ mol}^{-1}$$

$$K_P = 4.842 \times 10^{-3}$$

$$P = \left[\frac{\xi(4-2\xi)^2}{K_P(1-\xi)(3-2\xi)^2}\right]^{1/2}$$

$$= \left[\frac{(0.25)(3.5)^2}{4.842 \times 10^{-3}(0.75)(2.5)^2}\right]^{1/2}$$

$$= 11.6 \text{ bar}$$

(b)	CO	+	$2H_2$	=	CH_3OH	
Initial	1		1		0	
Equil	$1-\xi$		$1-2\xi$		ξ	Total = $2-2\xi$

$$K_P = \frac{\xi(2-2\xi)^2}{(1-\xi)(1-2\xi)^2\, P^2} \qquad P = \left[\frac{(0.25)(1.5)^2}{K_P(0.75)(0.5)^2}\right]^{1/2}$$

$$= 24.9 \text{ bar}$$

5.24 The following data apply to the reaction $Br_2(g) = 2Br(g)$.

T/K	1123	1173	1223	1273
$K_p/10^{-3}$	0.408	1.42	3.32	7.2

Determine by graphical means the enthalpy change when 1 mol of Br_2 dissociates completely at 1200 K.

SOLUTION

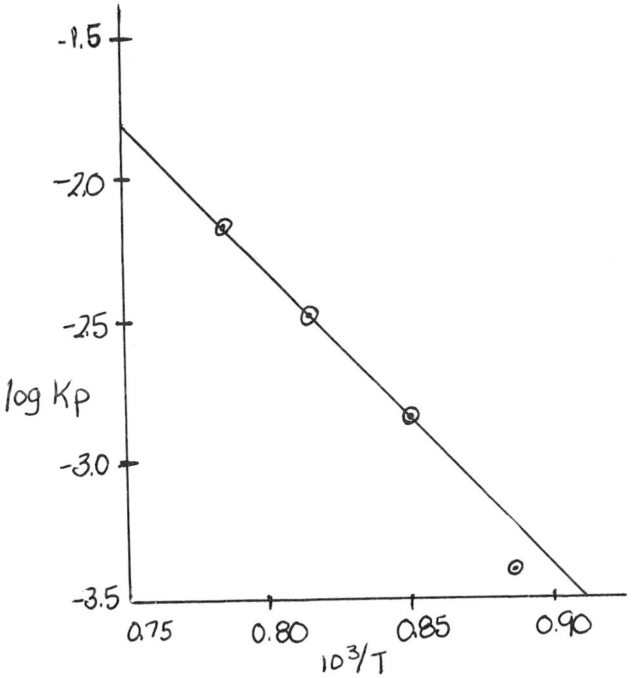

At T = 1200 K ($1/T = 0.833 \times 10^{-3}$) the slope is -10.4×10^3 K

$$H° = -2.303R(\text{slope})$$
$$= (2.303)(8.314 JK^{-1}mol^{-1})(10.4 \times 10^3 K)$$
$$= 199.2 \text{ kJ mol}^{-1}$$

5.25 The vapor pressure of water above mixtures of $CuCl_2 \cdot H_2O(cr)$ and $CuCl_2 \cdot 2H_2O(cr)$ is given as a function of temperature in the following table.

t/°C	17.9	39.8	60.0	80.0
P/bar	0.0049	0.0250	0.122	0.327

(a) Calculate $\Delta H°$ for the reaction

$$CuCl_2 \cdot 2H_2O(cr) = CuCl_2 \cdot H_2O(cr) + H_2O(g)$$

(b) Calculate $\Delta G°$ for the reaction at 60 °C.

(c) Calculate $\Delta S°$ for the reaction at 60 °C.

SOLUTION

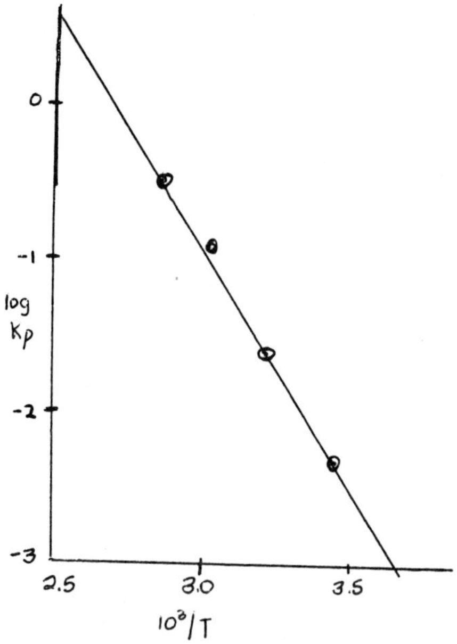

(a) $K_P = P$

slope = -3060 K

$\Delta H° = -2.303\ R(\text{slope})$

$= 57.3\ kJ\ mol^{-1}$

(b) $\Delta G° = -RT\ln K_P$

$= -(8.314)(333K)\ln 0.122$

$= 5.82\ kJ\ mol^{-1}$

(c) $\Delta S° = (\Delta H° - \Delta G°)/T$

$= 154.7\ J\ K^{-1}\ mol^{-1}$

5.26 The following reaction is nonspontaneous at room temperature and endothermic.

$$3C(\text{graphite}) + 2H_2O(g) = CH_4(g) + 2CO(g)$$

As the temperature is raised, the equilibrium constant will become equal to unity at some point. Estimate this temperature using data from Table A.2.

SOLUTION

At 1000 K

$$\Delta G° = 19.460 + 2(-200.297) - 2(-192.576)$$
$$= 4.018 \text{ kJ mol}^{-1}$$
$$= -(8.314\times10^{-3}\text{kJK}^{-1}\text{mol}^{-1})(1000\text{K})\ln K_P$$
$$K_P = 0.617$$
$$\Delta H° = -89.881 + 2(-112.010) - 2(-247.885)$$
$$= 181.869 \text{ kJ mol}^{-1}$$

$$\ln\frac{K_2}{K_1} \quad \frac{\Delta H°}{R}\left(\frac{1}{T_1} - \frac{1}{T_2}\right)$$

$$\ln\frac{1}{0.617} = \frac{181{,}869 \text{ J mol}^{-1}}{8.314 \text{ J K}^{-1}\text{ mol}^{-1}}\left(\frac{1}{1000 \text{ K}} - \frac{1}{T_2}\right)$$

$$T_2 = \frac{1}{\frac{1}{1000 \text{ K}} - \frac{8.314}{181{,}869}\ln\frac{1}{0.617}}$$
$$= 1023 \text{ K}$$

5.27 The measured density of an equilibrium mixture of N_2O_4 and NO_2 at 15 °C and 1.013 bar is 3.62 g L^{-1}, and the density at 75 °C and 1.013 bar is 1.84 g L^{-1}. What is the enthalpy change of the reaction $N_2O_4(g) = 2NO_2(g)$?

SOLUTION

At 15 °C

$$M = \frac{RT}{P}\frac{g}{V} = \frac{(0.08314)(288)(3.62)}{1.013}$$
$$= 85.57 \text{ g mol}^{-1}$$
$$\xi = \frac{M_1 - M_2}{M_2} = \frac{92.01 - 85.57}{85.57} = 0.0753$$
$$K = \frac{4\,\xi^2\,P}{1 - \xi^2} = \frac{4(0.0753)^2(1.013)}{1 - 0.0753^2} = 0.0231$$

At 75 °C

$$M = \frac{(0.08314)(348)(1.84)}{1.013} = 52.55 \text{ g mol}^{-1}$$

$$= \frac{92.01 - 52.55}{52.55} = 0.751 \qquad K = 5.24$$

$$\Delta H = \frac{RT_1T_2}{(T_2-T_1)} \ln\frac{K_2}{K_1} = \frac{(8.314)(288)(348)}{60} \ln\frac{5.24}{.0231}$$

$$= 75 \text{ kJ mol}^{-1}$$

5.28 The following reaction takes place in the presence of aluminum chloride.

Cyclohexane(ℓ) = Methylcyclopenthane(ℓ)

At 25 °C K_c = 0.143, and at 45 °C K_c = 0.193. From these data calculate: (a) $\Delta G°$ at 25 °C, (b) $\Delta H°$, and (c) $\Delta S°$.

SOLUTION

(a) $$\Delta G° = -RT\ln K$$

$$= -(8.314 \text{ J K}^{-1} \text{ mol}^{-1})(298.15 \text{ K})\ln 0.143$$

$$- 4820 \text{ J mol}^{-1}$$

(b) $$\Delta H° = \frac{RT_1T_2}{T_2-T_1} \ln\frac{K_2}{K_1} = \frac{(8.314)(298)(318)}{20} \ln\frac{0.193}{0.143}$$

$$= 11{,}800 \text{ J mol}^{-1}$$

(c) $$\Delta S° = \frac{\Delta H° - \Delta G°}{T} = \frac{11{,}800 - 4{,}820}{298.15}$$

$$= 23.41 \text{ J K}^{-1} \text{ mol}^{-1}$$

5.29 Mercuric oxide dissociates according to the reaction $2HgO(cr) = 2Hg(g) + O_2(g)$. At 420 °C the dissociation pressure is 5.16×10^4 Pa, and at 450 °C it is 10.8×10^4 Pa. Calculate (a) the equilibrium constants, and (b) the enthalpy of dissociation per mole of HgO.

SOLUTION

(a) $P_{Hg} = 2\ P_{O_2}$ $\quad P_{Hg} = \frac{2}{3}\ P$ $\quad P_{O_2} = \frac{1}{3}\ P$

$$K_{P420} = P_{Hg}\ P_{O_2} = (\tfrac{2}{3})^2\ (\tfrac{1}{3})\ P^3$$

$$= (\tfrac{4}{27})\ (\frac{5.16 \times 10^4\ \mathrm{Pa}}{1.013 \times 10^5\ \mathrm{Pa}})^3 = 0.0196$$

$$K_{P450} = (\tfrac{4}{27})\ (\frac{10.8 \times 10^4\ \mathrm{Pa}}{1.013 \times 10^5\ \mathrm{Pa}})^3 = 0.1794$$

(b) $\Delta H° = \frac{RT_1T_2}{T_2 - T_1} \ell n \frac{K_2}{K_1}$

$$= \frac{(8.314\mathrm{JK^{-1}mol^{-1}})(693\mathrm{K})(723\mathrm{K})}{30\ \mathrm{K}} \ell n \frac{0.1794}{0.0196}$$

$= 308\ \mathrm{kJ\ mol^{-1}}$ for the reaction as written

$= 154\ \mathrm{kJ\ mol^{-1}}$ of HgO(cr)

5.30 Show that for gaseous equilibria

$$\frac{\partial \ell n K_c}{\partial T} = \frac{\Delta U°}{RT^2}$$

SOLUTION

$$\frac{\partial \ell n K}{\partial (1/T)} = -\frac{\Delta H°}{R} \qquad \frac{\partial (1/T)}{\partial T} = -\frac{1}{T^2}$$

$$\frac{\partial \ell n K}{\partial T} = \frac{\Delta H°}{RT^2} \qquad K_c = (\frac{P°}{c°RT})^{\Sigma\nu_i} K_P$$

$$\ell n K_c = \Sigma\nu_i \ell n (\frac{P°}{c°RT}) + \ell n K_P$$

$$\frac{\partial \ell n K_c}{\partial T} = -\frac{\Sigma\nu_i}{T^2} + \frac{\Delta H°}{RT^2} = \frac{\Delta H° - RT\Sigma\nu_i}{RT^2}$$

$$= \frac{\Delta U°}{RT^2}$$

5.31 Calculate (a) K_P and (b) $\Delta G°$ for the following reaction at 20 °C.

$$CuSO_4 \cdot 4NH_3(cr) = CuSO_4 \cdot 2NH_3(cr) + 2NH_3(g)$$

The equilibrium pressure of NH_3 is 8.27 kPa.

SOLUTION

(a) $K_P = (P_{NH_3}/P°)^2 = \left(\frac{8270\ \text{Pa}}{101{,}325\ \text{Pa}}\right)^2 = 6.66 \times 10^{-3}$

(b) $\Delta G° = -RT\ell n K_P$

$= -(8.314\ JK^{-1}mol^{-1})(293.15K)\ell n 6.66 x 10^{-3}$

$= 12.2\ kJ\ mol^{-1}$

5.32 Air (20% O_2) will not oxidize silver at 200 °C and 1 bar pressure. Make a statement about the equilibrium constant for the reaction

$$2Ag(cr) + \frac{1}{2} O_2(g) = Ag_2O(cr)$$

SOLUTION

$$K_P = \left(\frac{P_{O_2}}{P°}\right)^{-1/2} = (0.2)^{-1/2} = 2.24 \quad \text{if}$$

Ag, O_2 and Ag_2O are in equilibrium. If they are not in equilibrium, a higher pressure of O_2 will be required and the value of K_P will be smaller than the above value: $K_P < 2.24$.

5.33 The dissociation of ammonium carbamate takes place according to the reaction

$$(NH_2)CO(ONH_4)(cr) = 2NH_3(g) + CO_2(g)$$

When an excess of ammonium carbamate is placed in a previously evacuated vessel, the partial pressure generated by NH_3 is twice the partial pressure of the CO_2, and the partial pressure of $(NH_2)CO(ONH_4)$ is negligible in comparison. Show that $K_P = (P_{NH_3})^2 P_{CO_2} = \frac{4}{27} P^3$ where P is the total pressure.

SOLUTION

$P = P_{NH_3} + P_{CO_2} = 3\ P_{CO_2}$ since $P_{NH_3} = 2\ P_{CO_2}$

$P_{CO_2} = P/3 \qquad P_{NH_3} = \frac{2}{3}\ P$

$$K_P = \left(\frac{P_{NH_3}}{P^\circ}\right)^2 \left(\frac{P_{CO_2}}{P^\circ}\right) = \left(\frac{2}{3}\ \frac{P}{P^\circ}\right)^2 \left(\frac{1}{3}\ \frac{P}{P^\circ}\right)$$

$$= \frac{4}{27}\ \left(\frac{P}{P^\circ}\right)^3$$

5.34 At 1000 K methane at 1 bar is in the presence of hydrogen. In the presence of a sufficiently high partial pressure of hydrogen, methane does not decompose to form graphite and hydrogen. What is this partial pressure?

SOLUTION

$CH_4(g) = C(\text{graphite}) + 2H_2(g)$

$\Delta G^\circ = -RT\ell nK = -19.46\ \text{kJ mol}^{-1} \qquad K = 10.39 = \frac{P_{H_2}^{\ 2}}{P_{CH_4}}$

$P_{H_2} = [(10.39\ \text{bar})(1\ \text{bar})]^{1/2} = 3.2\ \text{bar}$

5.35 A gaseous system contains CO, CO_2, H_2, H_2O, and C_6H_6 in chemical equilibrium. How many components are there? How many independent reactions? How many degrees of freedom are there?

SOLUTION

	CO	CO_2	H_2	H_2O	C_6H_6
C	1	1	0	0	6
O	1	2	0	1	0
H	0	0	2	2	6

Subtract the first row from the second and divide the third row by 2 to obtain

1	1	0	0	6
0	1	0	1	-6
0	0	1	1	3

Subtract the second row from the first

1	0	0	-1	12
0	1	0	1	-6
0	0	1	1	3

The rank of this matrix is 3, and so there are 3 independent components. The stoichiometric coefficients of 2 independent reactions are given by the last 2 columns.

$$CO_2 + H_2 = H_2O + CO$$
$$12CO + 3H_2 = C_6H_6 + 6CO_2$$

If the species are arranged in a different order in the matrix, a different pair of independent equations will be obtained.

$$v = c - p + 2$$

If the numbers of moles of 3 components and the temperature are specified, the system is completely determined; that is, the numbers of moles of the other two species and the pressure can be calculated--given values of the equilibrium constants. Alternatively, the mole fractions of 2 species and the temperature and pressure may be specified. We then have two mathematical expressions between 5 mole fractions, 4 of which are independent since $\Sigma y_i = 1$. If 2 mole fractions are known, the other two can be calculated from the two simultaneous equations.

5.36 Superheated steam is passed over coal, represented in these calculations by graphite, at

1000 K to produce CO, CO_2, H_2, and CH_4. Since there are 6 species and 3 element balance equations, there are 3 independent chemical reactions which may be written as follows:

$$C(graphite) + H_2O(g) = CO(g) + H_2(g)$$
$$3C(graphite) + 2H_2O(g) = CH_4(g) + 2CO(g)$$
$$2C(graphite) + 2H_2O(g) = CH_4(g) + CO_2(g)$$

What are the equilibrium constants for these reactions? If the total pressure is raised to 100 bar, one of these reactions will predominate. Neglecting the other two reactions, calculate the equilibrium mole fractions of the gases present and compare the results with Fig. 5.4.

SOLUTION

Using Table A.2 the following equilibrium constants are calculated for 1000 K.

$$C(graphite) + H_2O(g) = CO(g) + H_2(g) \qquad K_P = 2.5$$
$$3C(graphite) + 2H_2O(g) = CH_4(g) + 2CO(g) \qquad K_P = 0.61$$
$$2C(graphite) + 2H_2O(g) = CH_4(g) + CO_2(g) \qquad K_P = 0.35$$

At 100 bar the K_y values for these three reactions are 0.025, 0.0061, and 0.35, respectively. Thus the third reaction will predominate.

$$K_y = \frac{y_{CH_4}\, y_{CO_2}}{y_{H_2O}^2} = 0.35 \qquad y_{CH_4} = y_{CO_2} = y$$

$$1 = y_{H_2O} + y_{CH_4} + y_{CO_2} = y_{H_2O} + 2y$$

$$y_{H_2O} = 1 - 2y$$

Substituting in K_y and solving the quadratic equation for y yields

$y_{CH_4} = 0.27$ (0.2 in Fig. 5.4)

$y_{CO_2} = 0.27$ (0.24 in Fig. 5.4)

$y_{H_2O} = 0.46$ (0.36 in Fig. 5.4)

5.37 The equilibrium constant for the association of benzoic acid to a dimer in dilute benzene solutions is as follows at 43.9 °C.

$$2C_6H_5COOH = (C_6H_5COOH)_2 \qquad K_c = 2.7 \times 10^2$$

Molar concentrations are used in expressing the equilibrium constant. Calculate $\Delta G°$, and state its meaning.

SOLUTION

$$\Delta G° = -RT\ln K_c$$
$$= -(8.314\ JK^{-1}mol^{-1})(317.1K)\ln(2.7\times10^2)$$
$$= -14.8\ kJ\ mol^{-1}$$

This is the decrease in Gibbs energy when 2 mol of monomer at unit activity on the molar scale is converted to 1 mol of dimer at unity activity on the molar scale.

5.38 $\Delta G° > 17.2$ kJ mol^{-1}

5.39 0.696

5.40 1.87×10^{-4}, 1.64×10^{-4}, 1.90×10^{-4}, 0.364×10^{-4} As the pressure is increased, the gases behave more imperfectly.

5.41 (a) 17.0 kJ mol^{-1} (b) 0.795 bar

5.42 (a) 9.63 g (b) 1.215, 0.032, 1.457 bar

5.43 $k = 3^{3/2}/4Kp = 0.0164$

5.44 (a) 2.21×10^{-5} (b) 5.41×10^{-4}

5.45 10.2

5.46 (a) 5×10^{-3} bar (b) 1.79% (c) 0.0221 bar

5.47 (a) 19.70 bar (b) 35.6 bar

5.48 (a)

$t/°C$	25	45	65
ξ	0.1852	0.3775	0.6284
K_P	0.1451	0.6737	2.644

(b) 60.7 kJ mol^{-1} (c) 0.320 (d) 0.371

5.49 0.351 bar

5.50 3.74 bar

5.51 0.54 mol

5.52 $\xi = 0.800$ $K_P = 1.80$

5.53 Dissociation into atoms and solubility proportional to the partial pressure of atoms.

5.54 5.57×10^{-7} bar

5.55 6.9

5.56 Fe_2O_3

5.57 1.81×10^{-2}

5.58 2.99×10^{-3} 0.0273 0.0861

5.59 0.081 0.363

5.60 5.6×10^{-5} bar

5.61 0.53

5.62 2.66×10^{-6} bar

5.63 (a) 1.20×10^{10} 3.91×10^{-2}
(b) 8.25×10^{-2} 0.352

5.64 135.8 °C

5.65 153 kJ mol^{-1}

5.66 1.94×10^{-13} bar

5.67 6.88×10^{-24} 4.79×10^{-43}

5.68 452 kJ mol^{-1}

5.69 7.55 bar

5.70 (a) 9.75 g (b) 19.75 g

5.71 3.31×10^{-16} bar of O_2 is required to oxidize Ni, but the partial pressure of O_2 from partially dissociated H_2O in the presence of 3 mole percent hydrogen is only 9.7×10^{-18} bar.

5.72 3 components, 1 independent reaction, variance of 4. If C is present, the answers are 3, 2, and 3.

5.73 -324.3 kJ mol^{-1}

5.74 541

CHAPTER 6. Electrochemical Cells

6.1 How much work is required to bring two protons from an infinite distance of separation to 0.1 nm? Calculate the answer in joules using the protonic charge 1.602×10^{-19} C. What is the work in kJ mol^{-1} for a mole of proton pairs?

SOLUTION

$$\text{Potential } \phi = \frac{Q_2}{4\pi\varepsilon_0 r}$$

$$= \frac{(1.602\text{x}10^{-19}\ \text{C})(0.89875\text{x}10^{10}\ \text{N m}^2\ \text{C}^{-2})}{10^{-10}\ \text{m}}$$

$$= 14.398\ \text{J C}^{-1}$$

$$= Q_1\phi = (1.602\text{x}10^{-19}\ \text{C})(14.398\ \text{J C}^{-1})$$

$$= 2.307 \times 10^{-18}\ \text{J}$$

$$= (2.307 \times 10^{-18}\ \text{J})(6.022 \times 10^{23}\ \text{mol}^{-1})$$

$$= 1389.3\ \text{kJ mol}^{-1}$$

6.2 What is the electric field strength 0.5 nm from a proton?

SOLUTION

$$E = \frac{Q}{4\pi\varepsilon_0 r^2}$$

$$= \frac{1.602 \times 10^{-19}\ \text{C}}{4\pi(8.854\text{x}10^{-12}\ \text{C}^2\ \text{N}^{-1}\ \text{m}^{-2})(0.5\text{x}10^{-9}\ \text{m})^2}$$

$$= 5.759 \times 10^9\ \text{V m}^{-1}$$

6.3 A small dry battery of zinc and ammonium chloride weighing 85 g will operate continuously through a 4-Ω resistance for 450 min before its voltage falls below 0.75 V. The initial voltage is 1.60, and the effective voltage over the whole life of the battery is taken to be 1.00. Theoretically, how many kilometers above the earth could this battery be raised by the energy delivered under these conditions?

SOLUTION

$$I = \frac{E}{R} = \frac{IV}{4\Omega} = 0.25\ A$$

$$\text{Power} = I^2R = (0.25\ A)^2(4\Omega) = 0.25\ W$$

$$\text{Work} = (0.25\ W)(450 \times 60\ s) = 6.75 \times 10^3\ J$$

$$= (0.085\ kg)(9.80\ m\ s^{-2})h$$

$$h = \frac{6.75 \times 10^3\ J}{(0.085\ kg)(9.80\ m\ s^{-2})} = 8103\ m$$

$$= \frac{(8103 \times 10^5\ cm)}{(2.54\ cm\ in^{-1})(12\ in\ ft^{-1})(5280\ ft\ mile^{-1})}$$

$$= 5.04\ \text{miles}$$

6.4 What is the expression for the activity of Na_2SO_4 in terms of the mean ionic activity coefficient and the molality.

SOLUTION

$$a_{Na_2SO_4} = a_{Na^+}^2 \cdot a_{SO_4^=} \qquad \gamma_{Na^+} = \frac{a_{Na^+}}{2m}$$

$$= (2m\gamma_{Na^+})^2(\gamma_{SO_4^=}m) \qquad \gamma_{SO_4^=} = \frac{a_{SO_4^=}}{m}$$

$$= 4\ m^3\ \gamma_\pm^3$$

Where $\gamma_\pm = (\gamma_{Na^+}^2 \cdot \gamma_{SO_4^=})^{1/3}$

6.5 A solution of NaCl has an ionic strength of 0.24 mol kg^{-1}. (a) What is its concentration? (b) What concentration of Na_2SO_4 would have the same ionic strength? (c) What concentration of $MgSO_4$?

SOLUTION

(a) $I = \frac{1}{2}(m_1 z_1^2 + m_2 z_2^2)$

$0.24 = \frac{1}{2}(1^2 + 1^2)\ m$

$m = 0.24 \text{ mol kg}^{-1}$

(b) $0.24 = \frac{1}{2}(2m \times 1^2 + m2^2)$

$m = 0.08 \text{ mol kg}^{-1}$

(c) $0.24 = \frac{1}{2}(2^2 + 2^2)m$

$m = 0.06 \text{ mol kg}^{-1}$

6.6 For 0.002 mol kg^{-1} $CaCl_2$ at 25 °C use the Debye-Hückel limiting law to calculate the activity coefficients of Ca^{++} and Cl^-. What is the mean ionic activity coefficient for the electrolyte?

SOLUTION

$I = \frac{1}{2}(0.002 \times 2^2 + 0.004 \times 1^2) = 0.006$

$\log \gamma_i = -0.509\ z_i^2\ I^{1/2}$

$\log \gamma_{Ca^{2+}} = -0.509(2^2)(0.006)^{1/2}$

$\gamma_{Ca^{2+}} = 0.695$

$\log \gamma_{Cl^-} = -0.509(1^2)(0.006)^{1/2}$

$\gamma_{Cl^-} = 0.913$

$\gamma_\pm = (\gamma_+ \gamma_-^2)^{1/3} = (0.695 \times 0.913^2)^{1/3}$
$= 0.834$

6.7 The cell Pt| H_2(1 bar)|HBr(m)|AgBr|Ag has been studied by H. S. Harned, A. S. Keston, and J. G. Donelson [J. Am. Chem. Soc., 58, 989 (1936)].

The following table gives the electromotive forces obtained at 25 °C.

m	0.01	0.02	0.05	0.10
E	0.3127	0.2786	0.2340	0.2005

Calculate (a) E° and (b) the activity coefficient for a 0.10 mol kg^{-1} solution of hydrogen bromide.

SOLUTION

(a) Plot $E + 0.1183 \log m - 0.0602 \sqrt{m}$ versus m,

The intercept at m = 0 is E° = 0.0710 V.

(b) $E = 0.0710\ V - (0.05915\ V) \log \gamma_{\pm}^{2}\ m^{2}$

$$0.2005\ V = 0.0710\ V - (0.1183V)\log\gamma_{\pm} - (0.1183V)\log 0.1$$

$$\log \gamma_{\pm} = -\frac{0.2005 - 0.1183 - 0.0710}{0.1183}$$

$$\gamma_{\pm} = 0.804$$

6.8 According to Table A.1 what is the value of the equilibrium constant for the reaction

$$\tfrac{1}{2}H_2(g) + AgCl(cr) = Ag(cr) + H^+(ao) + Cl^-(ao)$$

at 25 °C and how is it defined?

SOLUTION

$\Delta G° = -131.228 - (-109.789) = -21.439$ kJ mol^{-1}

$K = e^{-\Delta G°/RT} = 5.701 \times 10^3$

$$= \frac{a(H^+)a(Cl^-)}{(P_{H_2}/P°)^{1/2}} = \frac{a(HCl)}{(P_{H_2}/P°)^{1/2}}$$

In writing the equilibrium constant expression the activities of the pure crystalline phases are taken equal to unity.

6.9 The electromotive force of the cell

$$Pb|PbSO_4|Na_2SO_4\cdot 10H_2O(sat)|Hg_2SO_4|Hg$$

is 0.9647 V at 25 °C. The temperature coefficient

is 1.74×10^{-4} V K^{-1}. (a) What is the cell reaction? (b) What are the values of ΔG, ΔS, and ΔH?

SOLUTION

(a) $Pb(cr) + Hg_2SO_4(cr) = PbSO_4(cr) + 2Hg(\ell)$

(b) $\Delta G = -nFE = -(2)(96{,}485 \text{ C mol}^{-1})(0.9647 \text{ V})$
$= -186.16 \text{ kJ mol}^{-1}$

$\Delta S = nF \left(\frac{\partial E}{\partial T}\right)_P$
$= (2)(96{,}485 \text{ C mol}^{-1})(1.74 \times 10^{-4} \text{ V K}^{-1})$
$= 33.58 \text{ J K}^{-1} \text{ mol}^{-1}$

$\Delta H = -nFE + nFT\left(\frac{\partial E}{\partial T}\right)_P$
$= -2(96{,}485 \text{ C mol}^{-1})(0.9647 \text{ V})$
$+ 2(96{,}485 \text{ C mol}^{-1})(298.15 \text{ K})(1.74 \times 10^{-4} \text{ V K}^{-1})$
$= -176.15 \text{ kJ mol}^{-1}$

6.10 (a) Write the reaction that occurs when the cell
$Zn|ZnCl_2(0.555 \text{ mol kg}^{-1})|AgCl|Ag$
delivers current and calculate (b) ΔG, (c) ΔS, and (d) ΔH at 25 °C for this reaction. At 25 °C $E = 1.015$ V and $(\partial E/\partial T)_P = -4.02 \times 10^{-4}$ V K^{-1}.

SOLUTION

(a) $Zn = Zn^{2+} + 2e^-$

$2AgCl + 2e^- = 2Cl^- + 2Ag$

$Zn + 2AgCl = 2Ag + ZnCl_2(0.555 \text{ mol kg}^{-1})$

(b) $\Delta G = -nFE = -2(96{,}485 \text{ C mol}^{-1})(1.015 \text{ V})$
$= -195.9 \text{ kJ mol}^{-1}$

(c) $\Delta S = nF\left(\frac{\partial E}{\partial T}\right)_P = 2(96{,}485 \text{ C mol}^{-1})(-4.02 \times 10^{-4} \text{ V K}^{-1})$
$= -77.6 \text{ J K}^{-1} \text{ mol}^{-1}$

(c) $\Delta H = \Delta G + T\Delta S = -195{,}900 + (298)(-77.6)$
$= -219.0 \text{ kJ mol}^{-1}$

6.11 We found in Section 6.7 that $Pt|H_2|HCl|AgCl|Ag$ $E° = 0.2224$ V at 25 °C. Using the value of ΔG_f° [Cl^-(ao)] obtained in Example 6.7, what is the value of ΔG_f°[AgCl(cr)]?

SOLUTION

$$H_2 = \frac{1}{2} H^4 + e^-$$

$$\frac{AgCl(cr) + e^- = Ag + Cl^-}{H_2 + AgCl(cr) = Ag + Cl^-} \qquad E° = 0.2224 \text{ V}$$

$$\Delta G° = -(96{,}485 \text{ C mol}^{-1})(0.2224 \text{ V})$$
$$= -21{,}458 \text{ J mol}^{-1}$$
$$= \Delta G_f^\circ[Cl^-(ao)] - \Delta G_f^\circ[AgCl(cr)]$$

$$\Delta G_f^\circ = [AgCl(cr)] = 21{,}458 - 131.263$$
$$= -109.805 \text{ kJ mol}^{-1}$$

6.12 Calculate the standard electromotive force of the cell $Li|LiCl(ai)|Cl_2(g)$, Pt at 25 °C using (a) electrode potentials and (b) standard Gibbs energies of formation.

SOLUTION

$Cl_2(g) + 2e^- = 2Cl^-(ao)$	1.360 V
$2Li = 2Li(ao) + 2e^-$	3.045 V
$2Li + Cl_2(g) = 2LiCl(2i)$	4.405 V

Using ΔG_f° values

$$\Delta G° = 2(-293.31 - 131.228) = -849.076 \text{ kJ mol}^{-1}$$

$$E = -\frac{\Delta G°}{nF} = \frac{849{,}076 \text{ J}}{2(96{,}485 \text{ C mol}^{-1})} = 4.400 \text{ V}$$

6.13 From standard electrode potentials in Table 6.1 what are the standard Gibbs energies of formation at 25 °C for $Cl^-(ao)$, $OH^-(ao)$ and $Na^+(ao)$?

SOLUTION

For $H_2(g)|HCl(ai)|CL_2(g)$

(1) $\frac{1}{2} H\ (g) = H^+(ao) + e^-$ $\qquad E° = 0$

(2) $\frac{1}{2} O_2(g) + e^- = Cl^-(ao)$ $\qquad E° = 1.3604\ V$

$$\frac{1}{2} H_2(g) + \frac{1}{2} Cl_2(g) = H^+(ao) + Cl^-(ao) \qquad E° = 1.3604\ V$$

$$\Delta G° = -nFE° = -(96{,}485\ C\ mol^{-1})(1.3604\ V) = -131.258\ kJ\ mol^{-1}$$

This is $\Delta G_f^\circ\ [Cl^-(ao)]$ since $\Delta G°$ for reaction 1 is zero by convention. Table A.1 gives $-131.228\ kJ\ mol^{-1}$.

For $H_2(g)|H^+(ao)||OH^-(ao)|O_2(g)$

$\frac{1}{2}H_2(g) = H^+(ao) + e^-$

$\frac{1}{4}O_2(g) + \frac{1}{2} H_2O(\ell) + e^- = OH^-(ao)$ $\qquad E° = 0.401\ V$

$$\frac{1}{2}H_2(g) + \frac{1}{4}O_2(g) + \frac{1}{2}H_2O(\ell) = H^+(ao) + OH^-(ao) \qquad E° = 0.401\ V$$

$$\Delta G° = -(96{,}485\ C\ mol^{-1})(0.401\ V) = -38.69\ kJ\ mol^{-1}$$

$$= \Delta G_f^\circ\ [OH^-(ao)] - \frac{1}{2}\ \Delta G_f^\circ\ [H_2O(\ell)]$$

$$\Delta G_f^\circ\ [OH^-(ao)] = -38.69 + \frac{1}{2}(-237.129) = -157.26\ kJ\ mol^{-1}$$

Table A.1 gives $-157.244\ kJ\ mol^{-1}$.

For $Na(cr)|Na^+(ao)||H^+(ao)|H_2(g)$

$Na(cr) = Na^+(ao) + e^-$ $\qquad E° = -2.714\ V$

$H^+(ao) + e^- = \frac{1}{2} H_2(g)$ $\qquad E° = 0$

$Na(cr) + H^+(ao) = Na^+(ao) + \frac{1}{2} H_2(g)$ $\quad E° = 2.714$ V

$\Delta G° = -(96{,}485 \text{ C mol}^{-1})(2.714 \text{ V}) = -261.86 \text{ kJ mol}^{-1}$

$= \Delta G_f°[Na^+(ao)]$

Table A.1 gives -261.905 kJ mol^{-1}.

6.14 What are the equilibrium constants for the following reactions at 25 °C?

(a) $H^+(ao) + Li(cr) = Li^+(ao) + \frac{1}{2}H_2(g)$

(b) $2H^+(ao) + Pb(cr) = Pb^{2+}(ao) + H_2(g)$

(c) $3H (ao) + Au(cr) = Au^{3+}(ao) + \frac{2}{3}H_2(g)$

SOLUTION

(a) $Li = Li^+ + e^-$ $\quad -3.045$ V

$H^+ + e^- = \frac{1}{2}H_2$ $\quad 0$

$H^+ + Li = Li^+ + \frac{1}{2} H_2$ $\quad E° = 3.045$ V

$K = e^{zFE°/RT}$

$= e^{(96{,}485\text{Cmol}^{-1})(3.045\text{V})/(8.314\text{JK}^{-1}\text{mol}^{-1}) \times (298\text{K})}$

$= 3.1 \times 10^{51}$

(b) $Pb = Pb^{2+} + 2e^-$ $\quad -0.126$ V

$2H^+ + 2e^- = H_2$ $\quad 0$

$2H^+ + Pb = Pb^{2+} + H_2$ $\quad E° = 0.126$ V

$K = e^{2(96{,}485\text{Cmol}^{-1})(0.126\text{V})/(8.314\text{JK}^{-1}\text{mol}^{-1}) \times (298\text{K})}$

$= 1.8 \times 10^4$

(c) $Au = Au^{3+} + 3e^-$ $\quad 1.50$ V

$3H^+ + 3e^- = \frac{3}{2}H_2$ $\quad 0$

$3H^+ + Au = Au^{3+} + \frac{3}{2}H_2$ $\quad E° = -1.50$ V

$K = e^{3(96{,}485\text{Cmol}^{-1})(-1.50\text{V})/(8.314\text{JK}^{-1}\text{mol}^{-1}) \times (298\text{K})}$

$= 7.8 \times 10^{-77}$

6.15 Devise a cell for which the cell reaction is

$$H_2O(\ell) = H^+(ao) + OH^-(ao)$$

Calculate $\Delta G°$ at 25 °C from electrode potentials. What is the value of the equilibrium constant at this temperature?

SOLUTION

$Pt|H_2(g)|H^+(ao)||OH^-(ao)|H_2(g)|Pt$

$$\tfrac{1}{2}H_2(g) = H^+(ao) + e^- \qquad E_L = 0$$

$$H_2O(\ell) + e^- = \tfrac{1}{2}H_2(g) + OH^-(ao) \qquad E_R = -0.8281\ V$$

$$H_2O(\ell) = H^+(ao) + OH^-(ao) \qquad E° = -0.8281\ V$$

$$\Delta G° = -nFE° = -(96{,}485\ C\ mol^{-1})(-0.8281\ V)$$

$$= 79.899\ kJ\ mol^{-1}$$

$$K = e^{-\Delta G°/RT} = e^{-79{,}899/(8.314)(298)}$$

$$= 1.003 \times 10^{-14}$$

6.16 Devise an electromotive force cell for which the cell reaction is $AgBr(cr) = Ag^+ + Br^-$ Calculate the equilibrium constant (usually called the solubility product) for this reaction at 25 °C.

SOLUTION

$Ag|Ag^+||Br^-|AgBr|Ag$

$$Ag = Ag^+ + e^- \qquad E° = 0.7991\ V$$

$$AgBr + e^- = Ag + Br^- \qquad E° = 0.095\ V$$

$$AgBr = Ag^+ + Br^- \qquad E° = -0.701\ V$$

$$\log K = \frac{E°}{0.05916} = \frac{-0.701\ V}{0.05916\ V} = -11.85$$

$$K = 10^{-11.85} = m^2\ \gamma_{\pm}^{\ 2}$$

6.17 Using data from Table A.1 calculate the solubility of AgCl(cr) in water at 298.15 K. The salt is completely dissociated in the aqueous phase.

SOLUTION

$AgCl(cr) = Ag^{+}(ao) + Cl^{-}(ao)$

$G° = 77.107 - 131.228 + 109.789$

$= 55.668 \text{ kJ mol}^{-1}$

$K = (a_{Ag^+})(a_{Cl^-}) = s^2$ where s is solubility

$s = e^{-55,668/(8.31441)(298.15)}$

$= 1.33 \times 10^{-5} \text{ mol kg}^{-1}$

6.18 When a hydrogen electrode and a normal calomel electrode are immersed in a solution at 25 °C a potential of 0.664 V is obtained. Calculate (a) the pH, and (b) the hydrogen-ion activity.

SOLUTION

(a) $E = E° - \frac{RT}{nF}\ell n a_{H^+} = E° + 0.0591 \text{ pH}$

$$\text{pH} = \frac{E - E°}{0.0591} = \frac{0.664 - 0.2802}{0.0591} = 6.49$$

(b) $a_{H^+} = 10^{-6.49} = 3.24 \times 10^{-7}$

6.19 Ammonia may be used as the anodic reactant in a fuel cell. The reactions occurring at the electrodes are

$$NH_3(g) + 3OH^-(ao) = \tfrac{1}{2}N_2(g) + 3H_2O(\ell) + 3e^-$$

$$O_2(g) + 2H_2O(\ell) + 4e^- = 4OH^-(ao)$$

What is the electromotive force of this fuel cell at 25 °C?

SOLUTION

$4NH_3 + 12OH^- = 2N_2 + 12H_2O + 12e^-$

$3\ O_2 + 6H_2O + 12e^- = 12OH^-$

$4NH_3 + 3\ O_2 = 2N_2 + 6H_2O$

Using Table A.1

$\Delta G° = 6(-237.129) - 4(-16.45) = -1356.97\ kJ\ mol^{-1}$

$= -nFE°$

$$E° = \frac{1{,}356{,}970\ J\ mol^{-1}}{(12)(96{,}485\ C\ mol^{-1})} = 1.1720\ V$$

6.20 Calculate the electromotive force of a methane-O_2 fuel cell at 25 °C.

SOLUTION

$CH_4(g) + 2H_2O(\ell) = CO_2(g) + 8H^+ + 8e^-$

$2\ O_2(g) + 8e^- + 8H^+ = 4H_2O(\ell)$

$CH_4(g) + 2\ O_2(g) = CO_2(g) + 2H_2O(\ell)$

$\Delta G° = -394.359 + 2(-237.129) - (-50.72)$

$= -817.90\ kJ\ mol^{-1}$

$= -nFE°$

$$E° = \frac{817.90 \times 10^3\ J\ mol^{-1}}{8(96{,}485\ C\ mol^{-1})} = 1.0596\ V$$

6.21 Calculate the electromotive force of $Li(\ell)|LiCl(\ell)|Cl_2(g)$ at 900 K for $P_{Cl_2} = 1$ bar. This high-temperature battery is attractive because of its high electromotive force and low atomic masses. Lithium chloride melts at 883 K and lithium at 453.69 K. [The $\Delta G_f°$ for $LiCl(\ell)$ at 900 K in JANAF Thermochemical Tables is $-335.140\ kJ\ mol^{-1}$.]

SOLUTION

$Li(\ell) + \frac{1}{2} Cl_2(g) = LiCl(\ell)$

$\Delta G° = -335{,}140 \text{ J mol}^{-1} = -FE°$

$$E° = \frac{-335{,}140 \text{ J mol}^{-1}}{-\ 96{,}485 \text{ C mol}^{-1}} = 3.474 \text{ V}$$

6.22 A membrane permeable only by Na^+ is used to separate the following two solutions:

α	0.10 mol kg^{-1} NaCl	0.05 mol kg^{-1} KCl
β	0.05 mol kg^{-1} NaCl	0.10 mol kg^{-1} KCl

What is the membrane potential at 25 °C and which solution has the highest positive potential?

SOLUTION

$$\phi^\beta - \phi^\alpha = -\frac{RT}{z_i F} \ell n \frac{a_i^{\ \beta}}{a_i^{\ \alpha}}$$

$$= -\frac{(8.314 \text{ J K}^{-1} \text{ mol}^{-1})(298 \text{ K})}{96{,}485 \text{ C mol}^{-1}} \ell n \frac{0.05}{0.10}$$

$$= 0.018 \text{ V}$$

The β phase is more positive because of the diffusion of Na^+ from α to β. Since the ionic strengths of the two solutions are the same, the activity coefficients of Na^+ in α and β are very nearly the same.

6.23 277.8 kJ mol^{-1}

6.24 461.1, 46.11, 5.77 kJ mol^{-1}

6.25 0.905

6.26 (a) $[(a_+)(a_-)]^{1/2}/m$

(b) $[(a_+)(a_-)^3]^{1/4}/3^{3/4}\ m$

(c) $[(a_+)(a_-)]^{1/2}/m$

6.27 (a) 0.1 (b) 0.3 (c) 0.4 (d) 0.4

6.28 (a) $Na|NaOH(m)|H_2,Pt$

At 25 °C $E = E° - 0.0591 \log(m^2 \gamma_\pm^2 P_{H_2}^{1/2})$

(b) $Pt|H_2|H_2SO_4(m)|Ag_2SO_4|Ag$

At 25 °C $E = E° - 0.0296 \log(4m^3 \gamma_\pm^3 P_{H_2}^{-1})$

6.29 $pK = 1.018 \sqrt{I} + \log \frac{m_1}{m_2} + \frac{(E-Eo)F}{2.303RT} + \log m_3$

6.30 -130.318 kJ mol^{-1} -125 J K^{-1} mol^{-1}

-167.580 kJ mol^{-1}

6.31 (a) -131.260 kJ mol^{-1} -167.127 kJ mol^{-1}

-120.3 J K^{-1} mol^{-1}

(b) $-131,260$ kJ mol^{-1} -167.127 kJ mol^{-1}

56.5 J K^{-1} mol^{-1}

6.33 (a) The electrode with the higher percent thallium is negative.

(b) -1456 J mol^{-1} (c) 0.030462 V

6.34 The equilibrium constants are 5.79×10^{-38}, 9.56×10^{45}, 2.01×10^8, 3.20×10^4, and 3.23.

6.35 0.828 V

6.36 (a) 0.403 V (b) $-38,890$ J mol^{-1}

6.37 1.30

6.38 49.0, 2.83×10^8, 8.02×10^{16}

6.39 1.34×10^{-5} mol L^{-1}

6.40 -0.152 V

6.41 79.885, 55.835 kJ mol^{-1}

-80.668 J K^{-1} mol^{-1} 1.008×10^{-14}

6.42 (a) 2.62 (b) 2.4×10^{-3}

6.43 (a) 1.229 V (b) 1.229 V (c) 1.229 V

6.44 0.60

6.45 15.3

CHAPTER 7. Equilibria of Biochemical Reactions

7.1 According to Table A.1, what are the values of $\Delta G°$, $\Delta H°$, and $\Delta S°$ at 298 K for

$$H_2O(\ell) = H^+(ao) + OH^-(ao)$$

Show that the same value of $\Delta S°$ is obtained from $\Delta G°$ and $\Delta H°$ as by using $\Delta S° = \Sigma \nu_i S_i°$

SOLUTION

$\Delta G° = -157.244 + 237.129 = 79.885 \text{ kJ mol}^{-1}$

$\Delta H° = -229.994 + 285.830 = 55.836 \text{ kJ mol}^{-1}$

$\Delta S° = -10.75 - 69.91 = -80.66 \text{ J K}^{-1} \text{ mol}^{-1}$ or

$$\Delta S° = \frac{\Delta H° - \Delta G°}{T} = \frac{(55{,}836 - 79{,}885) \text{ J mol}^{-1}}{298.15 \text{ K}}$$

$$= -80.66 \text{ J K}^{-1} \text{ mol}^{-1}$$

7.2 Using Table A.1 calculate $\Delta H°$, $\Delta G°$, and $\Delta S°$ for

$$CH_3CO_2H(aq) = H^+(ao) + CH_3CO_2^-(ao)$$

and compare the values in Table 7.1 for 298.15 K.

SOLUTION

$\Delta H° = -486.01 + 485.76 = -0.25 \text{ kJ mol}^{-1}$

$\Delta G° = -369.31 + 396.46 = 27.15 \text{ kJ mol}^{-1}$

$\Delta S° = 86.6 - 178.7 = -92.1 \text{ J K}^{-1} \text{ mol}^{-1}$

7.3 For the acid dissociation of acetic acid $\Delta H°$ is approximately zero at room temperature in H_2O. For the acidic form of aniline, which is approximately as strong an acid as acetic acid, $\Delta H°$ is approximately 21 kJ mol^{-1}. Calculate $\Delta S°$ for each of the following reactions.

$$CH_3CO_2H = H^+ + CH_3CO_2^- \qquad pK = 4.75$$
$$C_6H_5NH_3^+ = H^+ + C_6H_5NH_2 \qquad pK = 4.63$$

How do you interpret these entropy changes? What compensates for the increase in entropy expected from the increase in number of molecules in the reaction?

SOLUTION

For acetic acid

$$\Delta G° = -RT\ell nK$$
$$= RT\ 2.303\ pK$$
$$= (8.314\ J\ K^{-1}\ mol^{-1})(298\ K)(2.303)(4.75)$$
$$= 27.1\ kJ\ mol^{-1}$$
$$\Delta S° = (\Delta H° - \Delta G°)/T$$
$$= -(27.1 \times 10^3\ J\ mol^{-1})/(298\ K)$$
$$= -91\ J\ K^{-1}\ mol^{-1}$$

The increase in order is due to the hydration of the ions that are formed.

For aniline

$$\Delta G° = (8.314\ J\ K^{-1}\ mol^{-1})(298\ K)(2.303)(4.63)$$
$$= 26.4\ kJ\ mol^{-1}$$
$$\Delta S° = [(21 - 26.4) \times 10^3\ J\ mol^{-1}]/(298\ K)$$
$$= -18\ J\ K^{-1}\ mol^{-1}$$

The entropy change is much smaller than for acetic acid because there is no change in the number of ions.

7.4 Estimate pK_3 and pK_2 for H_3PO_4 at 25 °C and 0.1 mol L^{-1} ionic strength. The values at zero ionic strength are $pK_3 = 2.148$ $\quad pK_2 = 7.198$

SOLUTION

$$pK_I = pK_{I=0} - \frac{(2n + 1)AI^{1/2}}{1 + I^{1/2}}$$

A = 0.509 at 25 °C

n is defined by $HA^{-n} = H^{+} + A^{-(n+1)}$

For pK_3 of H_3PO_4, n = 0

$$pK_3 = 2.148 - \frac{(0.509)(0.1)^{1/2}}{1 + (0.1)^{1/2}}$$

$$= 2.148 - 0.122 = 2.026$$

For pK_2 of H_3PO_4, n = 1

$$pK_2 = 7.198 - \frac{(3)(0.509)(0.1)^{1/2}}{1 + (0.1)^{1/2}}$$

$$= 7.198 - 0.369 = 6.831$$

7.5 At what pH is the average net charge on a lysine molecule zero? That is, what is its isoelectric point?

$$pK_3 = 2.16(-CO_2H)$$
$$pK_2 = 9.18(\alpha - \overset{+}{N}H_3)$$
$$pK_1 = 10.79(\varepsilon - \overset{+}{N}H_3)$$

(Simply set up the equation for calculating (H^+), but do not attempt to calculate (H^+) because the equation is cubic.)

SOLUTION

The three acid dissociation constants are represented by

$$LH = L^- + H^+ \qquad K_1 = (L^-)(H^+)/(LH)$$
$$LH_2^+ = LH + H^+ \qquad K_2 = (LH)(H^+)/(LH_2^+)$$
$$LH_3^+ = LH_2^+ + H^+ \qquad K_3 = (LH_2^+)(H^+)/(LH_3^{2+})$$

The net charge z is given by

$$z = \frac{-(L^-) + (LH_2^+) + 2(LH_3^{2+})}{(L^-) + (LH) + (LH_2) + (LH_3^{2+})}$$

$$= \frac{-K_1(LH)/(H^+) + (LH)(H^+)/K_2 + 2(LH)(H^+)^2/K_2K_3}{\text{denominator}}$$

The net charge is zero when the numerator is zero so that

$$-\frac{K_1}{(H^+)} + \frac{(H^+)}{K_2} + \frac{2(H^+)^2}{K_2K_3} = 0$$

Multiplying each term by $K_2K_3(H^+)$ yields

$$-K_1K_2K_3 + K_3(H^+)^2 + 2(H^+)^3 = 0$$

7.6 At pH 7 and pMg 4 what value of pCa is required to put half the ATP in the form $CaATP^{-2}$? At 0.2 mol L^{-1} ionic strength and 25 °C the following constants are known.

$$HATP^{3-} = H^+ + ATP^{4-} \qquad pK = 6.95$$

$$MgATP^{2-} = Mg^{2+} + ATP^{4-} \qquad pK = 4.00$$

$$CaATP^{2-} = Ca^{2+} + ATP^{4-} \qquad pK = 3.60$$

SOLUTION

$$\frac{(CaATP^{2-})}{(ATP^{4-}) + (HATP^{3-}) + (MgATP^{2-}) + (CaATP^{2-})} = \frac{1}{2}$$

$$\frac{(CaATP^{2-})/(ATP^{4-})}{1 + \frac{(HATP^{3-})}{(ATP^{4-})} + \frac{(MgATP^{2-})}{(ATP^{4-})} + \frac{(CaATP^{2-})}{(ATP^{4-})}} = \frac{1}{2}$$

$$\frac{(Ca^{2+})/10^{-3.60}}{1 + 10^{-0.05} + 1 + (Ca^{2+})/10^{-3.60}} = \frac{1}{2}$$

$(Ca^{2+}) = 7.25 \times 10^{-4}$ mol L^{-1} pCa = 3.14

7.7 Show how the partition function method yields the number ν_A of A bound per P and the number ν_B of B

are bound at the same site so that

$$PA = P + A \qquad K_A = (P)(A)/(PA)$$

$$PB = P + B \qquad K_B = (P)(B)/(PB)$$

Show that

$$\frac{\partial \nu_A}{\partial \ell n(B)} = \frac{\partial \nu_B}{\partial \ell n(A)}$$

for these linked bindings.

SOLUTION

$$Z = 1 + (A)/K_A + (B)/K_B$$

$$\nu_A = \frac{\partial \ell n Z}{\partial \ell n(A)} = \frac{(A)\partial Z}{Z\partial(A)} = \frac{(A)/K_A}{1 + (A)/K_A + (B)/K_B}$$

$$\nu_B = \frac{(B)/K_B}{1 + (A)/K_A + (B)/K_B}$$

$$\frac{\partial \nu_A}{\partial \ell n(B)} = \frac{(B)\partial \nu_A}{\partial (B)} = \frac{(A)(B)/K_A K_B}{[\,1 + (A)/K_A + (B)/K_B]^2}$$

$$\frac{\partial \nu_B}{\partial \ell n(A)} = \frac{(A)\partial \nu_B}{\partial (A)} = \frac{(A)(B)/K_A K_B}{[1 + (A)/K_A + (B)/K_B]^2}$$

7.8 Will 0.01 mol L^{-1} creatine phosphate react with 0.01 mol L^{-1} adenosine diphosphate to produce 0.04 mol L^{-1} creatine and 0.02 mol L^{-1} adenosine triphosphate at 25 °C, pH 7, pMg 4? What concentration of ATP can be formed if the other reactants are maintained at the indicated concentrations?

SOLUTION

Creatine P + H_2O = Creatine + P	$\Delta G°' = -43.5$ kJmol^{-1}
ADP + P = ATP + H_2O	$\Delta G°' = 39.8$ kJmol^{-1}
Creatine P + ADP = Creatine + ATP	$\Delta G°' = -3.7$ kJmol^{-1}

$$\Delta G = \Delta G° + RT\ell n\frac{(\text{Creatine})(\text{ATP})}{(\text{Creatine P})(\text{ADP})}$$

$$= -3700 + (8.314)(298)\ell n\frac{(0.04)\ (0.02)}{(0.01)\ (0.01)}$$

$$= 1340\ \text{J mol}^{-1}$$

Therefore the answer to the first question is no.

$$K = e^{-\Delta G°/RT} = e^{1340/(8.314)(298)}$$

$$= 4.5 = \frac{(\text{Creatine})(\text{ATP})}{(\text{Creatine P})(\text{ADP})}$$

If the reactants are maintained at the indicated concentrations.

$$(\text{ATP}) = \frac{4.5\ (\text{Creatine P})(\text{ADP})}{(\text{Creatine})} = \frac{4.5\ (0.01)(0.01)}{0.04}$$

$$= 1.1 \times 10^{-2}\ \text{mol L}^{-1}$$

7.9 In a series of biochemical reactions the product in one reaction is a reactant in the next. This has the effect that spontaneous reactions drive nonspontaneous reactions. For example, reaction 2 follows reaction 1.

1. L-malate = fumarate + H_2O $\Delta G°' =$ 2.9 kJmol^{-1}
2. fumarate + ammonia = aspartate $\Delta G°' = -15.6$ kJmol^{-1}

The $\Delta G°'$ values are for pH 7 and 37 °C, and the state of ionization of the reactants is ignored. In reaction 1, the activity of H_2O is to be taken as 1. If the ammonia concentration is 10^{-2} mol L^{-1}, calculate (aspartate)/(L-malate) at equilibrium.

SOLUTION

L-malate + ammonia = aspartate + H_2O $\Delta G°' = -15.6$ kJmol^{-1}

$$\Delta G = \Delta G° + RT\ell n\frac{(\text{aspartate})}{(\text{L-malate})(\text{ammonia})}$$

$$0 = -15,600 + (8.314)(298)\ln\frac{(\text{aspartate})}{(\text{L-malate})(10^{-2})}$$

$$\frac{(\text{aspartate})}{(\text{L-malate})} = 1.3$$

7.10 Biochemistry textbooks give $\Delta G°' = -20.1\ \text{kJmol}^{-1}$ for the hydrolysis of ethyl acetate at pH 7 and 25 °C. Experiments in acid solution show that

$$\frac{(CH_3CH_2OH)(CH_3CO_2H)}{(CH_3CO_2CH_2CH_3)} = 14$$

where concentrations are in moles per liter. What is the value of $\Delta G°'$ obtained from this equilibrium quotient? The pK of acetic acid = 4.60 at 25 °C.

SOLUTION

$$K' = \frac{(CH_3CH_2OH)[(CH_3CO_2H) + (CH_3CO_2^-)]}{(CH_3CO_2CH_2CH_3)}$$

$$= \frac{(CH_3CH_2OH)(CH_3CO_2H)}{(CH_3CO_2CH_2CH_3)}\left[1 + \frac{(CH_3CO_2^-)}{(CH_3CO_2H)}\right]$$

$$= 14\ \text{mol L}^{-1}\left[1 + \frac{K_{CH_3CO_2H}}{(H^+)}\right]$$

At pH 7

$$K' = 14\left[1 + \frac{10^{-4.6}}{10^{-7}}\right] = 3530$$

$$\Delta G°' = -RT\ln K$$

$$= -(8.314\ \text{J K}^{-1}\ \text{mol}^{-1})(298.15\ \text{K})\ln 3540$$

$$= -20.3\ \text{kJ mol}^{-1}$$

7.11 The cleavage of fructose 1,6-diphosphate (FDP) to dihydroxyacetone phosphate (DHP) and glyceraldehyde 3-phosphate (GAP) is one of a series of reactions most organisms use to obtain energy. At 37 °C and pH 7, $\Delta G^{\circ\prime}$ for the reaction FDP = DHP + GAP is 23.97 kJ mol^{-1}. What is $\Delta G'$ in an erythrocyte in which (FDP) = 3×10^{-6} mol L^{-1}, (DHP) = 138×10^{-6} mol L^{-1}, and (GAP) = 18.5×10^{-6} mol L^{-1}?

SOLUTION

$$FDP + H_2O = DHP + GAP$$

$$\Delta G^{\circ\prime} = \Delta G^{\circ\prime} + RT\ell n\frac{(DHP)(GAP)}{(FDP)}$$

$$= 23{,}970 + (8.314)(310)\ell n\frac{(138\times10^{-6})(18.5\times10^{-6})}{(3 \times 10^{-6})}$$

$$= 5770 \text{ J mol}^{-1}$$

7.12 Fumarase catalyzes the reaction

fumarate + H_2O = L-malate At 25 °C and pH 7

K' = 4.4 = (L-malate)/(fumarate)

What is the value of K' at pH 4? Given:

For fumaric acid $K_1 = 10^{-4.18}$

For L-malic acid $K_1 = 10^{-4.73}$

SOLUTION

$$\frac{(\text{L-malate})}{(\text{fumarate})} = \frac{(M) + (HM)}{(F) + (HF)} = \frac{(M)[1 + (H^+)/K_{1M}]}{(F)[1 + (H^+)/K_{1F}]}$$

$$= 4.4\ \frac{[1 + 10^{-4}/10^{-4.73}]}{[1 + 10^{-4}/10^{-4.18}]} = 4.4\ \frac{6.37}{2.51} = 11.2$$

7.13 Given ΔG° = 49.4 kJ mol^{-1} for

$$ATP^{4-} + H_2O = AMP^{2-} + P_2O_4^{7-} + 2H^+$$

calculate $\Delta G°'$ at pH 7 and 25 °C and 0.2 mol L^{-1} ionic strength. See Table 7.4.

SOLUTION

$$ATP^{4-} + H_2O \rightleftharpoons AMP^{2-} + P_2O_7^{4-} + 2H^+$$

$$ATP^{4-} \underset{10^{-6.95}}{\rightleftharpoons} HATP^{3-} \qquad AMP^{2-} \underset{10^{-6.45}}{\rightleftharpoons} HAMP^{1-} \qquad P_2O_7^{4-} \underset{10^{-8.95}}{\rightleftharpoons} HP_2O_7^{3-} \underset{10^{-6.12}}{\rightleftharpoons} H_2P_2O_7^{2-}$$

$$K' = \frac{[(AMP^{2-}) + (HAMP^-)][(P_2O_7^{4-}) + (HP_2O_7^{3-}) + (H_2P_2O_7^{2-})]}{[(ATP^{4-}) + (HATP^{3-})]}$$

$$= \frac{\dfrac{(AMP^{2-})(P_2O_7^{4-})}{(ATP^{4-})}\left[1 + \dfrac{(HAMP^-)}{(AMP^{2-})}\right]\left[1 + \dfrac{(HP_2O_7^{3-})}{(P_2O_7^{4-})} + \dfrac{(H_2P_2O_7^{2-})}{(P_2O_7^{4-})}\right]}{\left[1 + \dfrac{(HATP^{3-})}{(ATP^{4-})}\right]}$$

$$= \frac{(AMP^{2-})(P_2O_7^{4-})(H^+)^2}{(ATP^{4-})} \cdot \frac{\left[1 + \dfrac{(H^+)}{K_{1AMP}}\right]\left[1 + \dfrac{(H^+)}{K_{1PP}} + \dfrac{(H^+)^2}{K_{1PP}K_{2PP}}\right]}{(H^+)^2 \left[1 + \dfrac{(H^+)}{K_{1ATP}}\right]}$$

$$= e^{\frac{-49,400}{(8.314)(298)}} \cdot \frac{\left[1 + \dfrac{10^{-7}}{10^{-6.45}}\right]\left[1 + \dfrac{10^{7}}{10^{-8.95}} + \dfrac{(10^{-7})^2}{10^{-8.95}\ 10^{-6.12}}\right]}{(10^{-7})^2 \left[1 + \dfrac{10^{-7}}{10^{-6.95}}\right]}$$

$$= 1.53 \times 10^7$$

$\Delta G°' = -RT\ell nK' = -(8.314 JK^{-1}mol^{-1})(298K)\ell n(1.53x10^{7})$
$= -41.0 \text{ kJ mol}^{-1}$

7.14 The hydrolysis of adenosine triphosphate ATP to adenosine diphosphate ADP and inorganic phosphate at pH 8 and 25 °C

$$ATP^{-4} + H_2O = ADP^{-3} + HPO_4^{-2} + H^+$$

has a standard enthalpy change of -13 kJ mol^{-1}. The standard enthalpy changes of acid dissociation of $HATP^{-3}$, $HADP^{-2}$, and $H_2PO_4^{-1}$ are -8, 0, and +8 kJ mol^{-1}, respectively. Calculate the standard enthalpy change for the reaction

$$HATP^{-3} + H_2O = HADP^{-2} + H_2PO_4^-$$

SOLUTION

$ATP^{4-} + H_2O = ADP^{3-} + HPO_4^{2-} + H^+$	$\Delta H° = -13 \text{ kJmol}^{-1}$
$HATP^{3-} = H^+ + ATP^{4-}$	- 8
$H^+ + ADP^{3-} = HADP^{2-}$	0
$H^+ + HPO_4^{2-} = H_2PO_4^{-1}$	- 8
$HATP^{3-} + H_2O = HADP^{2-} + H_2PO_4^{-1}$	$\Delta H° = -29 \text{ kJmol}^{-1}$

7.15 Ethyl acetate is hydrolyzed in an aqueous buffer at pH 7 in a calorimeter. The enthalpy of hydrolysis is measured in a calorimeter and does not correspond with what is calculated from the following standard enthalpies of formation from a chemical thermodynamic table.

	ΔH_f° (298 K)
acetic acid (l)	-484.5 kJ mol^{-1}
ethanol (l)	-277.0
ethyl acetate (l)	-479.0
H_2O (l)	-285.8

Please explain why. What additional information would you need to calculate the heat absorbed in this experiment?

SOLUTION

ethyl acetate (ℓ) + $H_2O(\ell)$ = ethanol (ℓ) + acetic acid (ℓ)

$\Delta H° = 3.3 \text{ kJ mol}^{-1}$

This heat absorption would be obtained in acidic solution (where the acetic acid is undissociated) if the reactants and products form ideal solutions. At pH 7 H^+ is produced and reacts with the buffer. Therefore, it is necessary to know the heat of acid dissociation of the buffer. To calculate an accurate value for the heat absorbed, ΔH_f° is needed for

acetic acid dissolved in water

ethanol dissolved in water

ethyl acetate dissolved in water

7.16 If n molecules of a ligand A combine with a molecule of protein to form PA_n without intermediate steps, derive the relation between the fractional saturation Y and the concentration of A.

SOLUTION

$$P + nA = PA_n$$

$$K = \frac{(P)(A)^n}{(PA_n)} \qquad (P)_o = (P) + (PA_n)$$

$$\frac{[(P)_o - (PA_n)](A)^n}{(PA_n)} = K$$

$$(P)_o(A)^n = (PA_n)[K + (A)^n]$$

$$Y = \frac{(PA_n)}{(P)_o} = \frac{1}{1 + K/(A)^n} = \frac{(A)^n/K}{1 + (A)^n/K}$$

This equilibrium represents a cooperative effect in that as soon as one ligand molecule is bound, the other (n - 1) ligand molecules are also bound.

7.17 The percent saturation of a sample of human hemoglobin was measured at a series of oxygen partial pressures at 20 °C, pH 7.1, 0.3 mol L^{-1} phosphate buffer, and 3×10^{-4} mol L^{-1} heme.

P_{O_2}/Pa	Percent Saturation
393	4.8
787	20
1183	45
2510	78
2990	90

Calculate the values of h and K_h in the Hill equation. (See problem 7.19.)

SOLUTION

log P	$\log\left(\frac{Y}{1 - Y}\right)$
2.594	-1.31
2.896	-0.602
3.074	-0.087
3.400	+0.550
3.476	+0.954

(Please see graph, top of page 127.)

$$\log\left(\frac{Y}{1 - Y}\right) = -\log K_h + n \log P$$

$$n = \text{slope} = 2.4$$

$$K_h = 4 \times 10^7$$

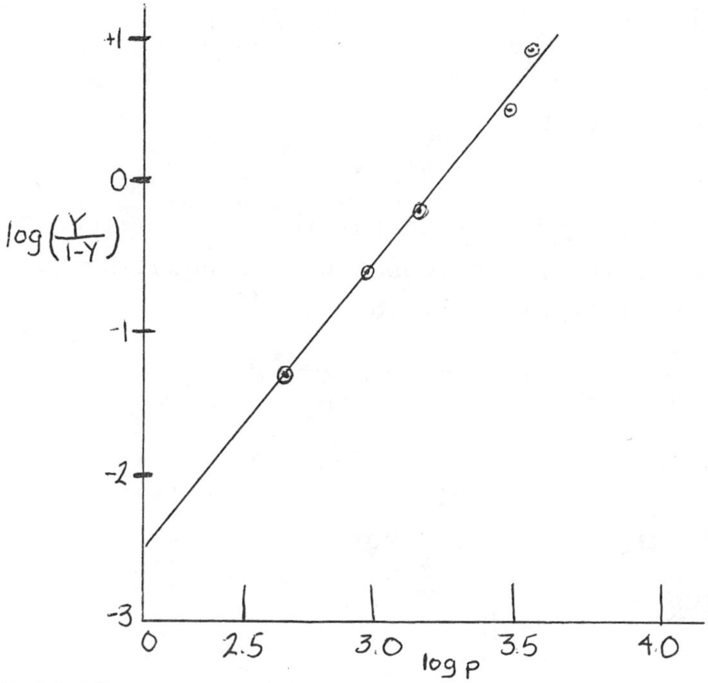

7.18 When myoglobin is in contact with air how many parts per million of CO are required to tie up 10% of the myoglobin? The partial pressure of oxygen required to half saturate myoglobin at 25 °C is 3.7 kPa. The partial pressure of CO required to half saturate myoglobin in the absence of oxygen is 0.009 kPa.

SOLUTION

$$\frac{P_{CO}/K_{CO}}{1 + P_{O_2}/K_{O_2} + P_{CO}/K_{CO}} = 0.1$$

$K_{O_2} = 3.7$ kPa $\qquad K_{CO} = 0.009$ kPa

$P_{O_2} = (0.2)(101.325 \text{ kPa}) = 20.3$ kPa

$$\frac{P_{CO}/0.009\text{ kPa}}{1 + 20.3\text{ kPa}/3.7\text{ kPa} + P_{CO}/0.009\text{ kPa}} = 0.1$$

$P_{CO} = 0.0065$ kPa

$$\frac{P_{CO}}{101.3\text{ kPa}} = 64\text{ ppm}$$

7.19 Since it is difficult to determine the values of the four dissociation constants in equation 7.74, the empirical Hill equation

$$Y = \frac{1}{1 + K_h/P_{O_2}^h}$$

is frequently used to characterize binding. Show that the Hill coefficient h may be obtained by plotting log[Y/(1 - Y)] versus log P_{O_2}.

<u>SOLUTION</u>

$$\frac{Y}{1 - Y} = \frac{P_{O_2}^h}{K_h}$$

$$\log \frac{Y}{1 - Y} = -\log K_h + h \log P_{O_2}$$

For a variety of binding systems relatively linear Hill plots are obtained for values of Y in the range 0.1 to 0.9, but deviations usually occur at the extremes unless h = 1. At the extremes the plot usually approaches a slope of unity.

7.20 Assuming that a protein molecule has n_α independent sites with intrinsic dissociation constant K_α and n_β independent sites with intrinsic dissociation constant K_β, what is the expression for the number of moles of A bound per mole of protein?

<u>SOLUTION</u>

$$\nu_A = \frac{n_\alpha\ (PA)}{(P) + (PA)} \quad \frac{n_\alpha\ (P'A)}{(P') + (P'A)}$$

$$= \frac{n_\alpha}{1 + (P)/(PA)} + \frac{n_\beta}{1 + (P')/(P'A)}$$

$$= \frac{n_\alpha}{1 + K_\alpha/(A)} + \frac{n_\beta}{1 + K_\beta/(A)}$$

7.21 $K_2 = 5.884 \times 10^{-14}$ $pK_2 = 13.230$ $pH = 6.615$

7.22 The pH of a solution of monosodium aspartate is $(pK_1 + pK_2)/2 = 6.9$. The pH of a solution of disodium aspartate is greater than 10.

7.24 4.710

7.25 (a) 4.335×10^{-7}
(b) 0.989

7.27 4.35

7.28 $$Z = 1 + \frac{(H^+)}{K_{1PP}} + \frac{(H^+)^2}{K_{1PP}K_{2PP}} + \frac{(Mg^{2+})}{K_{MgPP}} + \frac{(Mg^{2+})(H^+)}{K_{MgPP}K_{1PP}}$$

$$\nu_H = \left[\frac{(H^+)}{K_{1PP}} + \frac{2(H^+)^2}{K_{1PP}K_{2PP}} + \frac{(Mg^{2+})(H^+)}{K_{1PP}K_{MgPP}}\right]/Z$$

$$\nu_{Mg} = \left[\frac{(Mg^{2+})}{K_{MgPP}} + \frac{(H^+)(Mg^{2+})}{K_{1PP}K_{MgPP}}\right]/Z$$

7.29 0.063

7.30 $\Delta G = 9.2\ kJ\ mol^{-1}$ No

7.31 133 g

7.32 $3.88 \times 10^{-2}\ mol\ L^{-1}$

7.33 $-42.3\ kJ\ mol^{-1}$

7.34 $-30.33\ kJ\ mol^{-1}$

7.35 2.93×10^7

7.36

	pH 8, pMg 4	pH 8, pMg 8
$\Delta H°'$	$-26.57\ kJmol^{-1}$	$-19.7\ kJmol^{-1}$
$\Delta G°'$	$-42.7\ kJmol^{-1}$	$-44.4\ kJmol^{-1}$

7.37 (a) $K' = K_1[1 + K_{CH}/(H^+)]/[1 + K_{AH}/(H^+)]$
(b) $K_2K_{CH}/K_1K_{AH} = 1$

7.38 0.158 mol L^{-1} 7.39 25.6%

7.40 0.93, 2.47, 5.55, 14.8 kPa

7.41 $K_2 = 4K_1$
$K_2 < 4K_1$ positive cooperativity
$K_2 > 4K_1$ negative cooperativity

7.42 $K_1 = k/5$, $K_2 = k/2$, $K_3 = k$, $K_4 = 2k$, $K_5 = 5k$
1/32, 5/32, 10/32, 10/32, 5/32, 1/32

7.43 -10.75 J K^{-1} mol^{-1}

CHAPTER 8. Surface Thermodynamics

8.1 The surface tension of toluene at 20 °C is 0.0284 N m^{-1}, and its density at this temperature is 0.866 g cm^{-3}. What is the radius of the largest capillary that will permit the liquid to rise 2 cm?

SOLUTION

$$\gamma = \frac{1}{2} h\rho g r$$

$$\nu = \frac{2\gamma}{\rho g h} = \frac{2(0.284 \text{ N m}^{-1})}{(0.866\text{x}10^{3} \text{ kg m}^{-3})(9.80 \text{ m s}^{-2})(0.02 \text{ m})}$$

$$= 3.35 \times 10^{-4} \text{ m}$$

$$= 3.35 \times 10^{-2} \text{ cm}$$

8.2 If the surface tension of a soap solution is 0.050 N m^{-1}, what is the pressure inside (a) a soap bubble 2 mm in diameter and (b) a bubble 2 cm in diameter?

SOLUTION

The effective surface tension is 2γ because the soap film has 2 surfaces.

(a) $$\Delta P = \frac{2\gamma}{R} = \frac{4(0.050 \text{ N m}^{-1})}{0.002 \text{ m}} = 100 \text{ Pa}$$

(b) $$\Delta P = \frac{4(0.050 \text{ N m}^{-1})}{0.02 \text{ m}} = 10 \text{ Pa}$$

8.3 Calculate the vapor pressure of a water droplet at 25 °C that has a radius of 2 nm. The vapor pressure of a flat surface of water is 3167 Pa at 25 °C.

SOLUTION

$$\ell n\left(\frac{P}{P°}\right) = \frac{2V_m\gamma}{r\ RT}$$

$$\ell n\left(\frac{P}{3167\ Pa}\right) = \frac{2(18.016\times10^{-3}kgmol^{-1})(0.07197Nm^{-1})}{(2\times10^{-9}m)(8.314JK^{-1}mol^{-1})(298.15K)} \times \frac{1}{(10^3\ kgm^{-3})}$$

$$P = 5350\ Pa$$

8.4 The surface tensions of 0.05 and 0.127 mol kg^{-1} solutions of phenol in water are 67.7 and 60.1 mN m^{-1}, respectively, at 20 °C. What is the surface concentration $\Gamma_2^{(1)}$ in the range 0 to 0.05 mol kg^{-1} and 0.05 to 0.127 mol kg^{-1}, assuming that the phenol concentrations can be treated as activities? The surface tension of water at 20 °C is 72.7 mN m^{-1}.

SOLUTION

$$\Gamma_2^{(1)} = \frac{-a_2}{RT}\frac{\partial\gamma}{\partial a_2}$$

In the range 0-0.05 mol kg^{-1}

$$\frac{\partial\gamma}{\partial a_2} = \frac{67.7 - 72.7}{0.05} = -100 \times 10^{-3}\ N\ m^{-1}$$ (since the activity a_2 is taken as dimensionless)

$$\Gamma_2^{(1)} = -\frac{(0.025)(-100 \times 10^{-3}\ N\ m^{-1})}{(8.314\ J\ K^{-1}\ mol^{-1})(293\ K)}$$
$$= 1.03 \times 10^{-6}\ mol\ m^{-2}$$

In the range 0.05-0.127 mol kg^{-1}

$$\frac{\partial\gamma}{\partial a_2} = \frac{60.1 - 67.7}{0.077} = -99 \times 10^{-3}\ \text{N m}^{-1}$$

$$\Gamma_2^{(1)} = -\frac{[(0.05 + 0.127)/2](-99 \times 10^{-3}\ \text{N m}^{-1})}{(8.314\ \text{J K}^{-1}\ \text{mol}^{-1})(293\ \text{K})}$$

$$= 3.6 \times 10^{-6}\ \text{mol m}^{-2}$$

8.5 (a) The surface tension of water against air at 1 bar is given in the following table for various temperatures.

t/°C	20	22	25	28	30
γ/N m^{-1}	0.07275	0.07244	0.07197	0.07150	0.07118

Calculate the surface enthalpy at 25 °C in joules per square centimeter. (b) If a finely divided solid whose surface is covered with a very thin layer of water is dropped into a container of water at the same temperature, heat will be evolved. Calculate the heat evolution for 10 g of a powder having a surface area of 200 m^2 g^{-1} [W. D. Harkins and G. Jura, J.Am.Chem.Soc.,66, 1362 (1944)].

SOLUTION

(a) $h^\sigma = \gamma - T\left(\frac{\partial\gamma}{\partial T}\right)_P$

$$= 0.07197\ \text{Jm}^{-2} - (298.15\text{K})(-1.48\text{x}10^{-4}\text{Jm}^{-2}\text{K}^{-1})$$

$$= 0.1161\ \text{J m}^{-2}$$

$$= (0.1161\ \text{J m}^{-2})(10^{-4}\ \text{m}^2\ \text{cm}^{-2})$$

$$= 1.161 \times 10^{-5}\ \text{J cm}^{-2}$$

(b) $(10\text{g})(200\text{m}^2\text{g}^{-1})(10^4\text{cm}^2\text{m}^{-2})(1.161\text{x}10^{-5}\text{Jcm}^{-2}) = 233\ \text{J}$

8.6 The surface tension of liquid nitrogen at 75 K is 0.009 71 N m^{-1}, and the temperature coefficient of the surface tension is -2.3×10^{-4} N m^{-1} K^{-1}. What is the surface enthalpy?

SOLUTION

$$h^{\sigma} = \gamma - T\left(\frac{\partial \gamma}{\partial T}\right)_P$$

$$= 0.00971 \text{ N m}^{-1} - (75\text{K})(-2.3 \times 10^{-4} \text{ N m}^{-1} \text{ K}^{-1})$$

$$= 0.0269 \text{ J m}^{-2}$$

8.7 A solution of palmitic acid (M = 256 g mol^{-1}) in benzene contains 4.24 g of acid per liter. When this solution is dropped on a water surface, the benzene evaporates and the palmitic acid forms a monomolecular film of the solid type. If we wish to cover an area of 500 cm^2 with a monolayer, what volume of solution should be used? The area occupied by one palmitic acid molecule may be taken to be 21×10^{-20} m^2.

SOLUTION

The number of molecules in a cubic centimeter of solution is

$$\frac{(4.24 \text{ g L}^{-1}) \quad (6.022\text{x}10^{23} \text{ mol}^{-1})}{(256 \text{ g mol}^{-1}) \quad (1000 \text{ cm}^3 \text{ L}^{-1})} = 9.97 \times 10^{18} \text{ cm}^{-3}$$

The number of molecules required to cover 500 cm^2 is

$$\frac{(500 \text{ cm}^2)(10^{-2} \text{ m cm}^{-1})^2}{21 \times 10^{-20} \text{ m}^2} = 2.38 \times 10^{17}$$

$$\frac{2.38 \times 10^{17}}{9.97 \times 10^{18} \text{ cm}^{-3}} = 0.0239 \text{ cm}^3$$

8.8 A protein with a molar mass of 60,000 g mol^{-1} forms a perfect gaseous film on water. What area of film per milligram of protein will produce a pressure of 0.005 N m^{-1} at 25 °C?

SOLUTION

$$\sigma = \frac{kT}{\pi} \qquad \text{area} = \frac{gN_A\sigma}{M} = \frac{gN_AkT}{M\pi}$$

$$= \frac{(10^{-6}\ \text{kg})(6.022\text{x}10^{23}\ \text{mol}^{-1})(1.38\text{x}10^{-23}\ \text{J K}^{-1})(298\text{K})}{(60\ \text{Kg mol}^{-1})(0.005\ \text{N m}^{-1})}$$

$$= 82.5 \text{ x } 10^{-4}\ \text{m}^2$$

$$= 82.5\ \text{cm}^2$$

8.9 The acid $CH_3(CH_2)_{13}CO_2H$ forms a nearly perfect gaseous monolayer on water at 25 °C. Calculate the weight of acid per 100 cm^2 required to produce a film pressure of 10^{-3} N m^{-1}.

SOLUTION

$$\pi\mathcal{A} = \frac{N}{N_A}RT = \frac{g\ RT}{M}$$

$$g = \frac{\pi\mathcal{A}M}{RT}$$

$$= \frac{(10^{-3}\ \text{N m}^{-1})(10^{-2}\ \text{m}^2)(244 \text{ x } 10^{-3}\ \text{kg mol}^{-1})}{(8.314\ \text{J K}^{-1}\ \text{mol}^{-1})(298.15\ \text{K})}$$

$$= 9.8 \text{ x } 10^{-10}\ \text{kg} = 9.8 \text{ x } 10^{-7}\ \text{g}$$

8.10 What volume of oxygen, measured at 25 °C and 1 bar, is required to form an oxide film on 1 m^2 of a metal with atoms in a square array 0.1 nm apart?

SOLUTION

Assuming one oxygen atom combines with each metal atom, the number of moles of O_2 adsorbed per m^2 is

$$\frac{(1\ m^2)}{(2)(0.1 \times 10^{-9}\ m)^2}$$

The volume of oxygen is given by

$$\frac{(1\ m^2)(8.314\ J\ K^{-1}\ mol^{-1})(298\ K)}{(2)(0.1 \times 10^{-9} m)^2 (6.022 \times 10^{23} mol^{-1})(101\ 325\ Pa)}$$

$= 2 \times 10^{-6}\ m^3 \quad = \quad 2 \times 10^{-3}\ L$

8.11 The following table gives the number of milliliters (v) of nitrogen (reduced to 0 °C and 1 bar) adsorbed per gram of active carbon at 0 °C at a series of pressures.

P/Pa	524	1731	3058	4534	7497
$v/cm^3\ g^{-1}$	0.987	3.04	5.08	7.04	10.31

Plot the data according to the Langmuir isotherm, and determine the constants.

<u>SOLUTION</u>

$$\frac{1}{v} = \frac{1}{v_m} + \frac{1}{v_m KP} \qquad \text{Intercept} = 0.025\ g\ cm^{-3}$$

$$v_m = 1/(0.025\ g\ cm^{-3})$$
$$= 40\ cm^3\ g^{-1}$$

$$\text{Slope} = \frac{(1.00 - 0.025)\ g\ cm^{-3}}{1.9 \times 10^{-3}\ Pa^{-1}}$$

$$= 513\ g\ cm^{-3}\ Pa$$

$$= \frac{1}{v_m K} = \frac{1}{(40\ cm^3\ g^{-1})K}$$

$$K = [(513 g cm^{-3} Pa)(40 cm^3 g^{-1})]^{-1} = 4.8 \times 10^{-5}\ Pa^{-1}$$

(Please see graph page 137)

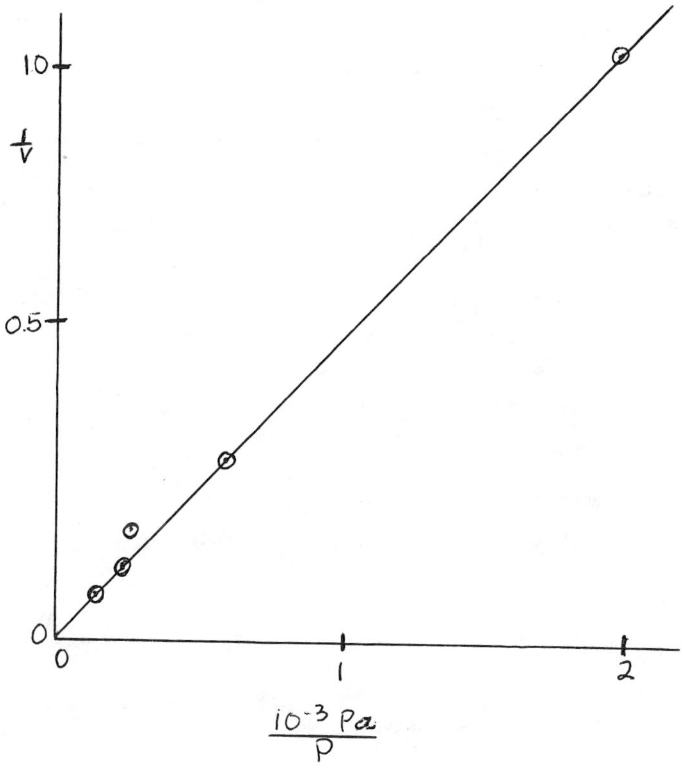

8.12 Hydrogen gas is adsorbed on a metal surface as atoms. Show that the fractional saturation θ of the surface is given by

$$\theta = \frac{K P_{H_2}^{1/2}}{1 + K P_{H_2}^{1/2}}$$

SOLUTION

$$H_2(g) = 2H(g) \qquad K_P = \frac{P_H^{\,2}}{P_{H_2}}$$

$$P_H = (K_P P_{H_2})^{1/2}$$

$$\theta = \frac{K_H P_H}{1 + K_H P_H} = \frac{K_H (K_P P_{H_2})^{1/2}}{1 + K_H (K_P P_{H_2})^{1/2}}$$

$$= \frac{KP_{H_2}^{1/2}}{1 + KP_{H_2}^{1/2}}$$

8.13 Calculate the surface area of a catalyst that adsorbs 103 cm^3 of nitrogen (calculated at 1.013 bar and 0 °C) per gram in order to form a monolayer. The adsorption is measured at -195 °C, and the effective area occupied by a nitrogen molecule on the surface is 16.2 x 10^{-20} m^2 at this temperature.

SOLUTION

The surface area is equal to the number of molecules times the effective area per molecule. The number of molecules is

$$\frac{PVN_A}{RT}$$

$$A = \frac{(1.013\ \text{bar})(0.103\text{L})(6.022\text{x}10^{23}\text{mol}^{-1})(16.2\text{x}10^{-20}\text{m}^2)}{(0.083\ \text{L bar K}^{-1}\ \text{mol}^{-1})(273\ \text{K})}$$

$$= 449\ \text{m}^2$$

8.14 The pressures of nitrogen required to cause the adsorption of 1.0 cm^3 g^{-1}(25 °C, 1.013 bar) of gas on P-33 graphitized carbon black are 24 Pa at 77.5 K and 290 Pa at 90.1 K. Calculate the enthalpy of adsorption at this fraction of surface coverage using the Clausius-Clapeyron equation.

SOLUTION

$$\Delta H = \frac{RT_1T_2\ell n(P_2/P_1)}{T_2 - T_1}$$

$$= \frac{(8.314\ \text{J K}^{-1}\ \text{mol}^{-1})(77.5\ \text{K})(90.1\ \text{K})\ell n(290/24)}{(90.1\ \text{K} - 77.5\ \text{K})}$$

$= 11.6\ \text{kJ mol}^{-1}$

8.15 Adsorption of nitrogen, at its boiling point -195.8 °C, is often used in estimating surface areas of solids. At this temperature nitrogen has a density of 0.808 g cm^{-3} and a surface tension of 8.85 mN m^{-1}. If the isotherm is of the type shown in Fig. 8.9b and hysteresis is encountered at about $P/P^{sat} = 0.5$, what does this imply about the radii of pores in the solid?

SOLUTION

$$r = \frac{2V_m\gamma}{\left(-\ln\frac{P}{P^\circ}\right)RT}$$

$$= \frac{2(14 \times 10^{-3}\text{kg mol}^{-1})(0.00885\ \text{N m}^{-1})}{(0.693)(8.314\text{JK}^{-1}\text{mol}^{-1})(77.4\text{K})(0.808\text{x}10^{3}\text{kgm}^{-3})}$$

$$= 0.69 \times 10^{-9}\ \text{m}$$

8.16 $0.0233\ \text{N m}^{-1}$

8.17 (a) 15.72 cm (b) 0.786 cm

8.18 (a) 15.2×10^{-10} m (b) 97

8.20 $4.04 \times 10^{-6}\ \text{mol m}^{-2}$

8.21 (a) A cooling effect because energy is required to move molecules to the surface.

(b) $0.1187\ \text{J m}^{-2}$

8.22 244 m

8.23 $$\pi = \frac{kT}{\sigma_o}\ln\left(\frac{1}{1-\theta}\right)$$

8.24 (a) $v_m = 38.2\ \text{mm}^3$ $K = 1.22 \times 10^{-3}\ \text{Pa}^{-1}$

(b) $30.4\ \text{mm}^3$

8.25 39.6 L

8.26 $$v = \frac{v_m(x_A K_A + x_B K_B)P}{1 + (x_A K_A + x_B K_B)P}$$

$$x_{A,ads} = \frac{\theta_A}{\theta_A + \theta_B} = \frac{x_A K_A}{x_A K_A + x_B K_B}$$

8.27 20.3 L at 25 °C and 1 bar

8.28 36.2 kJ mol^{-1}

8.29 29.5 kJ mol^{-1}

8.30 1.51 mm

PART TWO
QUANTUM CHEMISTRY

CHAPTER 9. Quantum Theory

9.1 A hollow box with an opening of 1 cm^2 area is heated electrically. (a) What is the total energy emitted per second at 800 K? (b) How much energy is emitted per second if the temperature is 1600 K? (c) How long would it take the radiant energy emitted from the box at this temperature, 1600 K, to melt 1000 g of ice?

SOLUTION

(a) $I = \sigma T^4 = (5.67 \times 10^{-8}\ J\ s^{-1}\ m^{-2}\ K^{-4})(800\ K)^4$

$= 2.32 \times 10^4\ J\ m^{-2}\ s^{-1}$

(b) $I = \sigma T^4 = (5.67 \times 10^{-8}\ J\ s^{-1}\ m^{-2}\ K^{-4})(1600\ K)^4$

$= 37.2 \times 10^4\ J\ m^{-2}\ s^{-1}$

(c) $\Delta H = (333\ J\ g^{-1})(1000\ g\ kg^{-1})$

$= 3.33 \times 10^5\ J$

$$t = \frac{3.33 \times 10^5\ J}{37.2\ J\ s^{-1}} = 8950\ s$$

9.2 At what temperature does M_λ for cavity radiation have a maximum at (a) 800 nm and (b) 400 nm? (c) What is λ_{max} for a cavity radiation at 25 °C?

SOLUTION

(a) $T = 2.898 \times 10^{-3}\ K\ m/\lambda_{max}$

$$= \frac{2.898 \times 10^{-3}\ K\ m}{800 \times 10^{-9}\ m} = 3623\ K$$

(b) $T = \dfrac{2.898 \times 10^{-3} \text{ K m}}{400 \times 10^{-9} \text{ m}} = 7245 \text{ K}$

(c) $\lambda_{max} = \dfrac{2.898 \times 10^{-3} \text{ K m}}{298 \text{ K}} = 9.72 \times 10^{-6} \text{ m}$

This wavelength is in the middle of the near infrared range.

9.3 Derive the value of the constant in the Wien displacement law (equation 9.7).

SOLUTION

$$M_\lambda = \frac{2\pi hc^2}{\lambda^4} \frac{1}{e^{hc/\lambda kT} - 1}$$

Setting $dM_\lambda/d\lambda = 0$ yields $\dfrac{xe^x}{e^x - 1} = 5$

where $x = \dfrac{hc}{kT\lambda}$

Substituting successive values of x shows that $x = 4.965$. Thus

$$\lambda_{max} T = \frac{hc}{4.965\ k}$$

$$= \frac{(6.626 \times 10^{-34} \text{ Js})(2.998 \times 10^8 \text{ ms}^{-1})}{(4.965)(1.3807 \times 10^{-23} \text{ J K}^{-1})}$$

$$= 2.898 \times 10^{-3} \text{ K m}$$

9.4 Calculate the wavelengths (in micrometers) of the first three lines of the Paschen series for atomic hydrogen.

SOLUTION

$$\lambda = \frac{1}{R} \frac{n_1^2\ n_2^2}{(n_2^2 - n_1^2)} = \frac{1}{R} \frac{9\ n_2^2}{(n_2^2 - 9)}$$

For $n_2 = 4$ $\lambda = \dfrac{(9)(16)}{(109677.58\ \text{cm}^{-1})(7)} = 1.8756\ \mu\text{m}$

For $n_2 = 5$ $\lambda = \dfrac{(9)(25)}{(109677.58\ \text{cm}^{-1})(16)} = 1.2822\ \mu\text{m}$

For $n_2 = 6$ $\lambda = \dfrac{(9)(36)}{(109677.58\ \text{cm}^{-1})(27)} = 1.0941\ \mu\text{m}$

9.5 In the Balmer series for atomic hydrogen what is the wavelength of the series limit?

SOLUTION

$$\lim_{n_2 \to \infty} \frac{1}{\lambda} = \lim_{n_2 \to \infty} R\left(\frac{1}{2^2} - \frac{1}{n_2^2}\right) = \frac{R}{4}$$

$$\lambda = \frac{4}{R} = \frac{4}{1.0967758 \times 10^7\ \text{m}^{-1}} = 364.7054\ \text{nm}$$

9.6 Calculate the wavelength of light emitted when an electron falls from the n = 100 orbit to the n = 99 orbit of the hydrogen atom. Such species are known as high Rydberg atoms. They are detected in astronomy and are more and more studied in the laboratory.

SOLUTION

$$E_{100} - E_{99} = -(1.0967 \times 10^7 \text{m}^{-1})(10^{-2} \text{mcm}^{-1})\left(\frac{1}{100^2} - \frac{1}{99^2}\right)$$

$$= 0.2227\ \text{cm}^{-1}$$

$$\lambda = 4.49\ \text{cm}$$

9.7 What potential difference is required to accelerate a singly charged gas ion in a vacuum so that it has (a) a kinetic energy equal to that of an average gas molecule at 25 °C; (b) an energy equivalent to 83.7 kJ mol^{-1}?

SOLUTION

(a) It has been shown earlier that the kinetic energy of an average gas molecule is (3/2)kT.

$$Ee = \frac{3}{2} kT$$

$$E = \frac{3(1.381 \times 10^{-23} \text{ J K}^{-1})(298 \text{ K})}{2(1.602 \times 10^{-19} \text{ C})}$$

$$= 0.0385 \text{ V}$$

(b) $$Ee = \frac{83{,}700 \text{ J mol}^{-1}}{6.022 \times 10^{23} \text{ mol}^{-1}} = 1.3896 \times 10^{-19} \text{ J}$$

$$E = \frac{1.3896 \times 10^{-19} \text{ J}}{1.602 \times 10^{-19} \text{ C}} = 0.867 \text{ V}$$

9.8 Calculate the velocity of an electron that has been accelerated by a potential difference of 1000 V.

SOLUTION

$$Ee = \frac{1}{2} m v^2$$

$$v = \left(\frac{2Ee}{m}\right)^{1/2}$$

$$= \left[\frac{2(1000 \text{ V})(1.602 \times 10^{-19} \text{ C})}{9.110 \times 10^{-31} \text{ kg}}\right]^{1/2}$$

$$= 1.875 \times 10^{7} \text{ ms}^{-1}$$

As higher accelerating potentials are used it becomes necessary to use the relativistic mass of the electron as discussed in the footnote in Section 1.12.

9.9 Electrons are accelerated by a 1000 V potential drop. (a) Calculate the de Broglie wavelength.

(b) Calculate the wavelength of the x-rays that could be produced when these electrons strike a solid.

SOLUTION

(a) $Ee = \frac{1}{2} m v^2$

$$v = \left(\frac{2Ee}{m}\right)^{1/2}$$

$$= \left[\frac{2(1000\ V)(1.602 \times 10^{-19}\ C)}{9.110 \times 10^{-31}\ kg}\right]^{1/2}$$

$$= 1.875 \times 10^{7}\ ms^{-1}$$

$$\lambda = \frac{h}{mv}$$

$$= \frac{6.626 \times 10^{-34}\ Js}{(9.110 \times 10^{-31}\ kg)(1.875 \times 10^{7}\ ms^{-1})}$$

$$= 0.0387\ nm$$

(b) $Ee = hc/\lambda \qquad \lambda = \frac{hc}{Ee}$

$$\lambda = \frac{(6.626 \times 10^{-34}\ Js)(2.998 \times 10^{8}\ ms^{-1})}{(1000\ V)(1.602 \times 10^{-19}\ C)}$$

$$= 1.24\ nm$$

9.10 An ultraviolet photon (λ = 58.4 nm) from a helium gas discharge tube is absorbed by a hydrogen molecule which is at rest. Since momentum is conserved, what is the velocity of the hydrogen molecule after absorbing the photon? What is the kinetic energy of the hydrogen molecule in J mol^{-1}?

SOLUTION

$$p = \frac{h}{\lambda} = \frac{6.626 \times 10^{-34}\ Js}{58.4 \times 10^{-9}\ m} = 1.135 \times 10^{-26}\ kg\ m\ s^{-1}$$

$$= mv = \frac{2(1.0079 \times 10^{-3} \text{ kg mol}^{-1})}{6.022 \times 10^{23} \text{ mol}^{-1}} v$$

$$v = \frac{(1.135 \times 10^{-26} \text{ kg m s}^{-1})(6.022 \times 10^{23} \text{ mol}^{-1})}{2(1.0079 \times 10^{-3} \text{ kg mol}^{-1})}$$

$$= 3.39 \text{ m s}^{-1}$$

$$E = \frac{1}{2} mv^2 N_A = 0.012 \text{ J mol}^{-1}$$

9.11 Calculate the de Broglie wavelength of a hydrogen atom with a translational energy corresponding to room temperature.

SOLUTION

$$\frac{1}{2} mv^2 = \frac{3}{2} kT \qquad v = \sqrt{\frac{3kT}{m}}$$

$$v = \sqrt{\frac{(3)(1.38\text{x}10^{-23} \text{ JK}^{-1})(298 \text{ K})(6.02 \times 10^{23} \text{ mol}^{-1})}{1.0079 \times 10^{-3} \text{ kg mol}^{-1}}}$$

$$= 2.71 \times 10^3 \text{ m s}^{-1}$$

$$\lambda = \frac{h}{mv} = \frac{(6.63\text{x}10^{-27} \text{ Js})(6.02\text{x}10^{23} \text{ mol}^{-1})}{(2.71\text{x}10^3 \text{ m s}^{-1})(1.0079\text{x}10^{-3} \text{ kg mol}^{-1})}$$

$$= 0.146 \text{ nm}$$

9.12 Since a wave function may be complex, it may be represented by $\psi = R + iI$ where $i = \sqrt{-1}$. Show that $\psi^*\psi$ is always real.

SOLUTION

$$\psi^*\psi = (R - iI)(R + iI)$$

$$= R^2 - i^2I = R^2 + I^2$$

9.13 Calculate the approximate quantum number corresponding with the translational energy of a 1-g bullet fired with a velocity of 300 m s^{-1} at a target 100 m away. What is the de Broglie wavelength?

SOLUTION

$$E = \frac{1}{2}\, m\, v^2 = \frac{1}{2}\,(10^{-3}\ \text{kg})(300\ \text{m s}^{-1})^2 = 45.00\ \text{J}$$

$$E = \frac{n^2\, h^2}{8\ ma^2}$$

$$n = (\frac{a}{h})(8\ m\ E)^{1/2}$$

$$= (\frac{100\ \text{m}}{6.626 \times 10^{-34}\ \text{Js}})[(8)(10^{-3}\ \text{kg})(45\ \text{J})]^{1/2}$$

$$= 9.05 \times 10^{34}$$

$$\lambda = \frac{h}{mv} = \frac{6.626 \times 10^{-34}\ \text{Js}}{(10^{-3}\ \text{kg})(300\ \text{m s}^{-1})} = 2.21 \times 10^{-33}\ \text{m}$$

9.14 (a) Calculate the energy levels for n = 1 and n = 2 for an electron in a potential well of width 0.5 nm with infinite barriers on either side. The energies should be expressed in kJ mol^{-1}. (b) If an electron makes a transition from n = 2 to n = 1, what will be the wavelength of the radiation emitted?

SOLUTION

(a) $$E = \frac{n^2\, h^2}{8\ ma^2}$$

$$= \frac{(6.626 \times 10^{-34}\ \text{J s})^2}{8(9.110 \times 10^{-31}\ \text{kg})(5 \times 10^{-10}\ \text{m})^2}$$

$$= 2.4096 \times 10^{-19}\ \text{J}$$

$$= (2.4096\text{x}10^{-19}\ \text{J})(6.022\text{x}10^{23}\ \text{mol}^{-1})$$

$$= 145.1\ \text{kJ mol}^{-1} \quad \text{for } n = 1$$

For n = 2

$$E = 4(145.1 \text{ kJ mol}^{-1})$$

$$= 580.4 \text{ kJ mol}^{-1}$$

(b) $\Delta E = (2.4096 \times 10^{-19} \text{ J})(4 - 1)$

$$\Delta E = \frac{hc}{\lambda}$$

$$(2.4096\times10^{-19}\text{J})(4-1) = \frac{(6.626\times10^{-34}\text{Js})(2.998\times10^{8}\text{ms}^{-1})}{\lambda}$$

$$\lambda = 274.7 \text{ nm}$$

9.15 For a hydrogen atom in a one-dimensional box calculate the value of the quantum number of the energy level for which the energy is equal to $\frac{3}{2}$ kT at 25 °C (a) for a box 1 nm long, and (b) for a box 1 cm long.

SOLUTION

(a) $E = \frac{3}{2} kT = \frac{3}{2}(1.38 \times 10^{-23} \text{ J K}^{-1})(298 \text{ K})$

$$= 6.17 \times 10^{-21} \text{ J}$$

$$n = \frac{a}{h}(8mE)^{1/2}$$

$$= \frac{10^{-9} \text{ m}}{6.63\times10^{-34}\text{Js}}\left[\frac{8(1.008\times10^{-3}\text{kgmol}^{-1})(6.17\times10^{-21}\text{J})}{6.02 \times 10^{23} \text{ mol}^{-1}}\right]^{1/2}$$

= 13.7 or approximately 14

(b) $n = 1.37 \times 10^{8}$

9.16 Calculate the degeneracies of the first three levels for a particle in a cubical box.

SOLUTION

$$E = \frac{h^2}{8 ma^2}(n_x^{\,2} + n_y^{\,2} + n_z^{\,2}) \quad n_x, n_y, n_z \text{ are } 1, 2, 3..$$

n_x	n_y	n_z	$E/\frac{h^2}{8\ ma^2}$	
1	1	1	3	Degeneracy = 1
2	1	1		
1	2	1	6	Degeneracy = 3
1	1	2		
2	2	1		
2	1	2	9	Degeneracy = 3
1	2	2		

9.17 Show that the function $\psi = 8e^{5x}$ is an eigenfunction of the operator d/dx. What is the eigenvalue?

SOLUTION

$$\frac{d\psi}{dx} = \frac{d}{dx}\ 8e^{5x}$$

$$= 40\ e^{5x} = 5\psi$$ Thus the eigenvalue is 5.

9.18 What are the results of operating on the following functions with the operator d/dx? (a) e^{-ax^2} (b) cos x, and (c) e^{ikx}. What eigenvalues are obtained?

SOLUTION

(a) $$\frac{de^{-ax^2}}{dx} = -2axe^{-ax^2}$$

(b) $$\frac{d\cos x}{dx} = -\sin x$$

(c) $$\frac{de^{ikx}}{dx} = ike^{ikx}$$

An eigenvalue, ik, is obtained only with the last function.

9.19 For the harmonic oscillator substitute the wave function for the ground state into the Schrödinger equation and obtain the expression for the ground state energy.

SOLUTION

$$\psi_0 = Ce^{-ax^2} \quad \text{where} \quad a = \frac{\pi}{h}\sqrt{km}$$

$$\frac{d\psi_0}{dx} = -2axCe^{-ax^2}$$

$$\frac{d^2\psi_0}{dx^2} = 4a^2x^2Ce^{-ax^2} - 2aCe^{-ax^2}$$

Substituting in the Schrödinger equation

$$\left(-\frac{\hbar^2}{2m}\frac{d^2}{dx^2} + \frac{1}{2}kx^2\right)\psi = E\psi$$

yields

$$E = \frac{1}{2}h\left(\frac{1}{2\pi}\sqrt{\frac{k}{m}}\right) = \frac{1}{2}h\nu_0$$

9.20 In the vibrational motion of HI, the iodine atom essentially remains stationary because of its large mass. Assuming that the hydrogen atom undergoes harmonic motion and that the force constant k is 317 N m^{-1}, what is the fundamental vibration frequency ν_0?

SOLUTION

$$m = \frac{1.0078 \times 10^{-3} \text{ kg mol}^{-1}}{6.022 \times 10^{23} \text{ mol}^{-1}} = 1.674 \times 10^{-27} \text{ kg}$$

$$\nu_0 = \frac{1}{2\pi}\sqrt{\frac{k}{m}} = \frac{1}{2\pi}\sqrt{\frac{317 \text{ kg s}^{-2}}{1.674 \times 10^{-27} \text{ kg}}}$$

$$= 6.93 \times 10^{13} \text{ s}^{-1}$$

9.21 Verify the normalization of ψ_1 for a harmonic oscillator.

SOLUTION

$$\psi_1 = \frac{(2\gamma)^{1/2}}{\pi^{1/4}} \gamma x e^{-\gamma^2 x^2/2}$$

$$\int_{\infty}^{\infty} \psi_1^{\,2}\, dx = \frac{2\gamma^3}{\pi^{1/2}} \int_{\infty}^{\infty} x^2 e^{-\gamma^2 x^2/2}\, dx = 1$$

The value of the integral is given in Table 16.1.

9.22 What are the reduced mass and moment of inertia of $^{23}Na^{35}Cl$? The equilibrium internuclear distance R_e is 236 pm. What are the values of L, L_z, and E for the state with J = 1?

SOLUTION

$$\mu = \frac{1}{\frac{1}{m_1} + \frac{1}{m_2}} = \frac{m_1\, m_2}{m_1 + m_2}$$

$$= \frac{(22.98977\text{x}10^{-3}\text{kgmol}^{-1})(34.96885\text{x}10^{-3}\text{kgmol}^{-1})}{[(22.98977+34.96885)\text{x}10^{-3}\text{kgmol}^{-1}](6.022045\text{x}10^{23}\text{mol}^{-1})}$$

$$= 2.30332 \times 10^{-26}\ \text{kg}$$

$$I = \mu R_e^{\,2} = (2.303 \times 10^{-26}\ \text{kg})(236 \times 10^{-12}\ \text{m})^2$$

$$= 1.283 \times 10^{-45}\ \text{kg m}^2$$

$$L = \sqrt{J(J+1)}\ \hbar = \frac{\sqrt{2}}{2\pi}(6.626 \times 10^{-34}\ \text{Js}) = 1.491\text{x}10^{-34}\ \text{Js}$$

$$L_z = 0 \text{ or } \pm \frac{1}{2\pi}(6.626 \times 10^{-34}\ \text{J s}) = 0 \text{ or} \pm 1.054\text{x}10^{-34}\ \text{Js}$$

$$E = \frac{\hbar^2}{2I} J(J+1) = \frac{(6.626\text{x}10^{-34}\text{Js})^2(2)}{4\pi^2(1.283\text{x}10^{-45}\text{kgm}^2)}$$

$$= 1.734 \times 10^{-23}\ \text{J}$$

9.23 857.7 watts

9.24 290 nm

9.25 1397 J min^{-1}

9.26 6.62×10^{-34} J s, 1.38×10^{-23} J K^{-1}

9.27 2.74×10^{14} s^{-1}, 1094 nm

9.28 4052, 2626 nm

9.29 1.325×10^{-27} kg m s^{-1}, 0.0113 m s^{-1}

9.30 5.93×10^{5} m s^{-1}

9.31 0.130 nm

9.32 0.00549 nm

9.33 $E = n^2h^2/8m\ell^2$

9.34 145.2, 582, 1305 kJ mol^{-1}

9.35

$n_xn_yn_z$	111	211	121	112	122	212	221	113	131	311
$E(8ma^2/h^2)$	3	6	6	6	9	9	9	11	11	11
Degeneracy	1		3			3			3	

9.36 -6a

9.37 1/k

9.38 $\Delta p = n\pi\hbar/\ell$, $\Delta p \Delta x = nh/2$

9.42

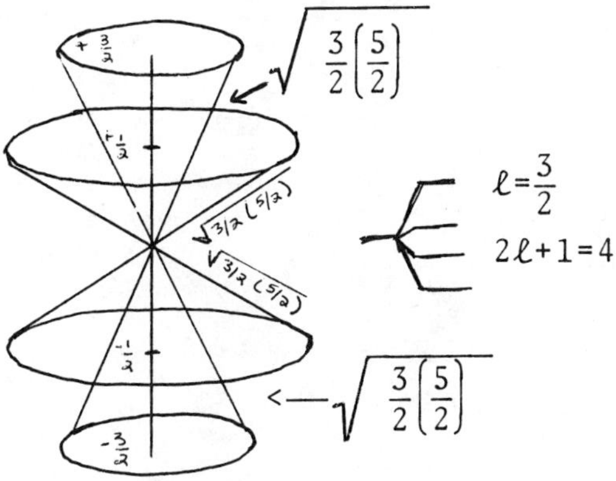

9.43

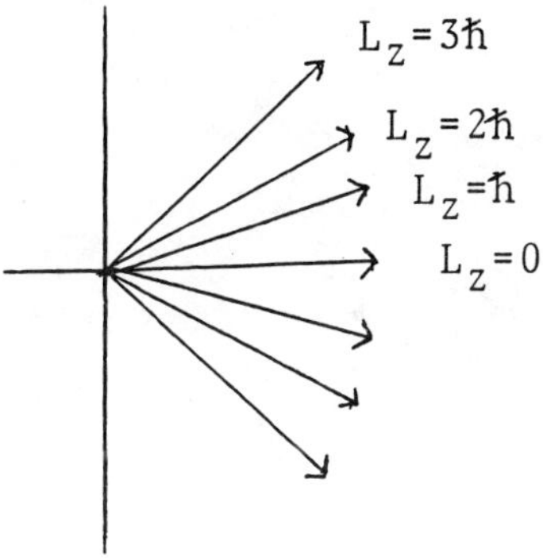

9.44 1.138×10^{-26} kg, 1.454×10^{-46} kg m^2, 2.583×10^{-34} J s, 1.442×10^{-21} J

CHAPTER 10. Atomic Structure

10.1 How much energy is required to remove a 4p electron from a hydrogen atom in eV and in kJ mol^{-1}?

SOLUTION

$$E = -(13.605\ 80\ V)/n^2$$
$$= -(13.605\ 80\ V)/16$$
$$= -0.850\ 36\ V$$
$$= (-0.850\ 36\ V)(96{,}485\ C\ mol^{-1})$$
$$= -82{,}047\ J\ mol^{-1}$$

10.2 What are the wavelengths of the first line in the Balmer series for H (relative atomic mass 1.007 825) and D (relative atomic mass 2.014 10)?

SOLUTION

$$\frac{1}{\lambda} = R_H(\frac{1}{2^2} - \frac{1}{3^2})$$

$$\lambda = \frac{1}{(1.0967758 \times 10^7\ m^{-1})(1/4 - 1/9)}$$

$$= 656.4696\ nm$$

$$m_H = \frac{1.007825 \times 10^{-3}\ kg\ mol^{-1}}{6.022045 \times 10^{23}\ mol^{-1}} = 1.673559 \times 10^{-27}\ kg$$

$$m_e = \underline{0.000911 \times 10^{-27}\ kg}$$
$$m_p = 1.672648 \times 10^{-27}\ kg$$

$$m_D = \frac{2.01410 \times 10^{-3}\ kg\ mol^{-1}}{6.022045 \times 10^{23}\ mol^{-1}} = 3.344545 \times 10^{-27}\ kg$$

$$m_e = \underline{0.000911 \times 10^{-27}\ kg}$$
$$m_d = 3.34363 \times 10^{-27}\ kg$$

$$R_H = \text{const.} \frac{m_e m_p}{m_e + m_p}$$

$$R_D = \text{const.} \frac{m_e m_d}{m_e + m_d}$$

$$R_D = R_H \frac{m_d(m_e + m_p)}{m_p(m_e + m_d)}$$

$$= (1.0967758\text{x}10^7\text{m}^{-1})\frac{(3.34363)(1.673559)}{(1.672648)(3.344545)}$$

$$= 1.097073 \text{ x } 10^7 \text{ m}^{-1}$$

$$\lambda_D = \frac{1}{(1.097073 \text{ x } 10^7)(1/4 - 1/9)}$$

$$= 656.292 \text{ nm}$$

10.3 Calculate the Bohr radius with the reduced mass of the hydrogen atom.

SOLUTION

$$\mu = \frac{1}{\frac{1}{m_e} + \frac{1}{m_p}}$$

$$= \frac{1}{\frac{1}{9.109534 \text{ x } 10^{-31} \text{ kg}} + \frac{1}{1.6726485 \text{ x } 10^{-27} \text{ kg}}}$$

$$= 9.104575 \text{ x } 10^{-31} \text{ kg}$$

$$a_0 = \frac{h^2(4\pi\varepsilon_0)}{4\pi^2 m_e e^2} = \frac{h^2\varepsilon_0}{\pi m_e e^2}$$

$$= \frac{(6.626176\text{x}10^{-34}\text{Js})^2(8.85418782\text{x}10^{-12}\text{C}^2\text{N}^{-1}\text{m}^{-2})}{\pi(9.104575\text{x}10^{-31}\text{kg})(1.6021892\text{x}10^{-19}\text{C})^2}$$

$$= 0.052\ 946\ 5 \text{ nm}$$

10.4 For a 1s level in a hydrogenlike atom the Schrödinger equation reduces to

$$\frac{1}{r^2}\frac{\partial}{\partial r}\left(r^2 \frac{\partial \psi}{\partial r}\right) + \frac{8\pi^2 m_e}{h^2}\left(E + \frac{Ze^2}{4\pi\varepsilon_0 r}\right)\psi = 0$$

Using the functional form for ψ_{1s} obtain the expression for the energy of the ground state.

SOLUTION

$$\psi_{1s} = \frac{1}{\sqrt{\pi}}\left(\frac{Z}{a_0}\right)^{3/2} e^{-Zr/a_0} = Ce^{-Zr/a_0}$$

$$\frac{\partial \psi_{1s}}{\partial r} = -\frac{ZC}{a_0} e^{-Zr/a_0}$$

$$\frac{\partial}{\partial r}\left(-\frac{ZCr^2}{a_0} e^{-Zr/a_0}\right) = \left(\frac{Z^2Cr^2}{a_0^2} - \frac{2ZCr}{a_0}\right)e^{-Zr/a_0}$$

Substituting in the Schrödinger equation and eliminating a_0 by use of

$$a_0 = \frac{h^2(4\pi\varepsilon_0)}{4\pi^2 m_e e^2}$$

yields

$$E = -\frac{2\pi^2 m_e{}^4 Z^2}{h^2(4\pi\varepsilon_0)^2}$$

10.5 What are the degeneracies of the following orbitals for hydrogenlike atoms? (a) n = 1, (b) n = 2, and (c) n = 3.

SOLUTION

(a) $n = 1 \quad \ell = 0 \quad m_\ell = 0 \qquad m_s = \pm\frac{1}{2}$

degeneracy = 2

(b) $n = 2$ $\quad \ell = 0 \quad m_\ell = 0 \quad m_s = \pm\frac{1}{2}$

$\ell = 1 \quad m_\ell = -1,0,+1 \quad m_s = \pm\frac{1}{2}$

degeneracy = 8

(c) $n = 3$ $\quad \ell = 0 \quad m_\ell = 0 \quad m_s = \pm\frac{1}{2}$

$\ell = 1 \quad m_\ell = -1,0,+1 \quad m_s = \pm\frac{1}{2}$

$\ell = 2 \quad m_\ell = -2,-1,0,+1,+2 \quad m_s = \pm\frac{1}{2}$

degeneracy = 18

[Note that the degeneracy is $2n^2$.]

10.6 Show that for a 1s orbital of a hydrogenlike atom the most probable distance from proton to electron is a_0/Z.

SOLUTION

$$\psi_{1s} \propto e^{-Zr/a_0}$$

Probability density of $r \propto r^2 \psi_{is}^2 = r^2 \times e^{-2Zr/a_0}$. Setting the derivative of the probability density equal to zero,

$$r^2 e^{-2Zr/a_0} \left(\frac{-2Z}{a_0}\right) + 2re^{-2Zr/a_0} = 0 \qquad r = \frac{a_0}{Z}$$

10.7 What is the average distance from an orbital electron to the nucleus for a 2s and 2p electron in (a) H and (b) Li^{2+}.

SOLUTION

(a) $$\langle r\rangle_{n,\ell} = \frac{n^2 a_0}{Z}\left\{1 + \frac{1}{2}\left[1 - \frac{\ell(\ell + 1)}{n^2}\right]\right\}$$

For H $$\langle r\rangle_{2,0} = \frac{2^2(52.92\ \text{pm})}{1}\cdot\frac{3}{2} = 317.5\ \text{pm}$$

$$\langle r\rangle_{2,1} = \frac{2^2(52.92\text{ pm})}{1}\left\{1 + \frac{1}{2}[1 - \frac{2}{4}]\right\}$$

$$= 264.6\text{ pm}$$

(b) For Li^{2+}

$$\langle r\rangle_{2,0} = \frac{2^2(52.92\text{ pm})}{3}\cdot\frac{3}{2} = 105.8\text{ pm}$$

$$\langle r\rangle_{2,1} = \frac{2^2(52.92\text{ pm})}{3}\left\{1 + \frac{1}{2}[1 - \frac{2}{4}]\right\}$$

$$= 88.2\text{ pm}$$

10.8 Use equation 10.30 to derive the expression for the average distance between the electron and proton in a hydrogenlike atom in the 1s level.

SOLUTION

$$\langle r\rangle = \int_0^\infty rP(r)dr$$

$$= \int_0^\infty r4\pi r^2\ \frac{Z^3}{\pi a_o{}^3}\ e^{-2Zr/a_o}\ dr$$

$$= \frac{3}{2}\frac{a_o}{Z} = 79\text{ pm}$$

10.9 For a 2p electron in a hydrogenlike atom what is the magnitude of the orbital angular momentum and what are the possible values of L_z?

SOLUTION

$n = 2 \qquad \ell = 1 \qquad m_\ell = -1,0,+1$

$$|\underset{\sim}{L}| = [\ell(\ell + 1)]^{1/2}\ \hbar$$

$$= \sqrt{2}\ (6.626 \times 10^{-34}\text{ J s})/2\pi$$

$$= 1.491 \times 10^{-34}\text{ J s}$$

$$L_Z = m_\ell \hbar = m_\ell(6.626 \times 10^{-34}\ \text{J s})/2\pi$$

$$= \begin{cases} + 1.054 \times 10^{-34}\ \text{J s} \\ 0 \\ - 1.054 \times 10^{-34}\ \text{J s} \end{cases}$$

10.10 Using data from Table A.2 at 0 K what is the ionization energy of H(g)?

$$H(g) = H^+(g) + e^-$$

SOLUTION

For H(g) at 0 K $\Delta H_f^\circ = 216.037$ kJ mol^{-1}

$H^+(g)$ at 0 K $\Delta H_f^\circ = 1528.085$ kJ mol^{-1}

$e^-(g)$ at 0 K $\Delta H_f^\circ = 0$

$\Delta H^\circ = 1312{,}048$ J mol^{-1}

$$= \frac{(1312{,}048\ \text{J mol}^{-1})}{(1.602\ 189\ 2 \times 10^{-19}\text{C})(6.022\ 045 \times 10^{23}\ \text{mol}^{-1})}$$

$= 13\ 598\ 53$ eV

10.11 What is the magnitude of the angular momentum for electrons in 3s, 3p, and 3d orbitals? How many radial and angular modes are there for each of these orbitals?

SOLUTION

	3s	3p	3d
$\lvert \underset{\sim}{L} \rvert = \sqrt{\ell(\ell + 1)}\ \hbar$	0	$\sqrt{2}\ \hbar$	$\sqrt{6}\ \hbar$
Radial modes	2	1	0
Angular modes	0	1	2

10.12 For the wave function

$$\psi = \begin{vmatrix} \psi_A(1) & \psi_A(2) \\ \psi_B(1) & \psi_B(2) \end{vmatrix}$$

show that (a) the interchange of two columns changes the sign of the wave function, (b) the interchange of two rows changes the sign of the wave function, and (c) the two electrons cannot have the same spin orbital.

SOLUTION

(a) The interchange of two columns yields

$$\begin{vmatrix} \psi_A(2) & \psi_A(1) \\ \psi_B(2) & \psi_B(1) \end{vmatrix} = \psi_A(2)\ \psi_B(1) - \psi_A(1)\ \psi_B(2)$$

This is the wave function for the atom with the two electrons interchanged, and it is the negative of the function given in the problem, as required by the anti-symmetrization principle.

(b) The interchange of two rows yields

$$\begin{vmatrix} \psi_B(1) & \psi_B(2) \\ \psi_A(1) & \psi_A(2) \end{vmatrix} = \psi_A(2)\ \psi_B(1) - \psi_A(1)\ \psi_B(2)$$

which is the negative of the wave function given in the problem. Thus the interchange of two columns or two rows is equivalent to the interchange of two electrons between orbitals.

(c) If two electrons are represented by the same spin orbitals, the determinant

$$\begin{vmatrix} \psi_A(1) & \psi_A(2) \\ \psi_A(1) & \psi_A(2) \end{vmatrix} = 0$$

is in accord with the principle that two electrons in an atom cannot have four identical quantum numbers.

10.13 What are the electron configurations for H^-, Li^+, O^{2-}, F^-, Na^+, and Mg^{2+}?

SOLUTION

$1s^2$, $1s^2$, $1s^22s^22p^6$, $1s^22s^22p^6$, $1s^22s^22p^6$, $1s^22s^22p^6$.

10.14 Calculate the ionization potentials for He^+, Li^{2+}, Be^{3+}, B^{4+}, and C^{5+}.

SOLUTION

$E = (13.61\ V)Z^2$

	Z	E/V
He^+	2	54.44
Li^{2+}	3	122.49
Be^{3+}	4	217.76
B^{4+}	5	340.25
C^{5+}	6	489.96

10.15 Calculate the ionization potential for H(g) from the energy given in Example 10.1.

SOLUTION

$$E = 13.60580\ V \left(\frac{\mu_H}{m_e}\right)$$

$$= 13.60580\ V \frac{9.104576 \times 10^{-31}\ kg}{9.109534 \times 10^{-31}\ kg}$$

$$= 13.59840\ V$$

10.16 Since the outer electron for Li is quite a bit further out than the two 1s electrons, this atom is something like a hydrogen atom in the 2s state. The first ionization potential of Li is

5.39 V. What ionization potential would be expected from this simple model of a Li atom? What effective nuclear charge Z' seen by the outer electron would give the correct first ionization potential?

SOLUTION

$$E = \frac{13.61\ V}{n^2} = 13.61\ V/4 = 3.40\ V$$

$$5.39\ V = (Z')^2 (13.61\ V)/4$$

$$Z' = 1.259$$

10.17 The first ionization potential of atomic hydrogen is 13.60 V. Calculate the wavelength of the light produced when a free electron without kinetic energy returns to the inner orbit.

SOLUTION

$$h\nu = hc/\lambda = Ee$$

$$\lambda = \frac{hc}{Ee} = \frac{(6.626 \times 10^{-34}\ J\ s)(2.998 \times 10^{8}\ m\ s^{-1})}{(13.60\ V)(1.602 \times 10^{-19}\ C)}$$

$$= 91.18\ nm$$

10.18 What is the electronic energy in Hartrees of He, Li, and Be with respect to the nuclei and free electrons? Ionization potentials are given in Table 10.3.

SOLUTION

The sums of the ionization potentials are

He	79.003 V
Li	203.481 V
Be	399.139 V

The ionization energies expressed in eV are given by the same numbers. The energies of the neutral atoms with respect to the nucleus and free electrons are the ionization energies with a negative sign. The energies of the neutral atoms are expressed in Hartrees (H) by dividing by 27.211 608 eV H^{-1}.

		Hartree-Fock (Fig. 10.9)
He	-2.903 H	-2.8617 H
Li	-7.478 H	-7.2364 H
Be	-14.668 H	-14.5730 H

10.19 The enthalpy of formation of $H^-(g)$ at 0 K is given as 143.264 kJ mol^{-1} in Table A.2. What is the electron affinity of H(g)?

SOLUTION

The electron affinity is the energy released in the reaction

$H(g) + e^- = H^-(g)$

$$\Delta H° = 143.264 - 216.037 = -72.773 \text{ kJ mol}^{-1}$$

$$= -\frac{(72{,}773 \text{ J mol}^{-1})}{(1.6022 \times 10^{-19} \text{ C})(6.022 \times 10^{23} \text{ mol}^{-1})}$$

$$= -0.7542 \text{ eV}$$

Thus the electron affinity of H(g) is 0.7542 eV.

10.20 1.153×10^{-18} J 7.197 eV 694.3 kJ mol^{-1}

10.21 983 kJ mol^{-1} 265

10.22 6 a_0

10.23 (a) 243 nm (b) 6.80 V

10.24 109,677.57 cm^{-1}, compared with the experimental value of 109,677.58 cm^{-1}.

10.25

$$\psi_{1s}|\psi_{2s} = (\text{const.}) \int_0^\infty r^2(2-\frac{r}{a_0})\ e^{-\frac{r}{a_0}}\ e^{-\frac{r}{2a_0}}\ dr$$

$$= (\text{const.}) \int_0^\infty (2x^2-x^3)\ e^{-\frac{3}{2}x}\ dx \quad \text{where } x = r/a_0$$

$$= (\text{const.})\left(\frac{32}{27} - \frac{32}{27}\right) = 0$$

10.26

$$\langle\psi_{1s}|\psi_{1s}\rangle = \frac{4}{\pi a_0^3} \int_0^\infty r^2 e^{-2r/a_0}\ dr$$

10.27 17.639 pm

10.28 0.7144, 0.6615, 0.5557 nm

10.29 0, 0, $\sqrt{2}\ \hbar$, $\sqrt{6}\ \hbar$

10.30

	4s	4p	4d	4f
$\lvert \underset{\sim}{L} \rvert = \sqrt{\ell(\ell + 1)}\ \hbar$	0	$\sqrt{2}\ \hbar$	$\sqrt{6}\ \hbar$	$\sqrt{12}\ \hbar$
Radial modes	3	2	1	0
Angular modes	0	1	2	3

10.31 $\psi = 1s\alpha(1)\ 1s\beta(2) - 1s\alpha(2)\ 1s\beta(1)$
Interchanging coordinates gives
$1s\alpha(2)\ 1s\beta(1) - 1s\alpha(1)\ 1s\beta(2) = -\psi$

10.32 The inner electrons (1s) screen the nucleus, but since the probability density of the 2s orbital is greater near the nucleus than that of the 2p, the 2s electrons are not screened as much as the 2p electrons, and therefore are bound more tightly (have a lower energy).

10.33 Those with outer electrons in s orbitals are spherically symmetrical: H, He, Li, Be, Na, and Mg.

10.34	Sc	$1s^2 2s^2 2p^6 3s^2 3p^6$	$3d4s^2$
	Sc^+	[argon core]	3d4s
	Sc^{2+}	[argon core]	3d
	Sc^{3+}	[argon core]	

10.35 230.1, 16.40 nm

10.36 54.4 V 164.0 nm 121.0 nm

10.37 13.6627, 13.5984 eV

CHAPTER 11. Symmetry

11.1 E, $C_2(z)$, $\sigma_v(xz)$, $\sigma_v'(yz)$

11.2 E, C_3, $3\sigma_v$

11.3 E, $C_4(z)$, $C_2(z)$, $S_4(z)$, $C_2'(x)$, $C_2'(y)$, $2C_2''$,
i, σ_h, $\sigma_v(xz)$, $\sigma_v(yz)$, $2\sigma_d$

11.4 E, $4C_3$, $4S_6$, $C_4(x)$, $C_4(y)$, $C_4(z)$, $3S_4$, $C_2(x)$, $C_2(y)$, $C_2(z)$, $6C_2'$, i, $3\sigma_h$, $6\sigma_d$

11.5 E, C_2

11.6 E, $C_3(z)$, $S_3(z)$, $3C_2$, σ_h, $3\sigma_v$

11.7 E

11.8 E, C_4, C_2, $2\sigma_v$, $2\sigma_d$

11.9 E, C_3, $3C_2$, i, $3\sigma_d$

11.10 E, $C_2(z)$, $C_2(y)$, $C_2(x)$, i, $\sigma(xy)$, $\sigma(xz)$, $\sigma(yz)$

11.11 E, C_∞, S_∞, σ_h, i, $\infty\sigma_v$, ∞C_2

11.12 E, $C_2(z)$, $C_2(y)$, $C_2(x)$, i, $\sigma(xy)$, $\sigma(xz)$, $\sigma(yz)$

11.13 E, $S_4(z)$, $C_2(z)$, $C_2'(x)$, $C_2'(y)$, $2\sigma_d$

11.14 E, $C_2(z)$, $\sigma_v(xz)$, $\sigma_v'(yz)$

11.15 E, $C_2(z)$, $C_2(y)$, $C_2(x)$, i, $\sigma(xy)$, $\sigma(xz)$, $\sigma(yz)$

11.16 E, $4C_3$, $S_4(x)$, $S_4(y)$, $S_4(z)$, $C_2(x)$, $C_2(y)$, $C_2(z)$

$6\sigma_d$

11.17 E, $C_3(z)$, $S_3(z)$, $3C_2$, σ_h, $3\sigma_v$

11.18 E, i

11.19 E, C_3, $3C_2$, i, $3\sigma_d$

11.20 Construct the operator multiplication table for the point group C_{2h}.

SOLUTION

	E	C_2^1	σ_h	i
E	E	C_2^1	σ_h	i
C_2^1	C_2^1	E	i	σ_h
σ_h	σ_h	i	E	C_2^1
i	i	σ_h	C_2^1	E

11.21 List the operators associated with the S^6 elements and their equivalents, if any. How many distinct operations are produced?

SOLUTION

S_6^1	S_6^2	S_6^3	S_6^4	S_6^5	S_6^6
	C_3^1	i	C_3^2		E

Thus an S_6 axis implies the existence of both a C_3 axis coincident with the S_6 axis and a center of symmetry (i). Only the S_6^1 and S_6^5 operations are distinct operations characteristic of only the S_6 axis.

11.22 The C_i point group has only the symmetry elements E and i. Write the matrices for these two symmetry elements and show that they follow the same multiplication table as E and i.

SOLUTION

$$E = \begin{bmatrix} 1 & 0 & 0 \\ 0 & 1 & 0 \\ 0 & 0 & 1 \end{bmatrix} \qquad i = \begin{bmatrix} -1 & 0 & 0 \\ 0 & -1 & 0 \\ 0 & 0 & -1 \end{bmatrix}$$

$$EE = \begin{bmatrix} 1 & 0 & 0 \\ 0 & 1 & 0 \\ 0 & 0 & 1 \end{bmatrix} = E \quad ii = \begin{bmatrix} 1 & 0 & 0 \\ 0 & 1 & 0 \\ 0 & 0 & 1 \end{bmatrix} = E$$

$$iE = \begin{bmatrix} -1 & 0 & 0 \\ 0 & -1 & 0 \\ 0 & 0 & -1 \end{bmatrix} = i \qquad Ei = \begin{bmatrix} -1 & 0 & 0 \\ 0 & -1 & 0 \\ 0 & 0 & -1 \end{bmatrix} = i$$

The multiplication table is

	E	i
E	E	i
i	i	E

11.23 O_h E, $4C_3$, $4S_6$, $3C_4$, $3S_4$, $3C_2$, $6C_2'$, i, $3\sigma_h$, $6\sigma_d$

11.24 D_{2d} E, S_4, C_2, $2C_2'$, $2\sigma_d$

11.25 C_s E, σ_h

11.26 D_{5h} E, C_5, S_5, $5C_2$, σ_h, $5\sigma_v$

11.27 C_{2v} E, C_2, σ_v, σ'_v

11.28 D_{6h} E, C_6, C_3, S_6, S_3, C_2, $3C_2'$, $3C_2''$, i, σ_h

$3\sigma_v$, $3\sigma_d$

11.29 D_{4d} E, S_8, C_4, C_2, $4C_2'$, $4\sigma_d$

11.30 O_h See 11.23

11.31 D_{3h} E, C_3, S_3, $3C_2$, σ_h, $3\sigma_v$

11.32 T_d E, $4C_3$, $3S_4$, $3C_2$, $6\sigma_d$

11.33 C_{2v} See 11.27

11.34 C_s See 11.25

11.35 D_{3h} See 11.31

11.36 C_3 E, C_3

11.37 C_{2v} See 11.27

11.38 C_{3v} E, C_3, $3\sigma_v$

11.39 T_d See 11.32

11.40 C_s See 11.25

11.41 D_{4h} E, C_4, C_2, S_4, $2C_2'$, $2C_2''$, i, σ_h, $2\sigma_v$, $2\sigma_d$

11.42 D_{2d} See 11.24

11.45

$$\begin{vmatrix} 1 & 0 & 0 \\ 0 & -1 & 0 \\ 0 & 0 & 1 \end{vmatrix} \begin{vmatrix} 1 & 0 & 0 \\ 0 & -1 & 0 \\ 0 & 0 & 1 \end{vmatrix} = \begin{vmatrix} 1 & 0 & 0 \\ 0 & 1 & 0 \\ 0 & 0 & 1 \end{vmatrix}$$

$$\sigma_v \cdot \sigma_v = E$$

CHAPTER 12. Molecular Electronic Structure

12.1 Given the equilibrium dissociation energy D_e for N_2 in Table 12.2 and the fundamental vibration frequency 2331 cm^{-1}, calculate the spectroscopic dissociation energy D_0 in kJ mol^{-1}.

SOLUTION

$$D_e = D_0 + \frac{1}{2} h\nu_0$$

$$\frac{1}{2}h\nu_0 = \frac{(6.626\times10^{-34}Js)(2.998\times10^{8}ms^{-1})(2.331\times10^{5}m^{-1})}{2(1000\ J\ kJ^{-1})} \times \frac{(6.022\times10^{23}mol^{-1})}{1}$$

$$= 13.93\ kJ\ mol^{-1}$$

$$D_0 = D_e - \frac{1}{2} h\nu_0$$

$$= 941.49\ kJ\ mol^{-1}$$

This is the thermodynamic dissociation energy at absolute zero. The standard change in enthalpy for dissociation at 25 °C calculated from Table A.1 is 945.408 kJ mol^{-1}.

12.2 Derive the values of the normalization constants given in equations 12.27 and 12.28.

SOLUTION

$$\psi_g = C(1s_A + 1s_B)$$

$$\int \psi_g^2 d\tau = C^2 \int (1s_A{}^2 + 2 1s_A 1s_B + 1s_B{}^2) d\tau$$

$$1 = C^2\left[1 + 2\int 1s_A 1s_B d\tau + 1\right]$$

$$C = \frac{1}{(2 + 2S)^{1/2}} \quad \text{where } S = \int 1s_A 1s_B d\tau$$

$$\psi_u = C'(1s_A - 1s_B)$$

$$\int \psi_u d\tau = (C')^2 \int (1s_A{}^2 - 2 1s_A 1s_B + 1s_B{}^2) d\tau$$

$$= (C')^2 [2 - 2S]$$

$$C' = \frac{1}{(2 - 2S)^{1/2}}$$

12.3 The internuclear repulsion $Z_\alpha Z_\beta e^2/4\pi\varepsilon_o R_{\alpha\beta}$ is omitted from the Hamiltonian operator $\hat{H}$ of a diatomic molecule. If ψ is an eigenfunction of $\hat{H}$ with eigenvalue E, what is the eigenvalue of the operator $\hat{H} + Z_\alpha Z_\beta e^2/4\pi\varepsilon_o R_{\alpha\beta}$?

SOLUTION

$$\left(\hat{H} + \frac{Z_\alpha Z_\beta e^2}{4\pi\varepsilon_o R_{\alpha\beta}}\right)\psi = E'\psi$$

$$\hat{H}\psi + \frac{Z_\alpha Z_\beta e^2}{4\pi\varepsilon_o R_{\alpha\beta}}\psi = E + \left(\frac{Z_\alpha Z_\beta e^2}{4\pi\varepsilon_o R_{\alpha\beta}}\right)\psi$$

$$E' = E + Z_\alpha Z_\beta e^2/4\pi\varepsilon_o R_{\alpha\beta}$$

12.4 Express the four VB wave functions for H_2 as Slater determinants.

SOLUTION

$$\psi_1 = \frac{1}{2^{\frac{1}{2}}}\begin{vmatrix} 1s_A(1)\alpha(1) & 1s_A(2)\alpha(2) \\ 1s_B(1)\beta(1) & 1s_B(2)\beta(2) \end{vmatrix} - \frac{1}{2^{\frac{1}{2}}}\begin{vmatrix} 1s_A(1)\beta(1) & 1s_A(2)\beta(2) \\ 1s_B(1)\alpha(1) & 1s_B(2)\alpha(2) \end{vmatrix}$$

$$\psi_2 = \frac{1}{2^{1/2}} \begin{vmatrix} 1s_A(1)\alpha(1) & 1s_A(2)\alpha(2) \\ 1s_B(1)\alpha(1) & 1s_B(2)\alpha(2) \end{vmatrix}$$

ψ_3 is the same as ψ_2 with β replacing α. ψ_4 is obtained from ψ_1 by replacing the minus sign with a plus sign.

12.5 Using data in Table A.1, calculate the electronegativities of Cl, Br, and I. The electronegativity of H is taken to be 2.1.

SOLUTION

$$|x_A - x_B| = 0.208 \left\{E(A\text{-}B) - \frac{1}{2}[E(A\text{-}A) + (B\text{-}B)]\right\}^{1/2}$$

For HCl

$$|x_{Cl} - x_H| = 0.0497 \left\{92.307 + 217.965 + 121.679 - \frac{1}{2}[2(217.965) + 2(121.679)]\right\}^{1/2}$$

$x_{Cl} - 2.1 = 1.0 \qquad x_{Cl} = 3.1$

For HBr

$$|x_{Br} - x_H| = 0.0497 \left\{36.401 + 217.965 + 111.884 - \frac{1}{2}[2(217.965) + 2 \times (111.884) - 30.907]\right\}^{1/2}$$

$x_{Br} - 2.1 = 0.7 \qquad x_{Br} = 2.8$

For HI

$$|x_I - x_H| = 0.0497 \left\{-26.49 + 217.965 + 106.838 - \frac{1}{2}[2(217.965) + 2(106.838) - 62.438]\right\}^{1/2}$$

$x_I - 2.1 = 0.2 \qquad x_I = 2.3$

12.6 For KF(g) the dissociation constant D_e is 5.18 eV and the dipole moment is 28.7×10^{-30} C m. Estimate these values assuming that the bonding is entirely ionic. The ionization potential K(g) is 4.34 eV, and the electron affinity of F(g) is 3.40 eV. The equilibrium internuclear distance in KF(g) is 0.217 nm.

SOLUTION

The work required to separate $K^+ F^-$ from the equilibrium internuclear distance to infinity is

$$\phi = \frac{Q_1 Q_2}{4\pi\varepsilon_0 R_e} = \frac{(1.602\times10^{-19}C)^2(0.8988\times10^{10} Nm^2C^{-2})}{217 \times 10^{-12} m}$$

$$= 1.063 \times 10^{-18} J$$

$$= \frac{1.063 \times 10^{-18} J}{1.602 \times 10^{-19} J eV^{-1}} = 6.64 eV$$

Since KF actually dissociates into neutral atoms, the ionization potential of the metal has to be subtracted and the electron affinity of the nonmetal has to be added to calculate D_e:

$$D_e = 6.64 - 4.34 + 3.40 = 5.70 eV$$

This is higher than the experimental value (5.18 eV) because there is some repulsion of K^+ and F^- due to inner electron clouds.

The dipole moment expected for the oversimplified structure $K^+ F^-$ is

$$(1.602\times10^{-19}C)(217\times10^{-12}m) = 34.7\times10^{-30}Cm$$

12.7 Show that the SI units are the same on the two sides of equation 12.57.

SOLUTION

$$\frac{\kappa-1}{\kappa+2}\frac{M}{\rho} = \frac{N_A}{3\varepsilon_0}\left(\frac{\mu^2}{3kT} + \alpha_e + \alpha_v\right)$$

μ = Cm

$$\alpha = \frac{\text{dipole moment}}{\text{electric field strength}} = \frac{\text{Cm}}{\text{Vm}^{-1}} = \frac{\text{Cm}^2}{\text{JC}^{-1}} = \frac{\text{C}^2\ \text{m}^2}{\text{gm}^2\text{s}^{-2}}$$

$$\varepsilon_o = \text{C}^2\ \text{N}^{-1}\ \text{m}^{-2}$$

Left side: $\dfrac{\text{kg mol}^{-1}}{\text{kg m}^{-3}} = \text{m}^3\ \text{mol}^{-1}$

Right side: $\dfrac{\text{mol}^{-1}}{\text{C}^2\text{N}^{-1}\text{m}^{-2}}\left(\dfrac{\text{C}^2\text{m}^2}{\text{gm}^2\text{s}^{-2}} + \dfrac{\text{C}^2\text{m}^2}{\text{gm}^2\text{s}^{-2}}\right) = \text{m}^3\ \text{mol}^{-1}$

12.8 The dipole moments of HCl, HBr, and HI are given in Table 12.4. Explain the relative magnitudes of these moments in terms of electronegativity.

<u>SOLUTION</u>

	$\mu/10^{-30}$ Cm	Electronegativity of halogen atom
HCl	3.44	3.0
HBr	2.64	2.8
HI	1.00	2.4

The electronegativities increase in going from I to Cl and so the negative charge on the halogen is greater in HCl than in HI and the dipole moment is larger.

12.9 Calculate the dipole moment HCl would have if it consisted of a proton and a chloride ion (considered to be a point charge) separated by 0.127 nm (the internuclear distance obtained from the infrared spectrum). The experimental value is 3.44×10^{-30} C m. How do you explain the difference?

<u>SOLUTION</u>

$\mu = qr = (1.602 \times 10^{-19}\ \text{C})(0.127 \times 10^{-9}\ \text{m})$

$\quad = 20.4 \times 10^{-30}\ \text{C m}$

$\mu_{expt} = 3.44 \times 10^{-30}$ C m

The charges are not completely separated in HCl.

12.10 The molar polarization $\mathcal{P}$ of ammonia varies with temperature as follows.

t/°C	19.1	35.9	59.9	113.9	139.9	172.9
$\mathcal{P}$/cm^{3}mol^{-1}	57.57	55.01	51.22	44.99	42.51	39.59

Calculate the dipole moment of ammonia.

SOLUTION

$$\mathcal{P} = \frac{N_A}{3\varepsilon_o}\left(\frac{\mu^2}{3kT} + \alpha_e + \alpha_v\right)$$

The slope of a plot of $\mathcal{P}$ versus 1/T is

$$\frac{N_A\mu^2}{9k\varepsilon_o} = \frac{(66 - 5)\ \text{cm}^3\ \text{mol}^{-1}}{4 \times 10^{-3}\ \text{K}^{-1}}$$

$$= 1.525 \times 10^4\ \text{cm}^3\ \text{mol}^{-1}\ \text{K}$$

$$= 1.525 \times 10^{-2}\ \text{m}^3\ \text{mol}^{-1}\ \text{K}$$

$$\mu = \left[\frac{(1.525\text{x}10^{-2}\text{m}^3\text{mol}^{-1}\text{K})\,9k\varepsilon_o}{N_A}\right]^{1/2}$$

$$= \left[\frac{(1.525\text{x}10^{-2}\text{m}^3\text{mol}^{-1}\text{K})(9)(1.381\text{x}10^{-23}\text{JK}^{-1})(8.854\text{x}10^{-12}\text{C}^2\text{N}^{-1}\text{m}^{-2})}{6.022 \times 10^{23}\ \text{mol}^{-1}}\right]$$

$= 5.28 \times 10^{-30}$ C m

(Please see graph, page 176)

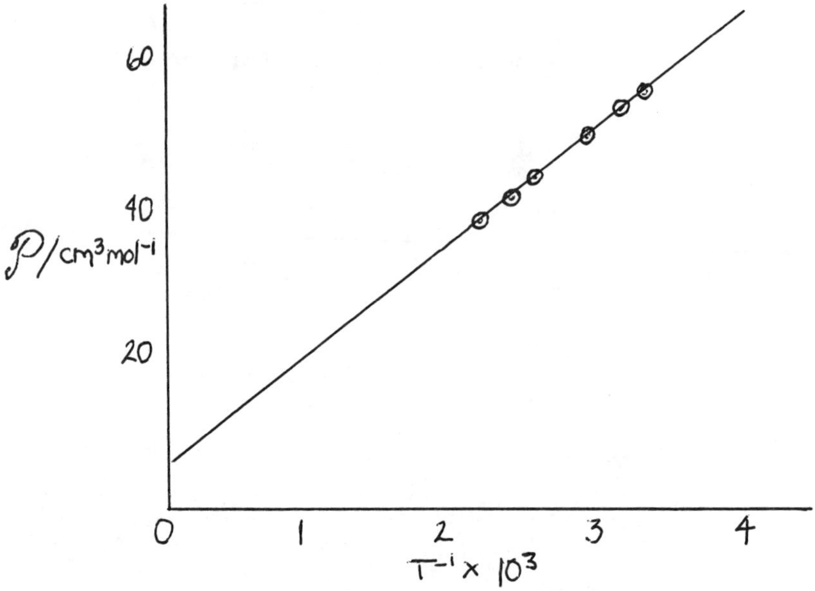

12.11 At 1 bar the dielectric constant of NH_3 gas is 1.007 20 at 292.2 K and 1.003 24 at 446.0 K. Calculate the dipole moment μ and the polarizability α.

SOLUTION

$$(\kappa-1)\ M/\rho = \frac{N_A}{\varepsilon_0}\left(\frac{\mu^2}{3kT} + \alpha\right)$$

Assuming NH_3 is a perfect gas $P = \rho RT/M$

At T = 292.2 K,

$M/\rho = RT/P = (8.314 JK^{-1}mol^{-1})(292.2K)/(101,325Pa)$

$= 23.98 \times 10^{-3}\ m^3\ mol^{-1}$

$(\kappa-1)M/\rho = (1.00720-1)(23.98\times10^{-3}m^3mol^{-1})$

$= 1.727 \times 10^{-4}\ m^3\ mol^{-1}$

At T = 446.0 K

$M/\rho = (8.314 JK^{-1}mol^{-1})(446.0\ K)/(101,325\ Pa)$

$$= 36.60 \times 10^{-3}\ m^3\ mol^{-1}$$

$$(\kappa-1)M/\rho = (1.00324-1)(36.60 \times 10^{-3}\ m^3\ mol^{-1})$$
$$= 1.186 \times 10^{-4}\ m^3\ mol^{-1}$$

We have the following two simultaneous equations

$$1.727 \times 10^{-4} m^3 mol^{-1} = \frac{N_A\mu^2}{3\varepsilon_o k(292.2K)} + \frac{N_A\alpha}{\varepsilon_o}$$

$$1.186 \times 10^{-4} m^3 mol^{-1} = \frac{N_A\mu^2}{3\varepsilon_o k(446.0K)} + \frac{N_A\alpha}{\varepsilon_o}$$

Taking the difference

$$0.541 \times 10^{-4} m^3 mol^{-1} = \frac{N_A\mu^2}{3\varepsilon_o k}\left(\frac{1}{292.2K} - \frac{1}{446.0K}\right)$$

$$\frac{N_A\mu^2}{3\varepsilon_o k} = 0.0458\ m^3 mol^{-1} K$$

$$\mu = \left[\frac{(0.0458 m^3 mol^{-1} K)(3)(8.854\times10^{-12} C^2 N^{-1} m^{-2})(1.3807\times10^{-23} JK^{-1})}{6.022 \times 10^{23}\ mol^{-1}}\right]^{1/2}$$

$$= 5.28 \times 10^{-30}\ C\ m$$

Solving the first equation for α

$$\alpha = \frac{\varepsilon_o}{N_A}\left[1.727\times10^{-4} m^3 mol^{-1} - \frac{N_A\mu^2}{3\varepsilon_o k(292.2k)}\right]$$

$$= \frac{8.854\times10^{-12} C^2 N^{-1} m^{-1}}{6.022\times10^{23} mol^{-1}}\left[1.727\times10^{-4} m^3 mol^{-1} - \frac{(6.022\times10^{23} mol^{-1})(5.28\times10^{-30} Cm)^2}{3(8.854\times10^{-12} C^2 N^{-1} m^{-2})(1.3807\times10^{-23} JK^{-1})(292.2K)}\right]$$

$$= 2.36 \times 10^{-40}\ C\ m/V\ m^{-1}$$

12.12 The equilibrium internuclear distance for NaCl(g) is 0.2361 nm. What dipole moment is expected? The actual value is 3.003×10^{-29} Cm. How do you explain the difference?

SOLUTION

$$\mu = (1.602 \times 10^{-19}\ C)(0.2361 \times 10^{-9}\ m)$$
$$= 3.783 \times 10^{-29}\ C\ m$$

The actual value is less because the ions polarize each other. The positive ion attracts the electrons of the negative ion toward itself, and the negative ion repels the electrons of the positive ion. The more polarizable the ions the larger the effect is.

12.13 A beam of sodium D light (589 nm) is passed through 100 cm of an aqueous solution of sucrose containing 10 g sucrose per 100 cm^3. Calculate I/I_0, where I_0 is the intensity that would have been obtained with pure water, given that M = 342.30 g mol^{-1} and dn/dc = 0.15 $g^{-1} cm^3$ for sucrose. The refractive index of water at 20 °C is 1.333 for the sodium D line.

SOLUTION

$$\tau = \frac{32\pi^3 n_0^2 (dn/dc)^2 Mc}{3N_A\lambda^4}$$

$$= \frac{32\pi^3(1.333)^2(0.15 g^{-1} cm^3)^2(342.30 g mol^{-1})(0.1 g cm^{-3})}{3(6.02 x 10^{23} mol^{-1})(5.89 x 10^{-5} cm)^4}$$

$$= 6.25 \times 10^{-5}\ cm^{-1}$$

$$\frac{I}{I_0} = e^{-\tau x} = e^{-(6.25 x 10^{-5} cm^{-1})(100 cm)}$$
$$= 0.9938$$

12.14 The Lennard-Jones parameters for nitrogen are ε/k = 95.1 K and σ = 0.37 nm. Plot the potential energy (expressed as V/k in K) for the interaction of two molecules of nitrogen.

SOLUTION

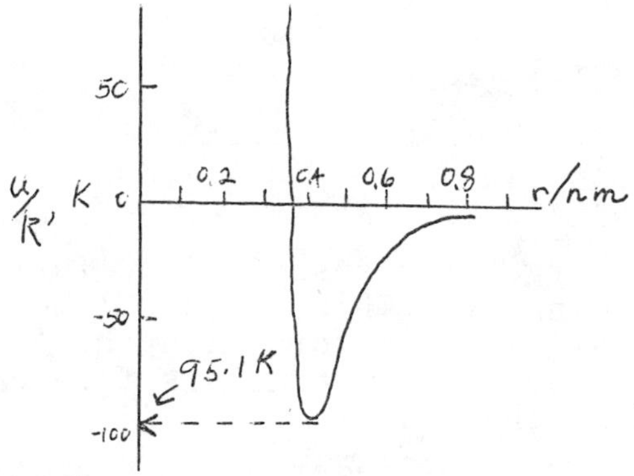

12.15 For Ne the parameters of the Lennard-Jones 6-12 potential are $\varepsilon/k = 35.6$ K and $\sigma = 0.275$ nm. Plot V in J mol^{-1} versus r and calculate the distance r_m where $dV/dr = 0$.

SOLUTION

$$r_m = 2^{1/6}\sigma = 2^{1/6}(0.275 \text{ nm}) = 0.309 \text{ nm}$$

$$\varepsilon = (8.314 \text{ J K}^{-1} \text{ mol}^{-1})(35.6 \text{ K}) = 296.0 \text{ J mol}^{-1}$$

$$V = 4\varepsilon\left[\left(\frac{\sigma}{r}\right)^{12} - \left(\frac{\sigma}{r}\right)^{6}\right]$$

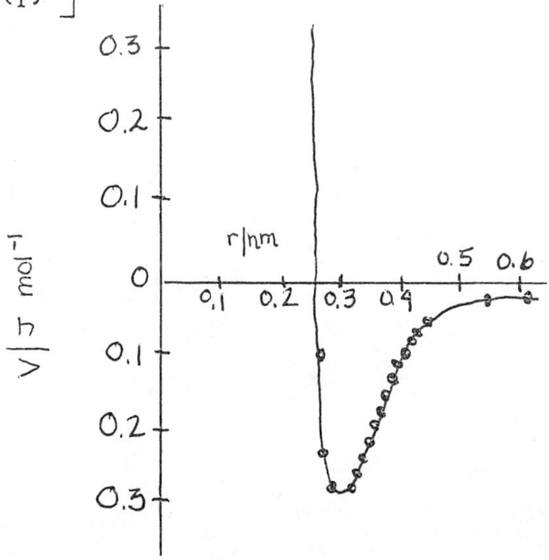

$$= 4(296.0 \text{ J mol}^{-1})\left[\left(\frac{2.75}{r}\right)^{12} - \left(\frac{2.75}{r}\right)^{6}\right]$$

12.16 494 kJ mol^{-1} 493.570 kJ mol^{-1} from Table A.2.

12.19 0.899 976 Hartrees <u>12.20</u> 0.375 Hartrees

12.21 -2(13.598 396 eV) = -27.196 792 eV
-27.196 792 - 4.7483 = -31.9451 eV

12.23 More charge is transferred in forming HCl than in forming HI. This makes HCl more stable than HI. This is in accord with the fact that more heat is evolved when HCl is formed from its elements than HI.

12.25 Because of the higher electronegativity of Cl, the Cl atom in CH_3Cl is more negative than the Br atom in CH_3Br.

12.26 6.408 x 10^{-29} C m

12.27 4.0 x 10^{-30} C m 0.88 x 10^{-40} C m/Vm^{-1}

12.28 5.29 x 10^{-30} C m

12.29 19,500 g mol^{-1} 0.9980

12.30

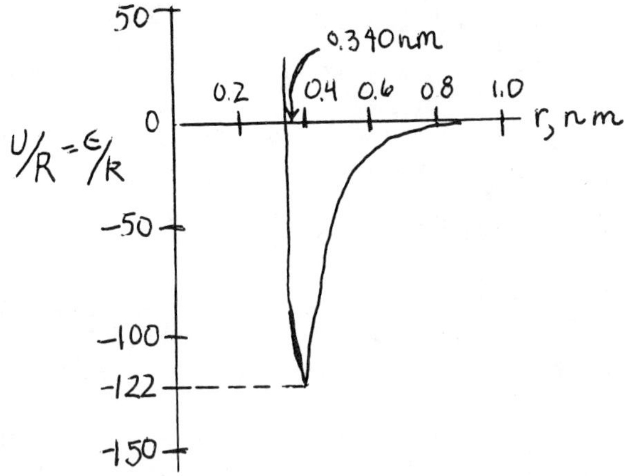

12.31 0.429 nm

CHAPTER 13. Molecular Spectroscopy

13.1 Since the energy of a molecular quantum state is divided by kT in the Boltzmann distribution, it is of interest to calculate the temperature at which kT is equal to the energy of photons of different wavelength. Calculate the temperature at which kT is equal to the energy of photons of wavelength 10^3 cm, 10^{-1} cm, 10^{-3} cm, 10^{-5} cm.

SOLUTION

$kT = hc/\lambda$

$$T = \frac{hc}{h\lambda} = \frac{(6.626\times10^{-34}\text{Js})(2.998\times10^{8}\text{ms}^{-1})}{(1.381\times10^{-23}\text{JK}^{-1})(10\text{m})}$$

$$= 1.44\times10^{-4}\text{ K} \quad \text{for} \quad \lambda = 10^3 \text{ cm}$$

For $\lambda = 10^{-1}$ cm, $T = 14.4$ K

For $\lambda = 10^{-3}$ cm, $T = 1440$ K

For $\lambda = 10^{-5}$ cm, $T = 144{,}000$ K

13.2 Most chemical reactions require activation energies ranging between 40 and 400 kJ mol^{-1}. What are the equivalents of 40 and 400 kJ mol^{-1} in terms of (a) nm, (b) wave numbers, and (c) electron volts?

SOLUTION

For 40 kJ mol^{-1}

(a) $\lambda = \frac{hc}{E}$

$$= \frac{(6.626\times10^{-34}\text{Js})(2.998\times10^{8}\text{ms}^{-1})(6.022\times10^{23}\text{mol}^{-1})}{(4\times10^{4}\text{ J mol}^{-1})}$$

$= 2.991 \times 10^{-6}$ m $= 2991$ nm

(b) $\tilde{\nu} = \frac{1}{\lambda} = \frac{1}{2.991 \times 10^{-4}\ \text{cm}} = 3343\ \text{cm}^{-1}$

(c) $\frac{(4 \times 10^{4}\ \text{J mol}^{-1})}{(6.022 \times 10^{23}\ \text{mol}^{-1})(1.602 \times 10^{-19}\text{J eV}^{-1})}$

$= 0.415$ eV

For 400 kJ mol^{-1}

(a) $\lambda = (2991\ \text{nm}) \frac{4\text{x}10^{4}\text{Jmol}^{-1}}{4\text{x}10^{5}\text{Jmol}^{-1}} = 299.1$ nm

(b) $\tilde{\nu} = \frac{1}{\lambda} = \frac{1}{2.991 \times 10^{-5}\ \text{cm}} = 33{,}430\ \text{cm}^{-1}$

(c) $(0.415\ \text{eV}) \frac{4\text{x}10^{5}\text{Jmol}^{-1}}{4\text{x}10^{4}\text{Jmol}^{-1}} = 4.15$ eV

13.3 Calculate the reduced mass and the moment of inertia of $D^{35}Cl$, given that $R_e = 0.1275$ nm.

SOLUTION

$$\mu = \frac{m_D\ m_{Cl}}{m_D + m_{Cl}}$$

$$= \frac{(2.01410\text{x}10^{-3}\text{kgmol}^{-1})(34.96885\text{x}10^{-3}\text{kgmol}^{-1})}{[(2.01410+34.96885)\text{x}10^{-3}\text{kgmol}^{-1}](6.022045\text{x}10^{23}\text{mol}^{-1})}$$

$= 3.1624 \times 10^{-27}$ kg

$I = \mu r^2 = (3.1624 \times 10^{-27}\ \text{kg})(1.275 \times 10^{-10}\ \text{m})^2$

$= 5.141 \times 10^{-47}$ kg m^2

13.4 Calculate the frequency in wave numbers and the wavelength in cm of the first rotational transition ($J = 0 \longrightarrow 1$) for $D^{35}Cl$.

SOLUTION

$$\tilde{\nu} = 2BJ = \frac{h}{4\pi c I} \quad \text{since } J = 1$$

The moment of inertia was calculated in the preceeding problem,

$$\tilde{\nu} = \frac{(6.626\text{x}10^{-34}\text{Js})}{4\pi^2(2.998\text{x}10^8\text{ms}^{-1})(5.141\text{x}10^{-47}\text{kgm}^2)}$$

$$= 1089\ \text{m}^{-1}$$

$$= (1089\ \text{m}^{-1})(10^{-2}\ \text{m cm}^{-1}) = 10.89\ \text{cm}^{-1}$$

$$\lambda = 1/\tilde{\nu} = 1/(10.89\ \text{cm}^{-1}) = 0.09183\ \text{cm}$$

13.5 The internuclear distance of $^{12}C^{16}O$ is 112.83 pm. Calculate the frequencies in wave numbers for the first four lines in the pure rotational spectrum.

SOLUTION

$$\mu = \frac{1/6.022045 \times 10^{23}\ \text{mol}^{-1}}{\dfrac{1}{12\text{x}10^{-3}\text{kg}} + \dfrac{1}{15.99491\text{x}10^{-3}\text{kg}}}$$

$$= 1.13852 \times 10^{-26}\ \text{kg}$$

$$I = (1.13852 \times 10^{-26}\ \text{kg})(112.83 \times 10^{-12}\ \text{m})^2$$

$$= 1.4494 \times 10^{-46}\ \text{kg m}^2$$

$$\tilde{\nu} = \frac{2hJ}{8\pi^2 Ic}$$

$$= \frac{2(6.626\text{x}10^{-34}\text{Js})(10^{-2}\text{mcm}^{-1})J}{8\pi^2(1.4494\text{x}10^{-46}\text{kgm}^2)(2.998\text{x}10^8\text{ms}^{-1})}$$

$$= (3.863\ \text{cm}^{-1})J$$

J	$\tilde{\nu}$
0 —> 1	3.863 cm^{-1}
1 —> 2	7.725 cm^{-1}
2 —> 3	11.588 cm^{-1}
3 —> 4	15.450 cm^{-1}

13.6 Considered classically, what is the average period of rotation of $H^{35}Cl$ molecules at room temperature? (The period of rotation is the time to rotate 2π radians.)

SOLUTION

If a molecule is considered to be a classical rotator, its angular momentum L is given by

$$\mu R_e^2 \omega$$

where ω is the angular velocity (radians per second). Setting this equal to the quantum mechanical expression for the angular momentum yields

$$\omega = \frac{[J(J+1)]^{1/2}\hbar}{\mu R_e^2}$$

But according to equation 13.10,

$$J(J+1)\hbar^2 = 2IE_r \quad \text{so that} \quad \omega = \left(\frac{2E_r}{\mu R_e^2}\right)^{1/2}$$

The average rotational energy of a molecule is kT. At room temperature

$$\omega = \left[\frac{2(1.381\text{x}10^{-23}\text{JK}^{-1})(298\text{K})}{(1.626\text{x}10^{-27}\text{kg})(1.275\text{x}10^{-10}\text{m})^2}\right]^{1/2}$$

$$= 1.765 \times 10^{13} \text{ s}^{-1}$$

$$\tau = \frac{2\pi}{\omega} = 3.56 \times 10^{-13} \text{ s}$$

13.7 The moment of inertia of $^{16}O{=}^{12}C{=}^{16}O$ is 7.167×10^{-46} kg m^2. (a) Calculate the CO bond length, R_{CO}, in CO_2. (b) Assuming that isotopic substitution does not alter R_{CO}, calculate the moments of inertia of

1. $^{18}O{=}^{12}C{=}^{18}O$ and 2. $^{16}O{=}^{13}C{=}^{16}O$

SOLUTION

(a) Since the CO_2 molecule is symmetrical, the carbon atom is on the axis of rotation and does not contribute to the moment of inertia.

$$I = 2mR_{CO}^2$$

$$R_{CO} = (I/2m)^{1/2}$$

$$= \left[\frac{(71.67\times10^{-47}\,kgm^2)(6.022045\times10^{23}\,mol^{-1})}{2(15.99491 \times 10^{-3}\ kg\ mol^{-1})}\right]^{1/2}$$

$$= 0.1162\ nm$$

(b) For $^{18}O\ ^{12}C\ ^{18}O$

$$I = 2mR_{CO}^2$$

$$= \frac{2(17.99916\times10^{-3}\,kgmol^{-1})(0.1162\times10^{-9}\,m)^2}{6.022045 \times 10^{23}\ mol^{-1}}$$

$$= 8.071 \times 10^{-46}\ kg\ m^2$$

For $^{16}O\ ^{13}C\ ^{16}O$ the moment of inertia is the same as for $^{16}O\ ^{12}C\ ^{16}O$.

13.8 The far infrared spectrum of HI consists of a series of equally spaced lines with $\Delta\tilde{\nu} = 12.8\ cm^{-1}$. What is (a) the moment of inertia, and (b) the internuclear distance?

SOLUTION

(a) $\Delta\tilde{\nu} = 2B = \dfrac{2h}{8\pi^2 cI}$

$$I = \frac{h}{8\pi^2 cB}$$

$$= \frac{(6.63 \times 10^{-34}\ J\ s)}{8\pi^2(2.998\times10^8\,ms^{-1})(6.4\,cm^{-1})(10^2\,cmm^{-1})}$$

$$= 4.37 \times 10^{-47}\ kg\ m^2$$

(b) $I = \frac{m_A m_B}{m_A + m_B} R_e^2$

$$R_e = \left[\frac{(4.37\text{x}10^{-47}\text{kgm}^2)(128\text{x}10^{-3}\text{kg})(6.02\text{x}10^{23}\text{mol}^{-1})}{(127 \text{ x } 10^{-3}\text{kg})(1 \text{ x } 10^{-3} \text{ kg})}\right]^{1/2}$$

$= 0.163$ nm

13.9 Show that for a rotational spectrum of a diatomic molecule the rotational quantum number (to the nearest integer value) for the maximum populated level is given by

$$J_{max} = \sqrt{\frac{kT}{2hc\tilde{B}}} - \frac{1}{2}$$

SOLUTION

$$\frac{dN_J}{dJ} = 2e^{\frac{-BhcJ(J+1)}{kT}} - (2J+1)\frac{Bhc}{kT}e^{\frac{-BhcJ(J+1)}{kT}}$$

At the maximum value of N_J this derivative is equal to zero so that the above expression is obtained. For $H^{35}Cl$ at 300 K

$$J_{max} = \sqrt{\frac{(1.3806\text{x}10^{-23})(300)}{2(6.626\text{x}10^{-34})(2.998\text{x}10^{8})(1044)}} - \frac{1}{2}$$

$= 2.7$, which rounds to 3

13.10 What are the relative populations of the first five rotational levels for $H^{35}Cl$ and $^{12}C\ ^{16}O$ at room temperature?

SOLUTION

$$N_J = K(2J+1)e^{-[hJ(J+1)B]/kT}$$

For $H^{35}Cl$, $B = 10.59\ \text{cm}^{-1}$

$$hB = (6.626\text{x}10^{-34}\text{Js})(2.9979\text{x}10^{8}\text{ms}^{-1})(1059\text{m}^{-1})$$
$$= 2.104 \text{ x } 10^{-22} \text{ J}$$

$$N_J/K = (2J+1)\ e^{-(2.104\text{x}10^{-22})J(J+1)/(1.3806\text{x}10^{-23})(300)}$$
$$= (2J+1)\ e^{-0.05080J(J+1)}$$

For $^{12}C\ ^{16}O$, $B = 1.9313\ \text{cm}^{-1}$

$$\frac{hB}{kT} = 9.26 \text{ x } 10^{-3}$$

$$N_J/K = (2J+1)\ e^{-(9.26\text{x}10^{-3})J(J+1)}$$

J	$H^{35}Cl$	$^{12}C^{16}O$
0	1	1
1	2.71	2.95
2	3.69	4.73
3	3.81	6.26
4	3.26	7.48

13.11 Derive the expression for the moment of inertia of a symmetrical tetrahedral molecule like CH_4 in terms of the bond length R and the masses of the four tetrahedral atoms. The easiest way to derive the expression is to consider an axis along one CH bond. Show that the same result is obtained if the axis is taken perpendicular to the plane defined by one group of three atoms HCH.

SOLUTION

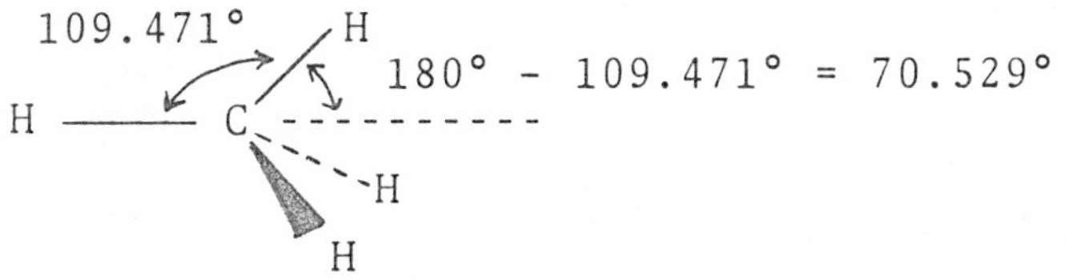

$\sin 70.529° = h/R$ where h is the perpendicular distance to the axis

$$I = \sum m_i h_i^{\,2} = 3mR^2 \sin^2(70.529°) = \frac{8}{3} mR^2$$

If the axis is taken perpendicular to the page through C

$$I = 2mR^2 + 2[R^2 \sin^2(90° - \frac{109.471°}{2})] = \frac{8}{3} mR^2$$

13.12 (a) What vibrational frequency in wave numbers corresponds to a thermal energy of kT at 25 °C? (b) What is the wavelength of this radiation?

SOLUTION

$$hc\tilde{\nu} = kT$$

$$= \frac{kT}{hc}$$

$$= \frac{(1.381 \times 10^{-23} JK^{-1})(298.15K)(10^{-2} m\ cm^{-1})}{(6.626 \times 10^{-34} JK^{-1})(2.998 \times 10^{8} ms^{-1})}$$

$$= 207\ cm^{-1}$$

$$= \frac{1}{\tilde{\nu}} = 4.83 \times 10^{-3}\ cm$$

13.13 Given the following fundamental vibration frequencies:

$H^{35}Cl$	2989 cm^{-1}	H^2D	3817 cm^{-1}
$^2D^{35}Cl$	2144 cm^{-1}	$^2D^2D$	2990 cm^{-1}

Calculate ΔH° for the reaction

$$H^{35}Cl(\nu = 0) + {}^2D^2D(\nu = 0) = {}^2D^{35}Cl(\nu = 0) + H^2D(\nu = 0)$$

SOLUTION

$$\Delta H° = E_{DCl} + E_{HD} - E_{HCl} - E_{D_2}$$

$$E = hc\tilde{\nu}/2 \quad \text{for} \quad v = 0$$

$$\Delta\tilde{\nu} = 2144 + 3817 - 2989 - 2990 = -18\ \text{cm}^{-1}$$

$$\Delta H° = -\frac{1}{2}\, hc\Delta\tilde{\nu}$$

$$= -\frac{1}{2}(6.63\text{x}10^{-34}\text{Js})\cdot(3\text{x}10^{8}\text{ms}^{-1})(-18\text{cm}^{-1})(100\text{cm m}^{-1})$$

$$= 1.79 \times 10^{-22}\ \text{J}$$

$$= (1.79 \times 10^{-22}\ \text{J})(6.022 \times 10^{23}\ \text{mol}^{-1})$$

$$= 108\ \text{J mol}^{-1}$$

13.14 The first three lines in the R branch of the fundamental vibration-rotation band of $H^{35}Cl$ have the following frequencies in cm^{-1}: 2906.25(0), 2925.78(1), 2944.89(2), where the numbers in parentheses are the J values for the initial level. What are the values of $\tilde{\omega}_0$, B_v', B_v'', B_e, and α ?

SOLUTION

According to equation 13.58

$$\tilde{\nu} = \tilde{\omega}_0(B_v' + B_v'')m + (B_v' - B_v'')m^2$$

$$= a + bm + cm^2 \quad \text{where} \quad m = J'' + 1$$

$$2906.25 = a + b + c$$

$$2925.78 = a + 2b + 4c$$

$$2944.89 = a + 3b + 9c$$

$$a = 2886.30\ \text{cm}^{-1} = \tilde{\omega}_0$$

$$b = 20.16 = B_v' + B_v''$$

$$c = -0.21 = B_v' - B_v''$$

$$B_v' = 9.98\ \text{cm}^{-1} = B_e - \frac{3}{2}\alpha$$

$$B_v'' = 10.19\ \text{cm}^{-1} = B_e - \frac{1}{2}\alpha$$

$B_e = 10.30\ cm^{-1}$

$\alpha = 0.21\ cm^{-1}$

13.15 The fundamental vibration frequency of $H^{35}Cl$ is $8.667 \times 10^{13}\ s^{-1}$. What would be the separation between the infrared absorption lines for $H^{35}Cl$ and $H^{37}Cl$ if the force constants of the bonds are assumed to be the same.

<u>SOLUTION</u>

$$\mu = \frac{m_H\ m_{Cl}}{m_H + m_{Cl}}$$

For $H^{35}Cl$

$$\mu_1 = \frac{(1.007825\text{x}10^{-3}\text{kgmol}^{-1})(34.96885\text{x}10^{-3}\text{kgmol}^{-1})}{(35.97668\text{x}10^{-3}\text{kgmol}^{-1})(6.022045\text{x}10^{23}\text{mol}^{-1})}$$

$$= 1.62668 \times 10^{-27}\ \text{kg}$$

For $H^{37}Cl$

$$\mu_2 = \frac{(1.007825\text{x}10^{-3}\text{kgmol}^{-1})(36.947\text{x}10^{-3}\text{kgmol}^{-1})}{(37.955\text{x}10^{-3}\text{kgmol}^{-1})(6.022045\text{x}10^{23}\text{mol}^{-1})}$$

$$= 1.62911 \times 10^{-27}\ \text{kg}$$

$$\nu_o = \frac{1}{2\pi}\left(\frac{k}{\mu}\right)^{1/2} \qquad \frac{\nu_1}{\nu_2} = \left(\frac{\mu_2}{\mu_1}\right)^{1/2} = \frac{\lambda_2}{\lambda_1}$$

$$\lambda_1 = \frac{(2.9979 \times 10^8\ \text{m s}^{-1})}{8.667 \times 10^{13}\ \text{s}^{-1}} = 3459.0 \times 10^{-9}\ \text{nm}$$

$$\lambda_2 = \lambda_1\left(\frac{\mu_2}{\mu_1}\right)^{1/2}$$

$$= (3459.0\ \text{nm})\left(\frac{1.62911\text{x}10^{-27}\text{kg}}{1.62668\text{x}10^{-27}\text{kg}}\right)^{1/2} = 3461.6\ \text{nm}$$

$$\lambda_2 - \lambda_1 = 2.6\ \text{nm}$$

13.16 Calculate the force constants k for $^{14}N_2$ in the ground state and in the $3\pi_g$ state. (See Table 13.4)

SOLUTION

$$\mu = \frac{m}{2} = \frac{14.0067 \times 10^{-3}\ \text{kg mol}^{-1}}{2(6.022045 \times 10^{23}\ \text{mol}^{-1})}$$

$$= 1.162\ 952 \times 10^{-26}\ \text{kg}$$

$$\nu = \frac{1}{2\pi}\left(\frac{k}{\mu}\right)^{1/2}$$

$$k = (2\pi\nu)^2\mu$$

$$= [2\pi(235\ 857\ \text{m}^{-1})(2.997\ 924 \times 10^8\ \text{m s}^{-1})]^2\mu$$

$$= 2295.41\ \text{N m}^{-1} \quad \text{for} \quad {}^1\Sigma_g^+$$

$$k = [2\pi(173\ 339\ \text{m}^{-1})(2.997\ 924 \times 10^8\ \text{m s}^{-1})]^2\mu$$

$$= 1239.81\ \text{N m}^{-1} \quad \text{for} \quad 3\pi_g$$

13.17 (a) What fraction of $H_2(g)$ molecules are in the v = 1 state at room temperature? (b) What fraction of $Cl_2(g)$ molecules are in the v = 1 state at room temperature?

SOLUTION

$$\frac{N_u}{N_\ell} = e^{-hc\tilde{\nu}/kT}$$

For H_2

$$\frac{N_u}{N_\ell} = \exp\left[-\frac{(6.626\text{x}10^{-34}\text{Js})(2.998\text{x}10^8\text{ms}^{-1})(440121\text{m}^{-1})}{(1.381\text{x}10^{-23}\text{JK}^{-1})(300\text{K})}\right]$$

$$= 6.84 \times 10^{-10}$$

For Cl_2

$$\frac{N_u}{N_\ell} = \exp\left[-\frac{(6.626\text{x}10^{-34}\text{Js})(2.998\text{x}10^8\text{ms}^{-1})(55970\text{m}^{-1})}{(1.381\text{x}10^{-23}\text{JK}^{-1})(300\text{K})}\right]$$

$$= 6.83 \times 10^{-2}$$

13.18 In Table 12.1 D_e for H_2 is given as 4.7483 eV or 458.135 kJ mol^{-1}. Given the vibrational parameters for H_2 in Table 13.4 calculate the value you would expect for ΔH_f° for H(g) at 0 K.

SOLUTION

$$D_0 = D_e - \frac{1}{2}\tilde{\omega}_e + \frac{1}{4}\tilde{\omega}_e x_e$$

$$= (4.7483 \text{ eV})(8065.478 \text{ cm}^{-1} \text{ eV}^{-1}) - \frac{1}{2}(4400.39) + \frac{1}{4}(120.815)$$

$$= 36{,}127 \text{ cm}^{-1} \quad \text{or} \quad 432.175 \text{ kJ mol}^{-1}$$

$$H_2(g) = 2H(g)$$

$$\Delta H_0^\circ = 2\Delta H_f^\circ(H) = 432.175 \text{ kJ mol}^{-1}$$

$$\Delta H_f^\circ(H) = 216.088 \text{ kJ mol}^{-1}$$

Table A.2 gives 216.037 kJ mol^{-1}, but of course this is for a mixture of hydrogen isotopes.

13.19 Calculate the wavelengths in (a) wave numbers, and (b) micrometers of the center two lines in the vibration-rotation spectrum of HBr for the fundamental vibration. The necessary data are to be found in Table 13.4.

SOLUTION

The reduced mass of $H^{80}Br$ is given in Table 13.4 as

$$\mu = \frac{0.99558 \times 10^{-3} \text{ kg mol}^{-1}}{6.022045 \times 10^{-23} \text{ mol}^{-1}}$$

$$= 1.65323 \times 10^{-27} \text{ kg}$$

$$I = \mu R^2 = (1.65323\times10^{-27}\text{kg})(1.4138\times10^{-10}\text{m})^2$$

$$= 3.30452 \times 10^{-47} \text{ kg m}^2$$

$$\tilde{B} = \frac{h}{8\pi^2 cI}$$

$$= \frac{(6.626 \times 10^{-34} Js)(10^{-2} m\ cm^{-1})}{8\pi^2 (2.998 x 10^8 ms^{-1})(3.3045 x 10^{-47} kgm^2)}$$

$$= 8.47\ cm^{-1}$$

(a) $\tilde{\nu}_P = \tilde{\nu}_o - 2BJ''$ where $J'' = 1,2,3,\ldots.$

$= 2649.67\ cm^{-1} - 2(8.47\ cm^{-1})$

$= 2632.73\ cm^{-1}$

$\tilde{\nu}_R = \tilde{\nu}_o + 2B + 2BJ''$ where $J'' = 0,1,2,\ldots.$

$= 2649.67\ cm^{-1} + 2(8.47\ cm^{-1})$

$= 2666.61\ cm^{-1}$

(b) $\lambda_P = 1/\tilde{\nu}_P = 3.79834 \times 10^{-4}$ cm

$= 3.79834\ \mu m$

$\lambda_R = 1/\tilde{\nu}_R = 3.75008 \times 10^{-4}$ cm

$= 3.75008\ \mu m$

13.20 How many normal modes of vibration are there for (a) SO_2(bent), (b) H_2O_2(bent), (c) HC≡CH(linear), and (d) C_6H_6?

SOLUTION

The number of normal modes of vibration is 3N-6 for a non-linear molecule and 3N-5 for a linear molecule.

(a) 3N-6 = 9-6 = 3
(b) 3N-6 = 12-6 = 6
(c) 3N-5 = 12-5 = 7
(d) 3N-6 = 36-6 = 30

13.21 List the numbers of translational, rotational, and vibrational degrees of freedom for (a) Ne, (b) N_2, (c) CO_2, and (d) CH_2O.

SOLUTION

	Molecule	Trans.	Rot.	Vib.	Total=3N
(a)	Ne	3	0	0	3
(b)	N_2	3	2	1	6
(c)	CO_2	3	2	4*	9
(d)	CH_2O	3	3	6**	12

*For linear molecules 3N-5
**For nonlinear molecules 3N-6

13.22 Acetylene is a symmetrical linear molecule. It has seven normal modes of vibration, two of which are doubly degenerate. These normal modes may be represented as follows:

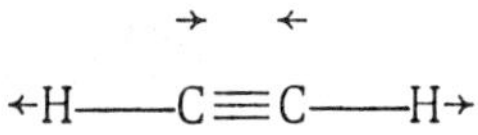

$$\overset{\rightarrow}{}\quad \overset{\leftarrow}{}$$
$\leftarrow$H——C≡C——H$\rightarrow$

$\tilde{\nu}_1 = 3374 \text{ cm}^{-1}$

$$\overset{\uparrow}{}\qquad\overset{\uparrow}{}$$
H——C≡C——H
$\downarrow\quad\downarrow$

$\tilde{\nu}_4 = 612 \text{ cm}^{-1}$

$$\overset{\rightarrow}{}\quad \overset{\leftarrow}{}$$
H$\rightarrow$—C≡C—$\leftarrow$H

$\tilde{\nu}_2 = 1974 \text{ cm}^{-1}$

$$\overset{\uparrow}{}\qquad\overset{\uparrow}{}$$
H——C≡C——H
$\downarrow\quad\downarrow$

$\tilde{\nu}_5 = 729 \text{ cm}^{-1}$

$$\overset{\leftarrow}{}$$
H $\rightarrow\leftarrow$ C≡C——H$\rightarrow$

$\tilde{\nu}_3 = 3287 \text{ cm}^{-1}$

(a) Which are the doubly degenerate vibrations?
(b) Which vibrations are infrared active?
(c) Which vibrations are Raman active?

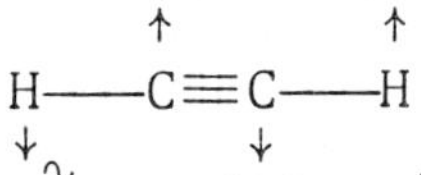

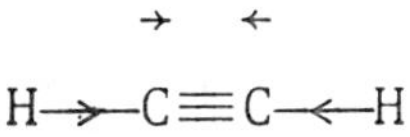

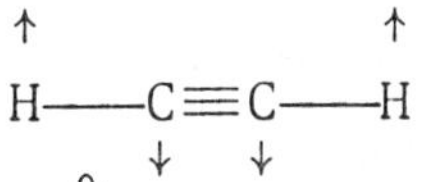

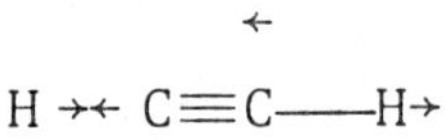

SOLUTION

(a) $\tilde{\nu}_4$ and $\tilde{\nu}_5$ since they can also take place in a plane perpendicular to the plane of the paper

(b) $\tilde{\nu}_3$ and $\tilde{\nu}_5$ because the dipole moment changes during the vibration

(c) $\tilde{\nu}_1$, $\tilde{\nu}_2$, and $\tilde{\nu}_4$ because the polarizability of the molecule changes during these vibrations

13.23 When CCl_4 is irradiated with the 435.8 nm mercury line, Raman lines are obtained at 439.9, 444.6, and 450.7 nm. Calculate the Raman frequencies of CCl_4 (expressed in wave numbers). Also calculate the wavelengths (expressed in μm) in the infrared at which absorption might be expected.

SOLUTION

$$\tilde{\nu}_{Raman} = \frac{1}{\lambda_{incid}} - \frac{1}{\lambda_{scatt}}$$

$$= \frac{1}{435.8\text{x}10^{-7}\text{cm}} - \frac{1}{439.9\text{x}10^{-7}\text{cm}} = 214\ \text{cm}^{-1}$$

$$\lambda = \frac{1}{\tilde{\nu}} = \frac{1}{214\ \text{cm}^{-1}} = 46.7\ \text{x}\ 10^{-6}\ \text{m}$$

For the remaining lines

$\tilde{\nu}_{Raman}/\text{cm}^{-1}$	312	454	759
$\lambda/\mu\text{m}$	32.0	22.0	13.2

13.24 The first several Raman frequencies of $^{14}N_2$ are 19.908, 27.857, 35.812, 43.762, 51.721, and

59.662 cm^{-1}. These lines are due to pure rotational transitions with J = 1,2,3,4,5, and 6. The spacing between the lines is $4B_e$. What is the internuclear distance?

SOLUTION

$$\mu = \frac{(14.00307\times10^{-3}kg)^2}{(28.00614\times10^{-3}kg)(6.022045\times10^{23}mol^{-1})}$$

$$= 1.162651 \times 10^{-26}kg$$

The average spacing between lines is 7.951 cm^{-1}, and so $\tilde{B}_e$ = (7.951 cm^{-1})(10^2 cm m^{-1})/4 = 198.78 m^{-1} since ΔJ = 2.

$$\tilde{B}_e = \frac{h}{8\pi^2 c\mu R_e^2}$$

$$R_e = \left(\frac{h}{8\pi^2 c\mu\tilde{B}_e}\right)^{1/2}$$

$$= \left[\frac{6.626176 \times 10^{-34}\ Js}{8\pi^2(2.99792458\times10^8 ms^{-1})(1.162651\times10^{-26}kg)(198.78m^{-1})}\right]$$

$$= 110\ pm$$

13.25 What Raman shifts are expected for the first four Stokes lines for CO_2?

SOLUTION

$I = 7.167 \times 10^{-46}$ kg m^2 (Problem 11.7)

$$B_e = \frac{h}{8\pi^2 Ic}$$

$$= \frac{(6.626\times10^{-34}Js)(10^{-2}m\ cm^{-1})}{8\pi^2(7.167\times10^{-46}kgm^2)(2.9979\times10^8 ms^{-1})}$$

$$= 0.3906\ cm^{-1} \qquad \Delta\tilde{\nu}_R = 2\tilde{B}_e(2J + 3)$$

J''	$\Delta\tilde{\nu}_R$
0	2.3436 cm^{-1}
1	3.9060
2	5.4684
3	7.0308

13.26 The spectroscopic dissociation energy of $H_2(g)$ is 4.4763 eV, and the fundamental vibrational frequency is 4395.24 cm^{-1}. What is the spectroscopic dissociation energy of $D_2(g)$ if it has the same force constant?

SOLUTION

Since deuterium atoms are more massive, the fundamental vibration frequency will be lower in D_2 than H_2. Therefore the zero point energy for D_2 is lower, and its dissociation energy is higher.

For H_2

$$\mu = \frac{(1.007825 \times 10^{-3}\ \text{kg mol}^{-1})^2}{2(1.007825\text{x}10^{-3}\text{kgmol}^{-1})(6.022045\text{x}10^{23}\text{mol}^{-1})}$$

$= 8.367797 \times 10^{-28}$ kg

For D_2

$$\mu = \frac{(2.01410 \times 10^{-3}\ \text{kg mol}^{-1})^2}{2(2.01410\text{x}10^{-3}\text{kgmol}^{-1})(6.022045\text{x}10^{23}\text{mol}^{-1})}$$

$= 16.722725 \times 10^{-28}$ kg

According to egn 11.54

$$\frac{\tilde{\omega}_{D_2}}{\tilde{\omega}_{H_2}} = \left[\frac{8.367797 \times 10^{-28}\ \text{kg}}{16.722725 \times 10^{-28}\ \text{kg}}\right]^{1/2}$$

$\tilde{\omega}_{H_2} = 4395.24\ cm^{-1}$ so

$\tilde{\omega}_{D_2} = 3109.10\ cm^{-1}$

$$\Delta\tilde{\omega} = 1286.1 \text{ cm}^{-1}$$

$$D^{\circ}_{D_2} = 4.4763 \text{ eV} + hc\Delta\tilde{\omega}/2$$

$$= 4.4763 \text{ eV} + \frac{(6.626\text{x}10^{-34}\text{Js})(2.9979\text{x}10^{8}\text{ms}^{-1})(1.286\text{x}10^{5}\text{m}^{-1})}{2(1.602 \text{ x } 10^{-19} \text{ J eV}^{-1})}$$

$$= 4.4763 \text{ eV} + 0.0797 \text{ eV} = 4.5560 \text{ eV}$$

13.27 According to the hypothesis of Franck, the molecules of the halogens dissociate into one normal atom and one excited atom. The wavelength of the convergence limit in the spectrum of iodine is 499.5 nm. (a) What is the energy of dissociation of iodine into one normal and one excited atom? (b) The lowest excitation energy of the iodine atom is 0.94 eV. What is the energy corresponding to this excitation? (c) Compute the heat of dissociation of the iodine molecule into two normal atoms, and compare it with the value obtained from thermochemical data, 144.4 kJ mol^{-1}.

SOLUTION

(a) $E = hcN_A/\lambda$

$$= \frac{(6.626\text{x}10^{-34}\text{Js})(2.998\text{x}10^{8}\text{ms}^{-1})(6.022\text{x}10^{23}\text{mol}^{-1})}{(4995 \text{ x } 10^{-10} \text{ m})(10^{3} \text{ J kJ}^{-1})}$$

$$= 239.5 \text{ kJ mol}^{-1}$$

(b) $(0.94 \text{ eV})(96{,}485 \text{ C mol}^{-1})$

$= 90.7 \text{ kJ mol}^{-1}$

(c)

$I_2 = I + I^*$	239.5
$I^* = I$	-90.7
$I_2 = 2I$	148.8 kJ mol^{-1}

13.28 A solution of a dye containing 0.1 g L^{-1} transmits 80% of the light at 435.6 nm in a glass cell 1 cm thick. (a) What percent of light will be absorbed by a solution containing 2g/100 cm^3 in a cell 1 cm thick? (b) What concentration will be required to absorb 50% of the light? (c) What percent of the light will be transmitted by a solution of the dye containing 1g/100 cm^3 in a cell 5 cm thick? (d) What thickness should the cell be in order to absorb 90% of the light with solution of this concentration?

SOLUTION

$$\log \frac{I_o}{I} = \varepsilon C \ell$$

$$\log \frac{100}{80} = \varepsilon(1 \times 10^{-2} \text{ g cm}^{-3})(1 \text{ cm})$$

$$\varepsilon = 9.69 \text{ cm}^2 \text{ g}^{-1}$$

(a) $$\log \frac{100}{I} = (9.69\text{cm}^2\text{g}^{-1})(2\text{x}10^{-2}\text{gcm}^{-3})(1 \text{ cm})$$

$$\log I = 2 - 0.1938$$

$$I = 64.0\%$$

Alternatively this calculation may be done using exponential notation.

$$I = I_o e^{-2.303\varepsilon C\ell}$$

$$= (100\%)e^{-2.303(9.69\text{cm}^2\text{g}^{-1})(2\text{x}10^{-2}\text{gcm}^{-3})(1 \text{ cm})}$$

$$= 64.0\%$$

(b) $$\log \frac{100}{50} = (9.69 \text{ cm}^2 \text{ g}^{-1})(1 \text{ cm})c$$

$$c = 0.0311 \text{ g cm}^{-3}$$

(c) $$\log \frac{100}{I} = (9.69 \text{ cm}^2 \text{ g}^{-1})(10^{-2} \text{ g cm}^{-3})(5 \text{ cm})$$ or

$$I = I_o e^{-2.303(9.69 cm^2 g^{-1})(10^{-2} g cm^{-3})(5\ cm)}$$
$$= 32.8\%$$

(d) $\log \frac{100}{10} = (9.69\ cm^{-2}\ g^{-1})(10^{-2} g\ cm^{-3})\ell$

$\ell = 10.32$ cm

13.29 The absorption coefficient α for a solid is defined by $I = I_o e^{-ax}$, where x is the thickness of the sample. The absorption coefficients for NaCl and KBr at a wavelength of 28 μm are 14 and 0.25 cm^{-1}. Calculate the percentage of this infrared radiation transmitted by 0.5 cm thicknesses of these crystals.

SOLUTION

For NaCl $\quad I = I_o e^{-ax}$

$= 100\ e^{-(14\ cm^{-1})(0.5\ cm)} = 0.091\%$

For KBr $\quad I = 100\ e^{-(0.25\ cm^{-1})(0.5\ cm)} = 88.2\%$

13.30 The following absorption data are obtained for solutions of oxyhemoglobin in pH 7 buffer at 575 nm in a 1-cm cell:

g/cm^3	Transmission, %
3×10^{-4}	53.5
5×10^{-4}	35.1
10×10^{-4}	12.3

The molar mass of hemoglobin is 64.0 kg mol^{-1}. (a) Is Beer's law obeyed? What is the molar absorption coefficient? (b) Calculate the percent transmission for a solution containing 10^{-4} g/cm^3.

SOLUTION

(a)

Grams per cm^3	Mol L^{-1}	I/I_0	$\log(I/I_0)$	ε
3×10^{-4}	4.69×10^{-6}	0.535	-0.272	5.80×10^{4}
5×10^{-4}	7.82×10^{-6}	0.351	-0.455	5.82×10^{4}
10×10^{-4}	15.64×10^{-6}	0.123	-0.910	5.82×10^{4}

Beer's law is obeyed and the molar absorption coefficient is $5.81 \times 10^4 (\text{mol L}^{-1})^{-1}\ \text{cm}^{-1}$.

(b) $\log (I/I_0) = -\varepsilon c \ell$

$= -[5.81\times10^4 (\text{molL}^{-1})^{-1}\text{cm}^{-1}](1.564\times10^{-6}\text{molL}^{-1})(1\ \text{cm})$

$= -0.091$

$I/I_0 = 0.81$ or 81% transmission

13.31 The protein metmyoglobin and imidazole form a complex in solution. The molar absorption coefficients in L mol^{-1} cm^{-1} of the metmyoglobin (Mb) and the complex (C) are as follows:

λ	$\dfrac{\varepsilon_{Mb}}{10^3\ \text{L mol}^{-1}\text{cm}^{-1}}$	$\dfrac{\varepsilon_{C}}{10^3\ \text{L mol}^{-1}\text{cm}^{-1}}$
500 nm	9.42	6.88
630 nm	3.58	1.30

An equilibrium mixture in a cell of 1-cm path length has an absorbance of 0.435 at 500 nm and 0.121 at 630 nm. What are the concentrations of metmyoglobin and complex?

SOLUTION

$$\log (I_0/I) = A = (\varepsilon_1 c_1 + \varepsilon_2 c_2)\ell$$

At 500 nm $\quad 0.435 = 9.42 \times 10^3\ c_1 + 6.88 \times 10^3\ c_2$

At 630 nm $\quad 0.121 = 3.58 \times 10^3\ c_1 + 1.30 \times 10^3\ c_2$

Solving these equations simultaneously

$c_1 = 2.17 \times 10^{-5}$ mol L^{-1} metmyoglobin

$c_2 = 3.37 \times 10^{-5}$ mol L^{-1} complex

13.32 (a) Calculate the energy levels for n = 1 and n = 2 for an electron in a potential well of width 0.5 nm with infinite barriers on either side. The energies should be expressed in J and in kJ mol^{-1}. (b) If an electron makes a transition from n = 2 to n = 1, what will be the wavelength of the radiation emitted?

SOLUTION

(a) $E = \dfrac{n^2 h^2}{8m_e a^2}$

$$= \frac{n^2(6.626 \times 10^{-34} \text{ J s})^2}{8(9.110 \times 10^{-31} \text{ kg})(5 \times 10^{-10} \text{ m})^2}$$

$= n^2\ 2.410 \times 10^{-19}$ J

$= 2.410 \times 10^{-19}$ J for n = 1

$= 9.640 \times 10^{-19}$ J for n = 2

$$= \frac{n^2(2.410 \times 10^{-19} \text{ J})(6.022 \times 10^{23} \text{ mol}^{-1})}{(10^3 \text{ J kJ}^{-1})}$$

$= n^2\ 145.14$ kJ mol^{-1}

$= 145.14$ kJ mol^{-1} for n = 1

$= 580.57$ kJ mol^{-1} for n = 2

(b) $E_2 - E_1 = \dfrac{hc}{\lambda}$

$\lambda = \dfrac{hc}{E_2 - E_1}$

$$= \frac{(6.626 \times 10^{-34} \text{ J s})(2.998 \times 10^8 \text{ ms}^{-1})}{(9.640 - 2.410) \times 10^{-19} \text{ J}} =$$

275 nm

13.33 The life times of vibrationally excited states of molecules of a liquid are limited by the collision rates in the liquid. If one in ten collisions deactivates a vibrationally excited state, what is the broadening of vibrational lines if a molecule undergoes 10^{13} collisions per second?

SOLUTION

$$\Delta E = hc\Delta\tilde{\nu} \geqslant h/2\pi\Delta t$$

$$\Delta\tilde{\nu} \geqslant (2\pi c\Delta t)^{-1} = [2\pi(3\times10^{10}\text{cms}^{-1})(10^{-12}\text{s})]^{-1}$$

$$\Delta\tilde{\nu} \geqslant 5.3 \text{ cm}^{-1}$$

13.34 The absorption spectrum for benzene in Fig. 13.13 shows maxima at about 180, 200, and 250 nm. Estimate the integrated absorption coefficients using $\varepsilon_{max}\Delta\tilde{\nu}_{1/2}$ and assuming that the width at half maximum is 5000 cm^{-1} in each case. What are the three oscillator strengths?

SOLUTION

At 180 nm

$$\varepsilon_{max}\Delta\tilde{\nu}_{1/2} = (50.1 \times 10^3 \text{L mol}^{-1} \text{ cm}^{-1})(5000 \text{ cm}^{-1})$$
$$= 2.50 \times 10^8 \text{ L mol}^{-1} \text{ cm}^{-2}$$

$$f = (4.33 \times 10^{-9}\text{molL}^{-1}\text{cm}^2)(2.50\text{x}10^8\text{Lmol}^{-1}\text{cm}^{-2})$$
$$= 1.08$$

At 200 nm

$$\varepsilon_{max}\Delta\tilde{\nu}_{1/2} = (7000 \text{ L mol}^{-1} \text{ cm}^{-1})(5000 \text{ cm}^{-1})$$
$$= 3.5 \times 10^7 \text{ L mol}^{-1} \text{ cm}^{-2}$$

$$f = (4.33\text{x}10^{-9}\text{molL}^{-1}\text{cm}^2)(3.5\text{x}10^7\text{Lmol}^{-1}\text{cm}^{-2})$$
$$= 0.152$$

At 250 nm

$$\varepsilon_{max}\,\Delta\tilde{\nu}_{1/2} = (100\ \text{L mol}^{-1}\ \text{cm}^{-1})(5000\ \text{cm}^{-1})$$

$$= 5 \times 10^5\ \text{L mol}^{-1}\ \text{cm}^{-2}$$

$$f = (4.33\text{x}10^{-9}\text{molL}^{-1}\text{cm}^{2})(5\text{x}10^{5}\text{Lmol}^{-1}\text{cm}^{-2})$$

$$= 0.0022$$

13.35 A sample of oxygen gas is irradiated with $MgK_{\alpha_1\alpha_2}$ radiation of 0.99 nm (1253.6 eV). A strong emission of electrons with velocities of 1.57×10^7 m s^{-1} is found. What is the binding energy of these electrons?

SOLUTION

$$\frac{1}{2}\,mv^2 = h\nu - I$$

$$I = h\nu - \frac{1}{2}\,mv^2$$

$$= 1253.6\text{eV} - \frac{1}{2}\,\frac{(9.109\text{x}10^{-31}\text{kg})(1.57\text{x}10^{7}\text{ms}^{-1})^2}{(1.602 \times 10^{-19}\ \text{J eV}^{-1})}$$

$$= 553\ \text{eV}$$

13.36 What is the difference in refractive index for left (n_l) and right (n_r) circularly polarized radiation if the measured rotation ϕ for a one decimeter path length is 10^{-3} degrees at a wavelength of 600 nm.

SOLUTION

$$n_l - n_r = \frac{\phi\lambda}{1800} = \frac{(10^{-3}\ \text{deg dm}^{-1})(6 \times 10^{-6}\ \text{dm})}{1800\ \text{deg}}$$

$$= 0.3 \times 10^{-11}$$

13.37 When α-D-mannose ($[\alpha]_D^{20} = +29.3°$) is dissolved in water, the optical rotation decreases as β-D-mannose is formed until at equilibrium $[\alpha]_D^{20} = +14.2°$). This process is referred to as mutarotation. As expected, when β-D-mannose ($[\alpha]_D^{20} = -17.0°$) is dissolved in water, the optical rotation increases until $[\alpha]_D^{20} = +14.2°$ is obtained. Calculate the percentage of α form in the equilibrium mixture.

SOLUTION

$[\alpha]_\alpha = 29.3°$

$[\alpha]_\beta = -17.0°$

$$[\alpha]_{mixture} = +14.2° = 29.3°\ f_\alpha - 17.0°\ f_\beta$$
$$= 29.3°\ f_\alpha - 17.0°(1 - f_\alpha)$$
$$= -17.0° + 36.3\ f_\alpha$$

$$f_\alpha = \frac{31.2°}{46.3°} = 0.67 = \text{fraction of } \alpha \text{ form}$$

13.38 (a) 3.10 nm (b) 414 nm

13.39 (a) 1.13852×10^{-26} kg (b) 1.4492×10^{-46} kgm^2

13.40 21.1, 42.2, 63.3, 189.9 cm^{-1}
474, 237, 158, 53 μm

13.41 (a) 4.61×10^{-48} (b) 6.15×10^{-48}
(c) 6.92×10^{-48} (d) 9.22×10^{-48} kg m^2

13.42 0.4515 kJ 0.0265 cm

13.43 113 pm

13.44 $1.629\,143 \times 10^{-27}$ kg 2.648×10^{-47} kg m^2

13.45 $\nu = Jh/4\pi^2 I$

13.48 12,168.5, 24,337.0, 36,505.6 Mc

13.49 9.10, 14.39 K

13.50 0.1855, 0.1737 eV

13.51 8.731×10^{-20} J 52.59 kJ 0.5449 eV

13.52 (a) 26.288, (b) 3.381 kJ mol^{-1}

13.53 6.216×10^{13} s^{-1} 4.83 μm

13.54 4.09×10^{8} N m^{-1}

13.55 8.9×10^{-7} 0.358

13.56 5.25 eV

13.57 90,530 cm^{-1} 11.22 eV

13.58 4592.8, 4605.9, 4632.1, 4645.2 cm^{-1}
2.1773, 2.1154, 2.1588, 2.1528 μm

13.59 1870 cm^{-1} 8.45 cm^{-1}

13.60

Molecule	Translational	Rotational	Vibrational
Cl_2	3	2	3N - 5 = 1
H_2O	3	3	3N - 6 = 3
HC≡CH	3	3	3N - 5 = 7

13.61

Molecule	Translational	Rotational	Vibrational
NNO	3	2	3N - 5 = 4
NH_3	3	3	3N - 6 = 6

13.62 0.07519 nm

13.63 109.7 pm

13.64 -77.8 kJ mol^{-1}

13.65 (a) 15,885 cm^{-1} (b) 1.9695 eV

13.66 25.2%

13.67 $(7.7 \pm 0.7) \times 10^{-5}$ g cm^{-3}

13.68 2.6 μg cm^{-3}

13.69 (a) 3.24×10^{-3} mol L^{-1} 5.12×10^{-3} mol L^{-1}
(b) 2.95×10^{4} (mol L^{-1})$^{-1}$

13.70 0.522 nm

13.71 (a) $\lambda = 8m\ell^2ck^2/(2k + 1)h$
(b) 207,589,1230 nm

13.72 0.0053 cm^{-1}

13.73 16×10^{4} L mol^{-1} cm^{-2} 7×10^{-4}

13.74 5.20 eV

13.75 0.418

13.76 0.636

CHAPTER 14. Magnetic Resonance Spectroscopy

14.1 Calculate the magnetic flux density to give a precessional frequency for fluorine of 60 MHz.

SOLUTION

$$B = \frac{h\nu}{g_N \mu_N}$$

$$= \frac{(6.6262 \times 10^{-34}\ \mathrm{J\ s})(6 \times 10^{7}\ \mathrm{s^{-1}})}{(5.257)(5.0508 \times 10^{-27}\ \mathrm{J\ T^{-1}})} = 1.4973\ \mathrm{T}$$

14.2 Frequencies used in nuclear magnetic resonance spectrometers are of the order of 60 to 300 MHz. Calculate the corresponding energies in kilojoules per mole.

SOLUTION

$$E = h\nu$$

$$= \frac{(6.626\mathrm{x}10^{-34}\mathrm{Js})(6\mathrm{x}10^{7}\mathrm{s^{-1}})(6.022\mathrm{x}10^{23}\mathrm{mol^{-1}})}{(1000\ \mathrm{J\ kJ^{-1}})}$$

$$= 2.394 \times 10^{-5}\ \mathrm{kJ\ mol^{-1}}$$

$$E = (2.394\mathrm{x}10^{-5}\mathrm{kJmol^{-1}})(300\ \mathrm{MHz})/(60\ \mathrm{MHz})$$

$$= 11.970 \times 10^{-5}\ \mathrm{kJ\ mol^{-1}}$$

14.3 What magnetic flux density is required for proton magnetic resonance at 220 MHz?

SOLUTION

$$\frac{220\ \mathrm{MHz}}{60\ \mathrm{MHz}} \times 1.4092\ \mathrm{T} = 5.1670\ \mathrm{T}$$

14.4 (a) What are the energy levels for a ^{23}Na nucleus in a magnetic field of 2 T? (b) What is the absorption frequency?

SOLUTION

$$E = |g_N|\mu_N m_I B$$
$$= (1.478)(5.0508 \times 10^{-27}\ J\ T^{-1})(2T)m_I$$
$$= (1.493 \times 10^{-26}\ J)m_I$$

Since $I = \frac{3}{2}$, $m_I = -\frac{3}{2}, -\frac{1}{2}, +\frac{1}{2}, +\frac{3}{2}$
and
$E = -2.240\text{x}10^{-26}$, $-0.7465\text{x}10^{-26}$, $0.7465\text{x}10^{-26}$, and $2.240\text{x}10^{-26}$ J

$$\nu = \frac{\Delta E}{h} = \frac{1.493 \times 10^{-26}\ J}{6.6262 \times 10^{-34}\ J\ s} = 22.53\ MHz$$

14.5 The gyromagnetic ratio γ_N for a nucleus is defined by $\mu_N = \gamma_N \hbar I$
What is the value of γ_N for H?

SOLUTION

$\mu_N = g_N\mu_N I$ Therefore, $\gamma_N = g_N\mu_N/\hbar$

$$= \frac{2\pi(5.585)(5.0508 \times 10^{-27}\ J\ T^{-1})}{(6.6262 \times 10^{-34}\ J\ s)}$$
$$= 2.675 \times 10^{8}\ s^{-1}\ T^{-1}$$

14.6 What is the ratio of the number of ^{31}P spins in the lower state to the number in the upper state in a magnetic field of 1.724 T at room temperature? (Given: $g_N = 2.263$.)

SOLUTION $\quad \frac{N_\alpha}{N_\beta} = 1 + \frac{g_N\mu_N B}{kT}$

$$= 1 + \frac{(2.263)(5.0508\text{x}10^{-27}\,\text{JT}^{-1})(1.724\text{T})}{(1.3807\text{x}10^{-23}\ \text{JK}^{-1})(298.15\text{K})}$$

$$= 1 + 4.79 \text{ x } 10^{-6}$$

14.7 In a magnetic field of 2 T, what fraction of the protons have their spin lined up with the field at room temperature?

SOLUTION

$$\frac{N_\ell}{N_h} = 1 + \frac{g_N \mu_N B}{kT}$$

$$= 1 + \frac{(5.585)(5.05\text{x}10^{-27}\,\text{JT}^{-1})(2\text{T})}{(1.38\text{x}10^{-23}\,\text{JK}^{-1})(298\text{K})}$$

$$= 1.00001372$$

$$\frac{N_\ell}{N_\ell + N_h} = \frac{1}{1 + N_h/N_\ell} = \frac{1}{1 + 1/(N_h/N_\ell)}$$

$$= \frac{1}{1 + 1/1.00001372} = 0.50000343$$

14.8 Using information from Tables 14.2 and 14.3, sketch the spectrum you would expect for ethyl acetate ($CH_3CO_2CH_2CH_3$).

SOLUTION

$$\underset{\substack{\uparrow \\ \delta \approx 2}}{CH_3} - \underset{\substack{\| \\ O}}{C} - O - \underset{\substack{\uparrow \\ \delta \approx 3.6}}{CH_2} - \underset{\substack{\uparrow \\ \delta \approx 1}}{CH_3}$$

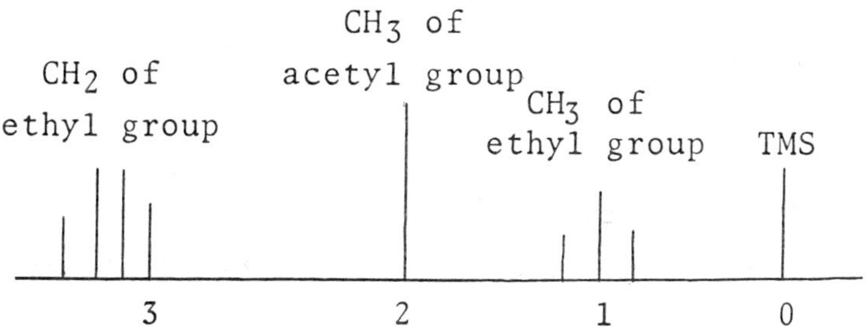

14.9 Chemical shifts δ are generally expressed in ppm, but it is also convenient to express them in Hz because coupling constants are expressed in Hz. What shifts in frequency correspond with a chemical shift of δ = 1 for protons in a spectrometer with (a) B = 1.41 T and (b) 7.00 T?

SOLUTION

(a) $\nu_i = |g_i| \mu_N (1 - \sigma_i) B/h$

$$\Delta\nu_i = |g_i| \mu_N \sigma_i B/h$$

$$= \frac{(5.585)(5.051 \times 10^{-27} \text{ J T}^{-1})(10^{-6})(1.41 \text{ T})}{6.626 \times 10^{-34} \text{ J s}}$$

$$= 60 \text{ Hz}$$

(b) $\Delta\nu_i = (60 \text{ Hz})(7.00 \text{ T})/(1.41 \text{ T}) = 298 \text{ Hz}$

14.10 The proton resonance pattern of 2,3-dibromothiophene shows an AB-type spectrum with lines at 405.22, 410.85, 425.07, and 430.84 Hz measured from tetramethylsilane at 1.41 T [K. F. Kuhlmann and C. L. Braun, J. Chem. Ed., 46, 750 (1969]. (a) What is the coupling constant J?

(b) What is the difference in the chemical shifts of the A and B hydrogens? (c) At what frequencies would the lines be found at 2 T?

SOLUTION

(a) The average spacing of the two doublets is

$J = 5.70 \pm 0.07$ Hz.

(b) $$\nu_o\delta = \left[(a - d)(b - c)\right]^{1/2}$$

$$= \left[(25.62)(14.22)\right]^{1/2}$$

$$= 19.09 \text{ Hz}$$

$$\delta = (19.09\text{Hz})/(60\text{x}10^6\text{Hz}) = 0.318 \text{ x } 10^{-6}$$

$$= 0.318 \text{ ppm}$$

(c) At 2T $$\nu_o^1 = \frac{2}{1.41} \quad 60 \text{ x } 10^6 \text{ Hz}$$

$$= 85.1 \text{ x } 10^6 \text{ Hz}$$

The center of the spectrum shifts by ν_o^1/ν_o

$$418.00 \; \frac{85.1 \text{ x } 10^6}{60 \text{ x } 10^6} = 592.86$$

Distance of b and c from center

$$= \frac{[(\nu_o\delta)^2+J^2]^{\frac{1}{2}}-J}{2} = \frac{1}{2}\left\{[(85.1\text{x}0.318)^2+5.70^2]^{\frac{1}{2}} - 5.40\right\}$$

$$= 10.98$$

$a = 592.86 - 10.98 - 5.70 = 576.18$ Hz

$b = 592.86 - 10.98 = 581.88$ Hz

$c = 592.86 + 10.98 = 603.84$ Hz

$d = 592.86 + 10.98 + 5.70 = 609.54$ Hz

14.11. Are protons a and b magnetically equivalent in

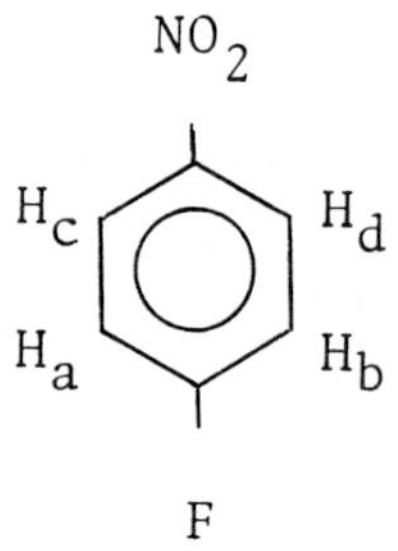

SOLUTION

They are not magnetically equivalent (although they are chemically equivalent) because the coupling H_a — H_c is different from the coupling H_a — H_d.

14.12 At room temperature the chemical shift of cyclohexane protons is an average of the chemical shifts of the axial and equatorial protons. Explain.

SOLUTION

At room temperature the rate of conversion of cyclohexane from boat to chair forms is so fast that the protons are at the average local magnetic field.

14.13 Calculate the precessional frequency of electrons in a 1.5-T field.

SOLUTION

$$\nu = \frac{B g_e \mu_B}{h}$$

$$= \frac{(1.5T)(2.0023)(9.2742\text{x}10^{-24}JT^{-1})}{6.6262 \times 10^{-34} \text{ J s}}$$

$$= 4.2037 \times 10^{10} \text{ s}^{-1} = 42{,}037 \text{ MHz}$$

14.14 Line separations in ESR may be expressed in G or MHz. Show how the conversion factor $1\ T = 2.80 \times 10^4$ MHz is obtained.

SOLUTION

The resonance frequency for electrons in a 1 gauss field is given by

$$\nu = \frac{g_e B \mu_B}{h}$$

$$= \frac{(2.00)(9.274 \times 10^{-24}\ J\ T^{-1})(1\ T)}{6.626 \times 10^{-34}\ J\ s}$$

$$= 2.80 \times 10^{10}\ s^{-1} = 2.80 \times 10^4\ MHz$$

14.15 Sketch the ESR spectrum expected for p-benzosemiquinone radical ion.

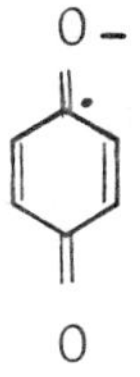

The four hydrogens are magnetically equivalent.

SOLUTION

			1		1				1H
		1		2		1			2H
	1		3		3		1		3H
1		4		6		4		1	4H

Thus there will be five equally spaced lines with relative intensities of 1, 4, 6, 4, 1.

14.16 An unpaired electron in the presence of two protons gives the following four-line ESR spectrum: $\Delta B/10^{-4}$ T = 0, 1, 3, 4. What are the two coupling constants in T and in MHz?

SOLUTION

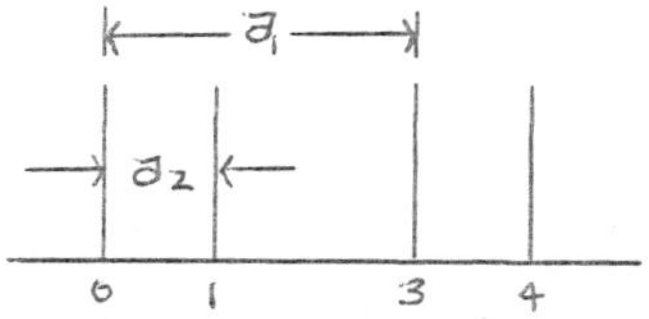

$a_1 = 3 \times 10^{-4}$ T

$a_2 = 1 \times 10^{-4}$ T

Multiplying by the factor in problem 14.14

$a_1 = (3\times10^{-4}\text{T})(2.80\times10^{4}\text{MHzT}^{-1}) = 8.40$ MHz

$a_2 = (1\times10^{-4}\text{T})(2.80\times10^{4}\text{MHzT}^{-1}) = 2.80$ MHz

14.17 Sketch the ESR spectrum for an unpaired electron in the presence of three protons for the following cases: (a) the protons are not equivalent, (b) the protons are equivalent, and (c) two protons are equivalent and the third is different.

SOLUTION

(a)

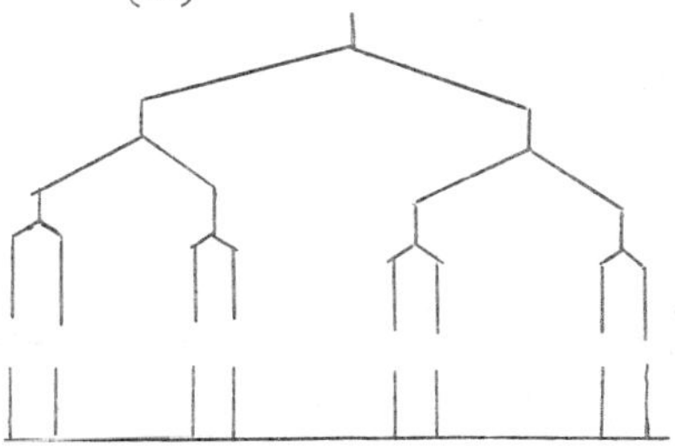

(b)

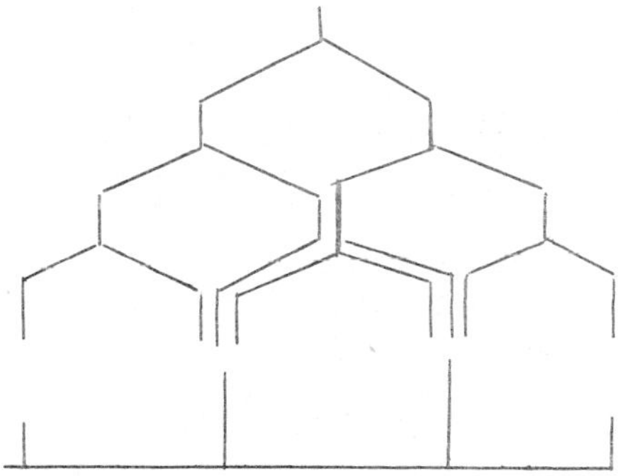

(c)

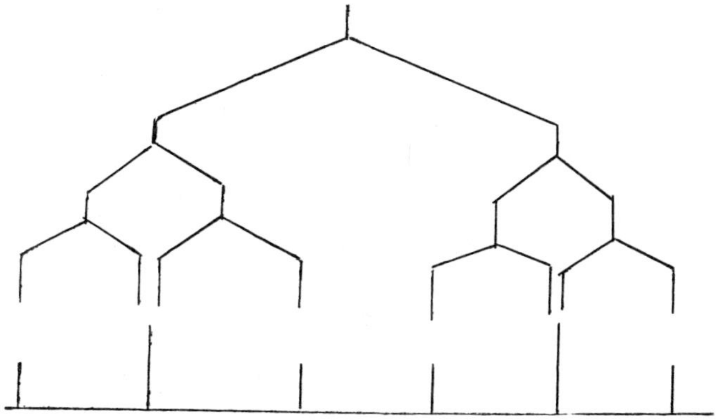

14.18 Sketch the ESR spectrum of CH_3.

SOLUTION

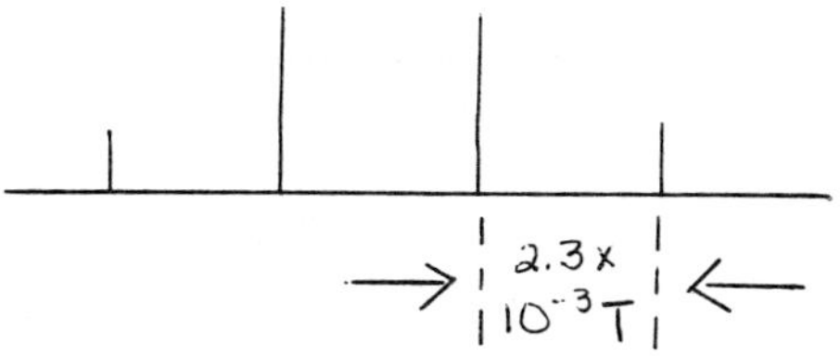

14.19 1.410×10^{-27} J T^{-1}

14.20 (a) 0.4669 T (b) 0.1248 T

14.21 498 MHz

14.22 0.4991 T

14.23 10.706 MHz

14.24 0.50000171

14.25 7.270

14.26 4257.2×10^{4} Hz T^{-1}

14.27 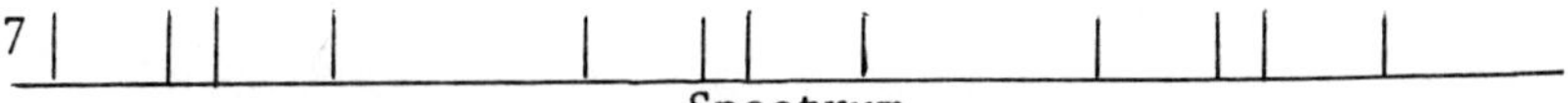

Spectrum

14.28 Assuming rapid exchange of OH proton

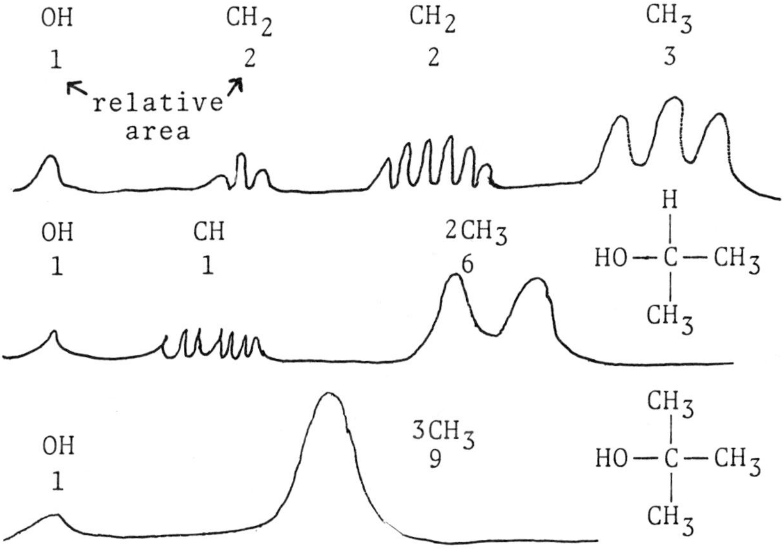

14.29 The proton resonance would be split into three lines by the deuteron and would occur at 42.6 MHz at 1 T. The deuteron resonance would be split into two lines by the proton and would occur at 6.5 MHz at 1 T.

14.30 The H_2 resonance will be split into a triplet with a coupling constant of 0.5-4 Hz. The H_5 resonance will be split into a triplet with a coupling constant of 6-9 Hz. The H_4-H_6 resonance will produce an AB-type spectrum.

14.31 9230 MHz

14.32 The deuteron has a spin of +1, 0, or -1. Therefore the following combinations of spins are possible:

Spin comb.	Total spin	No. of ways		Spectrum
all +	+3	1		
2+,10	+2	3		
2+,1-	+1	3	6	
20,1+	+1	3		
+, -, and 0	0	3! = 6	7	
all 0	0	1		
2-,1+	-1	3	6	
20,1-	-1	3		
2-,10	-2	3		
all -	-3	1		

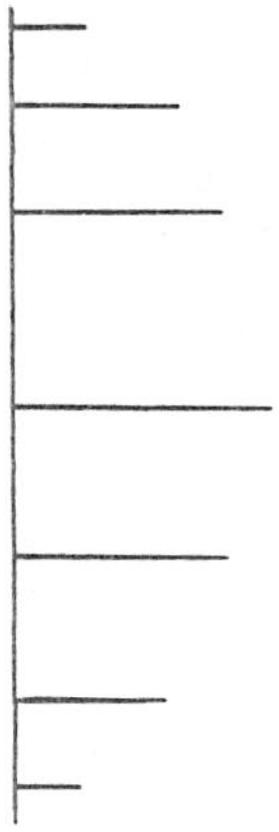

14.33

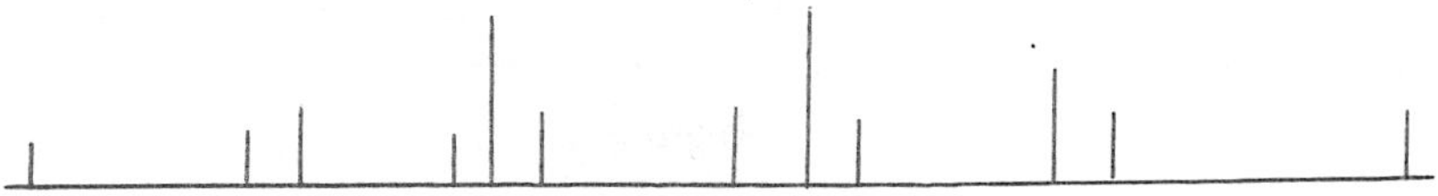

See M. Bersohn and J. C. Baird, An Introduction to Electron Paramagnetic Resonance, Benjamin, New York, 1966, p. 93.

14.34

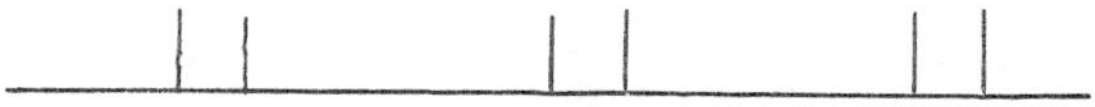

CHAPTER 15. Statistical Mechanics

15.1 (a) How many different ways can two distinguishable balls be placed in two boxes? (b) How many different ways can two distinguishable balls be placed in three boxes? (c) What are the answers to (a) and (b) if the balls are indistinguishable?

SOLUTION

(a) The number of ways two can be placed in the first box and none in the other is

$W = \frac{N!}{N_1!N_2!} = \frac{2!}{2!\ 0!} = 1.$ The number of ways one can be placed in each box is $\frac{2!}{1!\ 1!} = 2.$

The number of ways two can be placed in the second box and none in the other is

$\frac{2!}{0!\ 2!} = 1.$ Thus the total number of different ways that two distinguishable balls can be placed in two boxes is 1 + 2 + 1 = 4.

(b) $W = \Sigma W_i = \sum \frac{N!}{N_1!N_2!N_3!}$

$$= \frac{2!}{1!1!0!} + \frac{2!}{0!1!1!} + \frac{2!}{1!0!1!} + \frac{2!}{2!0!0!} + \frac{2!}{0!2!0!} + \frac{2!}{0!0!2!}$$

$$= 2 + 2 + 2 + 1 + 1 + 1 = 9$$

(c)

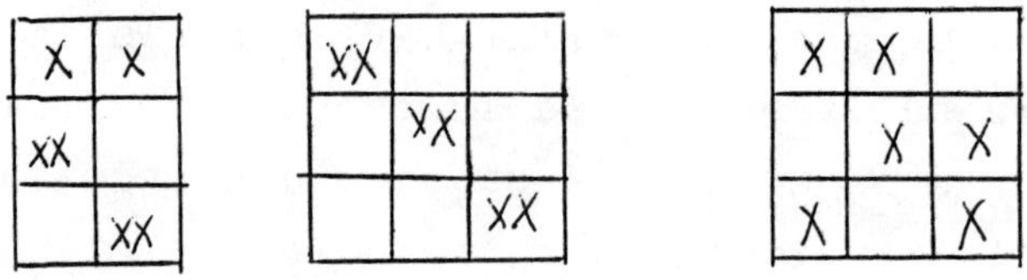

3 6

15.2 (a) How many different ways can four distinguishable balls be placed in two boxes? (b) How many different ways are there if the balls are indistinguishable?

<u>SOLUTION</u>

(a) $W = \Sigma W_i$

$$W_1 = \frac{4!}{2!2!} = 6$$

$$W_2 = \frac{4!}{1!3!} = 4$$

$$W_3 = \frac{4!}{3!1!} = 4$$

$$W_4 = \frac{4!}{0!4!} = 1$$

$$W_5 = \frac{4!}{4!0!} = 1 \qquad W = 16$$

(b)

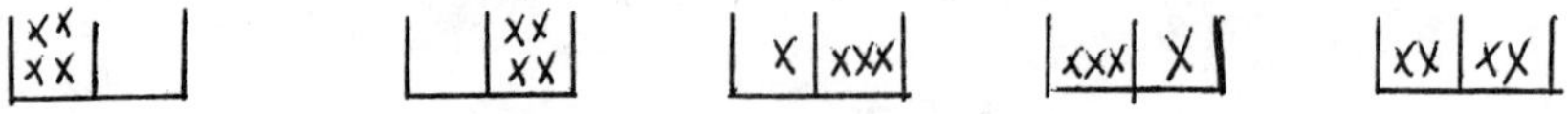

15.3 Using the Boltzmann distribution calculate the ratio of populations at 25 °C of energy levels separated by (a) 1000 cm^{-1}, and (b) 10 kJ mol^{-1}.

SOLUTION

(a) $\exp\left(-\frac{hc\tilde{\nu}}{kT}\right) = \exp\left[-\frac{(6.626\times10^{-34}\,Js)(2.998\times10^{8}\,ms^{-1})(10^{5}\,m^{-1})}{(1.380\times10^{-23}\,JK^{-1})(298.15K)}\right]$

$= 0.0080$

(b) $\exp\left[-\frac{10^{4}\ J\ mol^{-1}}{(8.314\ J\ K^{-1}\ mol^{-1})(298\ K)}\right]$

$= 0.0177$

15.4 The difference of 4.60 J K^{-1} mol^{-1} between the third law entropy of CO and the statistical mechanical value can be attributed to the randomness of orientation of CO molecules in its crystals at absolute zero. If half of the molecules are oriented CO and half OC, calculate the entropy of a crystal at absolute zero using equation 15.7 and Stirling's approximation.

SOLUTION

$$W = \frac{N!}{(N/2)!\ (N/2)!}$$

Stirling's approximation $\ell n N! \cong N\ell n N - N$

$$\ell n W = \ell n N! - \ell n \frac{N}{2}! - \ell n \frac{N}{2}!$$

$$= N\ell n N - N - 2\,\frac{N}{2}\ell n\frac{N}{2} + 2\,\frac{N}{2}$$

$$= N\ell n N - N - N\ell n\frac{N}{2} + N$$

$$= N\ell n 2$$

$$S = k\ell n W$$

$$= kN\ell n 2$$

$$= (8.314\ J\ K^{-1}\ mol^{-1})(2.303)(0.301)$$

$$= 5.76\ J\ K^{-1}\ mol^{-1}$$

15.5 Calculate the translational partition function for $H_2(g)$ at 1000 K and 1 bar.

SOLUTION

$$V = \frac{RT}{P} = \frac{(1\text{mol})(8.314\text{JK}^{-1}\text{mol}^{-1})(1000\text{K})}{10^5\ \text{N m}^{-2}}$$

$$= 0.08314\ \text{m}^3$$

$$m = 2(1.0078\text{x}10^{-3}\text{kgmol}^{-1})/(6.022\text{x}10^{23}\text{mol}^{-1})$$

$$= 3.347\ \text{x}\ 10^{-27}\ \text{kg}$$

$$q_{tr} = \left(\frac{2\pi mkT}{h^2}\right)^{3/2} V$$

$$= \left[\frac{2\pi(3.347\text{x}10^{-27}\text{kg})(1.3806\text{x}10^{-23}\text{JK}^{-1})(1000\text{K})}{(6.626\ \text{x}\ 10^{-34}\ \text{J s})^2}\right]^{3/2} \text{x}\ (0.08314\ \text{m}^3)$$

$$= 1.414\ \text{x}\ 10^{30}$$

15.6 Calculate the entropy of one mole of H-atom gas at 1000 K and (a) 1 bar, and (b) 1000 bar.

SOLUTION

$$S° = R\ln\left[\frac{(2\pi mkT)^{3/2}V\ e^{5/2}}{h^3N_A}\right] + R\ln 2$$

$$m = \frac{1.0079\ \text{x}\ 10^{-3}\ \text{kg mol}^{-1}}{6.022045\ \text{x}\ 10^{23}\ \text{mol}^{-1}}$$

$$= 1.67385\ \text{x}\ 10^{-27}\ \text{kg}$$

$$V = \frac{RT}{P} = \frac{(8.314\ \text{J K}^{-1}\ \text{mol}^{-1})(1000\ \text{K})}{10^5\ \text{N m}^{-2}}$$

$$= 0.08314\ \text{m}^3\ \text{mol}^{-1}\quad \text{at}\ \ 1\ \text{bar}$$

$$V = 8.314\ \text{x}\ 10^{-5}\ \text{m}^3\ \text{mol}^{-1}\ \text{at}\ 1000\ \text{bar}$$

$$k = 1.380662\ \text{x}\ 10^{-23}\ \text{J K}^{-1}$$

$$T = 1000\ \text{K}$$

$$e = 2.7182818$$

$h = 6.626176 \times 10^{-34}$ J K^{-1}

$N_A = 6.022045 \times 10^{23}$ mol^{-1}

(a) At 1000 K and 1 bar

$S° = 139.871$ J K^{-1} mol^{-1}

(b) At 1000 K and 1000 bar

$S° = 82.437$ J K^{-1} mol^{-1}

15.7 Calculate the entropy of neon at 25 °C and 1 bar.

SOLUTION

$$m = \frac{20.179 \times 10^{-3} \text{ kg mol}^{-1}}{6.022\ 045 \times 10^{23} \text{ mol}^{-1}}$$

$$= 3.350\ 86 \times 10^{-26} \text{ kg}$$

$$V = \frac{RT}{P} = \frac{(8.314 \text{ J K}^{-1}\text{mol}^{-1})(298.15 \text{ K})}{10^5 \text{ N m}^{-2}}$$

$$= 2.479 \times 10^{-2} \text{ m}^3 \text{ mol}^{-1}$$

$$S° = R\left\{\frac{5}{2} + \ln\left[\left(\frac{2\pi mkT}{h^2}\right)^{3/2} \frac{V}{N_A}\right]\right\}$$

$$= 146.328 \text{ J K}^{-1} \text{ mol}^{-1}$$

15.8 Calculate S° and $C_p°$ for argon ($M = 39.948$ g mol^{-1}) at 25 °C and 1 bar.

SOLUTION

$$m = \frac{39.948 \times 10^{-3} \text{ kg mol}^{-1}}{6.022045 \times 10^{23} \text{ mol}^{-1}}$$

$$= 6.63363 \times 10^{26} \text{ kg}$$

$$V = \frac{RT}{P} = \frac{(8.314\ 41 \text{ J K}^{-1} \text{ mol}^{-1})(298.15 \text{ K})}{10^5 \text{ N m}^{-2}}$$

$$= 2.478\ 94 \times 10^{-2} \text{ m}^3 \text{ mol}^{-1}$$

$$S° = R\left\{\frac{5}{2} + \ell n\left[\left(\frac{2\pi mkT}{h^2}\right)^{3/2} \frac{V}{N_A}\right]\right\}$$

$$= 154.844 \text{ J K}^{-1} \text{ mol}^{-1}$$

$$C_P^\circ = \frac{5}{2} R = 20.786 \text{ J K}^{-1} \text{ mol}^{-1}$$

15.9 What are the translational partition functions of hydrogen atoms and hydrogen molecules at 500 K in a volume of 4.157×10^{-2} m^3? (This is the molar volume of a perfect gas at this temperature and a pressure of 1 bar.)

<u>SOLUTION</u>

$$q_t = \left(\frac{2\pi mkT}{h^2}\right)^{3/2} V$$

For H

$$q_t = \left[\frac{2\pi(1.008\text{x}10^{-3}\text{kgmol}^{-1})(1.3806\text{x}10^{-23}\text{JK}^{-1})(500\text{K})}{(6.022045\text{x}10^{23}\text{mol}^{-1})(6.626\text{x}10^{-34}\text{Js})^2}\right] \text{x } (4.157 \text{ x } 10^{-2}\text{m}^3)$$

$$= 8.84 \text{ x } 10^{-28}$$

For H_2, $m_{H_2} = 2m_H$

$$q_{tH_2} = 2^{3/2} \qquad q_{tH} = 2.50 \text{ x } 10^{29}$$

15.10 Calculate the translational partition functions for H, H_2, and H_3 at 1000 K and 1 bar. What are the rotational partition functions of H_2 and H_3 (linear) at 1000 K? The internuclear distances in H_3 are 0.094 nm.

SOLUTION

$$V = \frac{RT}{P} = \frac{(8.31441\ J\ K^{-1}\ mol^{-1})(1000\ K)}{10^5\ N\ m^{-2}}$$

$$= 0.0831441\ m^3\ mol^{-1}$$

$$q_t = \frac{(2\pi mkT)^{3/2}\ V}{h^3}$$

$$= \frac{[2\pi(1.0079\times10^{-3}kg/6.022\times10^{23}mol^{-1})(1.38\times10^{-23}JK^{-1})(10^3K)]^{3/2}\ V}{(6.62\ \times\ 10^{-34}J\ s)^3}$$

$$= 6.026\ \times\ 10^{30}\ V = 5.00\ \times\ 10^{29}$$

For H_2 $\quad q_t = 6.026\ \times\ 10^{30}\ V\ 2^{3/2} = 1.42\ \times\ 10^{30}$

For H_3 $\quad q_t = 6.026\ \times\ 10^{30}\ V\ 3^{3/2} = 2.60\ \times\ 10^{30}$

$$q_r = \frac{8\pi^2 IkT}{2h^2}$$

For H_2

$$q_r = \frac{8\pi^2(4.6054\times10^{-48}kgm^2)(1.38066\times10^{-23}JK^{-1})(1000K)}{2(6.62618\ \times\ 10^{-34}\ J\ s)^2}$$

$$= 5.72$$

For H_3

$$I = \frac{m_1\ m_3}{m_1 + m_3}\ R^2$$

$$= \frac{(1.0079\times10^{-3}kgmol^{-1})^2(1.88\times10^{-10}m)^2}{2(1.0078\times10^{-3}kgmol^{-1})(6.02205\times10^{23}mol^{-1})}$$

$$= 2.96\ \times\ 10^{-47}\ kg\ m^2$$

(Note that m_2 is on the axis of rotation, and does not contribute to the moment of inertia.)

$$q_t = 5.72\ \frac{29.6\ \times\ 10^{-48}\ kg\ m^2}{4.60\ \times\ 10^{-48}\ kg\ m^2} = 36.8$$

15.11 What is the rotational contribution to C_P° and S° of CH_4 at 298.15 K?

SOLUTION

$$(C_P^\circ)_r = \frac{3}{2}R = \frac{3}{2}(8.31441) = 12.472 \text{ J K}^{-1}\text{ mol}^{-1}$$

$$S_r^\circ = R\ell n\left[\frac{\pi^{1/2}}{\sigma}\left(\frac{T^3\ e^3}{\Theta_a\ \Theta_b\ \Theta_c}\right)^{1/2}\right]$$

$$= 8.31441\ell n\left[\frac{\pi^{1/2}}{12}\left(\frac{298.15^3 \times 2.7183^3}{435.6}\right)^{1/2}\right]$$

$$= 42.366 \text{ J K}^{-1}\text{ mol}^{-1}$$

15.12 Calculate the ratio of the number of HBr molecules in state v = 2, J = 5 to the number in state v = 1, J = 2 at 1000 K. Assume that all of the molecules are in their electronic ground states. (Θ_v = 3700 K, Θ_r = 12.1 K)

SOLUTION

$$\frac{N(v=2,J=5)}{N(v=1,J=2)} = \frac{g_{2vib}e^{\frac{-\varepsilon_{2vib}}{kT}}\ g_{5rot}e^{\frac{-\varepsilon_{5rot}}{kT}}}{g_{1vib}e^{\frac{-\varepsilon_{1vib}}{kT}}\ g_{2rot}e^{\frac{-\varepsilon_{2rot}}{kT}}}$$

g_{vib} = 1 for all v

g_{rot} = 2J + 1 = 5 for all g_{2rot}

= 11 for all g_{5rot}

$\varepsilon_{vib} = vh\nu$

Given $\Theta_v = h\nu/k = 3700$ K for HBr

$$\varepsilon_{vib} = vk\Theta_v$$

$$\frac{\varepsilon_{vib}}{kT} = \frac{v\Theta_v}{T}$$

$$\varepsilon_{rot} = \frac{J(J+1)h^2}{8\pi^2 I}$$

Given $\Theta_r = \dfrac{h^2}{8\pi^2 Ik} = 12.1$ K for HBr

$$\varepsilon_{rot} = J(J+1)\Theta_r k$$

$$\frac{\varepsilon_{rot}}{kT} = \frac{J(J+1)\Theta_r}{T}$$

$$\frac{N(v=2,J=5)}{N(v=1,J=2)} = \frac{e^{-(2)(3.7)} \times 11 \times e^{-(5)(5+1)(12.1)/10^3}}{e^{-3.7} \times 5 \times e^{-2(2+1)(12.1)/10^3}}$$

$$= 0.0407$$

15.13 Derive the expression for the vibrational contribution to the internal energy

$$U = \frac{RTx}{e^x - 1} \qquad \text{where } x = h\nu/kT.$$

SOLUTION

$$q_v = \frac{1}{1 - e^{-h\nu/kT}}$$

$$U = RT^2 \frac{\partial \ln q}{\partial T} \qquad \ln q_v = -\ln(1 - e^{-h\nu/kT})$$

$$\frac{\partial \ln q_v}{\partial T} = \frac{e^{-h\nu/kT}}{1 - e^{-h\nu/kT}}\left(\frac{h\nu}{kT^2}\right) = \frac{x}{T(e^x - 1)}$$

where $x = h\nu/kT$

$$U = \frac{RTx}{e^x - 1}$$

15.14 Using

$$\left(\frac{\partial G}{\partial T}\right)_P = -S$$

and the contribution of vibration to the Gibbs energy for a diatomic molecule in a perfect gas $G = RT\ell n(1 - e^{-x})$ derive the expression for the corresponding contribution to the entropy.

SOLUTION

$$x = \frac{h\nu}{kT} \qquad \frac{dx}{dT} = -\frac{h\nu}{kT^2}$$

$$\frac{\partial G}{\partial T} = R\ell n(1 - e^{-x}) + \frac{RT}{(1 - e^{-x})}\frac{d}{dT}(1 - e^{-x})$$

$$= R\ell n(1 - e^{-x}) + \frac{RT}{(1 - e^{-x})}\left[-e^{-x}\frac{d(-x)}{dT}\right]$$

$$= R\ell n(1 - e^{-x}) - \frac{Rxe^{-x}}{(1 - e^{-x})}$$

$$S = -R\ell n(1 - e^{-x}) + \frac{Rx}{e^x - 1}$$

$$= R\left[\frac{x}{e^x - 1} - \ell n(1 - e^{-x})\right]$$

15.15 By use of series expansions show the vibrational contribution to C_V° for a diatomic molecule approaches R as $T \longrightarrow \infty$.

SOLUTION

$$(C_V^\circ)_v = R\left(\frac{\Theta_v}{T}\right)^2 \frac{e^{\Theta_v/T}}{(e^{\Theta_v/T} - 1)^2}$$

$$e^x = 1 + x + \frac{x^2}{2!} + \ldots$$

$$e^{\Theta_v/T} = 1 + \frac{\Theta_v}{T} + \ldots$$

$$(C_V^\circ)_v = R\left(\frac{\Theta_v}{T}\right) \frac{\left[1 + \frac{\Theta_v}{T} + \ldots\right]}{\left[\frac{\Theta_v}{T}\right]}$$

As $T \longrightarrow \infty \qquad (C_V^\circ)_v \longrightarrow R$

15.16 Calculate the temperature at which 10% of the molecules in a system will be in the first excited electronic state if this state is 400 kJ mol^{-1} above the ground state.

SOLUTION

$$\frac{Ni}{N} = \frac{e^{-E_i/RT}}{\Sigma e^{-E_i/RT}}$$

$$\frac{N1}{N} = \frac{e^{-E_1/RT}}{e^{0/RT} + e^{-E_1/RT}}$$

$$0.1 = \frac{1}{e^{E_1/RT} + 1}$$

$$e^{E_1/RT} = \frac{0.9}{0.1} = 9 \qquad \frac{E_1}{RT} = \ell n 9$$

$$T = \frac{E_1}{R\ell n 9} = \frac{(400{,}000 \text{ J mol}^{-1})}{(8.314 \text{ J K}^{-1} \text{ mol}^{-1})\ell n 9} = 22{,}000 \text{ K}$$

15.17 Calculate the fraction of hydrogen atoms that at equilibrium at 1000 °C would have n = 2.

SOLUTION

Since the fraction will be very small, it is given by the ratio of the number with N = 2 to the number N = 1.

$$\text{Fraction} = \frac{e^{-E_2/kT}}{e^{-E_1/kT}} = e^{-(E_2 - E_1)/kT}$$

From Example 10.1 $\quad E = \dfrac{-2.179907 \times 10^{-18}\ \text{J}}{h^2}$

$$\text{Fraction} = \exp\left[-\frac{(-2.179907\text{x}10^{-18}\text{J})(-0.75)}{(1.380662\text{x}10^{-23}\text{JK}^{-1})(1273\text{K})}\right]$$

$$= 4 \times 10^{-41}$$

15.18 The ground state of Cl(g) is twofold degenerate. The first excited state is 875.4 cm^{-1} higher in energy and is twofold degenerate. What is the value of the electronic partition function at 25 °C? At 1000 K?

SOLUTION

$$q = g_0\ e^{-0} + g_1\ e^{-\varepsilon_1/kT}$$

$$= 2 + 2e^{-hc\tilde{\nu}/kT}$$

$$= 2 + 2\exp\left[\frac{-(6.626176\text{x}10^{-34}\text{Js})(2.99792\text{x}10^{8}\text{ms}^{-1})(8.754\text{x}10^{4}\text{m}^{-1})}{(1.380662\text{x}10^{-23}\text{JK}^{-1})(298.15\text{K})}\right]$$

$$= 2 + 2\ e^{-4.224} = 2.029$$

At 1000 K

$$q = 2 + 2\ e^{-1.259} = 2.568$$

15.19 Calculate the values of D_0 in Table 13.4 from data in Table A.2 for $H_2(g)$, $O_2(g)$, $Cl_2(g)$, HCl(g), and CO(g).

SOLUTION

$H_2(g) = 2H(g)$

$\Delta H_o^\circ = 2(216.037) = 432.074 \text{ kJ mol}^{-1} = 4.4781 \text{ eV}$

$O_2(g) = 2O(g)$

$\Delta H_o^\circ = 2(246.785) = 493.570 \text{ kJ mol}^{-1} = 5.1155 \text{ eV}$

$Cl_2(g) = 2Cl(g)$

$\Delta H_o^\circ = 2(119.608) = 239.216 \text{ kJ mol}^{-1} = 2.479 \text{ eV}$

$HCl(g) = H(g) + Cl(g)$

$\Delta H_o^\circ = 216.037 + 119.608 + 92.127 = 427.772 \text{ kJ mol}^{-1}$
$= 4.4336 \text{ eV}$

$CO(g) = C(g) + O(g)$

$\Delta H_o^\circ = 709.506 + 246.785 + 113.805 = 1070.096 \text{ kJ mol}^{-1}$
$= 11.0909 \text{ eV}$

15.20 Calculate the entropy of nitrogen gas at 25 °C and 1 bar pressure. The equilibrium separation of atoms is 0.1095 nm and the vibrational wave number is 2330.7 cm^{-1}.

SOLUTION

$$m = \frac{2(14.0067 \times 10^{-3} \text{ kg mol}^{-1})}{6.022045 \times 10^{23} \text{ mol}^{-1}} = 4.65181 \times 10^{-26} \text{ kg}$$

$$\frac{kT}{P^\circ} = \frac{(1.380662 \times 10^{-23} \text{ J K}^{-1})(298.15 \text{ K})}{10^5 \text{ N m}^{-2}}$$

$$= 4.11644 \times 10^{-26} \text{ m}^3$$

$$S_t^\circ = R\left\{\frac{5}{2} + \ell n\left[\left(\frac{2\pi mkT}{h^2}\right)^{3/2}\frac{kT}{P^\circ}\right]\right\}$$

$$\ell n\left[\left(\frac{2\pi(4.65181\text{x}10^{-26}\text{kg})(1.380662\text{x}10^{-23}\text{JK}^{-1})(298.15\text{K})}{(6.626176 \text{ x } 10^{-34} \text{ J s})^2}\right)^{\frac{3}{2}}\text{x}(4.11644\text{x}10^{-26}\text{m}^3)\right]$$

$$= 15.59136$$

$$S_t^\circ = (8.314\ 41 \text{ JK}^{-1}\text{mol}^{-1})(\tfrac{5}{2} + 15.59136)$$

$$= 150.419 \text{ J K}^{-1} \text{ mol}^{-1}$$

$$\Theta_r = \frac{h^2}{8\pi^2 Ik}$$

$$\mu = \frac{m_N{}^2}{2m_N} = \frac{m_N}{2} = \frac{14.0067 \text{ x } 10^{-3} \text{ kg mol}^{-1}}{2(6.022045 \text{ x } 10^{23} \text{ mol}^{-1}}$$

$$= 1.16295 \text{ x } 10^{-26} \text{ kg}$$

$$I = \mu R^2 = (1.16295 \text{ x } 10^{-26} \text{ kg})(1.095 \text{ x } 10^{-10} \text{ m})^2$$

$$= 1.39441 \text{ x } 10^{-46} \text{ kg m}^2$$

$$\Theta_r = \frac{(6.626176 \text{ x } 10^{-34} \text{ J s})^2}{8\pi^2(1.39441\text{x}10^{-46}\text{kgm}^2)(1.380662\text{x}10^{-23}\text{JK}^{-1})}$$

$$= 2.88841 \text{ K}$$

$$S_r^\circ = R\ell n\left(\frac{eT}{\sigma\Theta_r}\right)$$

$$= (8.31441 \text{ JK}^{-1}\text{mol}^{-1})\ell n\left[\frac{(2.71828)(298.15\text{K})}{(2)(2.88841\text{K})}\right]$$

$$= 41.104 \text{ J K}^{-1} \text{ mol}^{-1}$$

$$x = \frac{h\nu}{kT} = \frac{hc\tilde{\nu}}{kT} = \frac{(6.626176\text{x}10^{-34}\text{Js})(2.997925\text{x}10^{8}\text{ms}^{-1})(2.3307\text{x}10^{5}\text{m}^{-1})}{(1.380662 \text{ x } 10^{-23}\text{J K}^{-1})(298.15 \text{ K})}$$

$$= 11.247$$

$$S_v^\circ = R\left[\frac{x}{e^x - 1} - \ln(1 - e^{-x})\right]$$

$$= (8.31441\ JK^{-1}mol^{-1})\left[\frac{11.247}{e^{11.247} - 1} - \ln(1-e^{-11.247})\right]$$

$$= 1.21 \times 10^{-3}\ J\ K^{-1}\ mol^{-1}$$

$$S^\circ = S_t^\circ + S_r^\circ + S_v^\circ$$

$$= 150.419 + 41.104 + 0 = 191.524\ J\ K^{-1}\ mol^{-1}$$

15.21 Calculate C_P for CO_2 at 1000 K. Compare the actual contributions to C_P from the various normal modes with the classical expectations.

SOLUTION

$$C_P^\circ = \frac{5}{2}R + R + \sum_{i=1}^{4}\frac{Rx_i^2\ e^{x_i}}{(e^{x_i} - 1)^2}$$

$$x_i = \frac{hc\tilde{\nu}_i}{kT} = (1.438 \times 10^{-5}\ m)\tilde{\nu}$$

ν_i/m^{-1}	1.3512×10^5	6.722×10^4	6.722×10^4	2.3964×10^5
x_i	1.943	0.967	0.967	3.446

$$C_P^\circ = \frac{7}{2}(8.31441) + 6.127 + 2(7.695) + 3.357$$

$$= 53.974\ J\ K^{-1}\ mol^{-1}$$

Classically $C_P^\circ = C_{Pt}^\circ + C_{Pr}^\circ + C_{Pv}^\circ$

$$= \frac{5}{2}R + R + 4R$$

$$= \frac{15}{2}R = 62.355\ J\ K^{-1}\ mol^{-1}$$

15.22 Calculate the statistical mechanical values of C_P°, S°, H°, and G° for H(g) at 3000 K.

SOLUTION

$$(C_P^\circ)_t = \frac{5}{2}(8.314\ JK^{-1}mol^{-1}) = 20.786\ JK^{-1}mol^{-1}$$

$$H_t^\circ = \frac{5}{2}(8.31441\ JK^{-1}mol^{-1})(3000K)$$

$$= 62.358\ kJ\ mol^{-1}$$

$$\left(\frac{2\pi mkT}{h^2}\right)^{3/2}\frac{kT}{P^\circ} =$$

$$\left[\frac{2\pi(1.0080x10^{-3}kgmol^{-1})(1.380662x10^{-23}JK^{-1})(3000K)}{(6.626176x10^{-34}Js)^2(6.022045x10^{23}mol^{-1})}\right]^{3/2}$$

$$x\ \frac{(1.380662x10^{-23}JK^{-1})(3000K)}{10^5\ N\ m^{-2}} = 1.29443\ x\ 10^7$$

$$S_t^\circ = (8.314\ 41\ JK^{-1}mol^{-1})(2.5 + \ln 1.294\ 43\ x\ 10^7)$$

$$= 156.944\ JK^{-1}\ mol^{-1}$$

$$G_t^\circ = -(8.314\ 41\ JK^{-1}mol^{-1})(3000\ K)\ln 1.294\ 43\ x\ 10^7)$$

$$= -408.474\ kJ\ mol^{-1}$$

For the purposes of this calculation $(C_P^\circ)_e = 0$ and $H_e^\circ = 0$.

$$S^\circ = S_t^\circ + S_e^\circ$$

$$= (156.944 + 8.314\ln 2)JK^{-1}mol^{-1} = 162.707\ J\ K^{-1}\ mol^{-1}$$

$$G^\circ = G_t^\circ + G_e^\circ$$

$$= [-408.474 + (8.314x10^{-3})(3000)\ln 2]kJmol^{-1}$$

$$= -425.763\ kJ\ mol^{-1}$$

15.23 Calculate the statistical mechanical values of C_P°, S°, H°, and G° for $H_2(g)$ at 3000 K.

SOLUTION

$$(C_P^\circ)_t = \frac{5}{2}R = 20.786\ J\ K^{-1}\ mol^{-1}$$

$$H_t^\circ = \frac{5}{2} RT = 62.358 \text{ kJ mol}^{-1}$$

$$\left(\frac{2\pi mkT}{h^2}\right)^{3/2} \frac{kT}{P^\circ} = 3.661\ 20 \times 10^7$$

$$S_t^\circ = R(2.5 + \ell n 3.661\ 20 \times 10^7)$$
$$= 165.589 \text{ J K}^{-1} \text{ mol}^{-1}$$

$$G_t^\circ = -(8.31441 \text{ JK}^{-1}\text{mol}^{-1})(3000\text{K})\ell n 3.66120 \times 10^7$$
$$= -434.409 \text{ kJ mol}^{-1}$$

$$(C_p^\circ)_r = R = 8.314 \text{ J K}^{-1} \text{ mol}^{-1}$$

$$H_r^\circ = RT = 24.943 \text{ kJ mol}^{-1}$$

$$S_r^\circ = R\ell n\left(\frac{eT}{\sigma\Theta_r}\right)$$
$$= (8.31441)\ell n\left|\frac{(2.7183)(3000 \text{ K})}{(2)(87.547 \text{ K})}\right|$$
$$= 31.936 \text{ J K}^{-1} \text{ mol}^{-1}$$

$$G_r^\circ = -RT\ell n\left(\frac{T}{\sigma\Theta_r}\right)$$
$$= -70.865 \text{ kJ mol}^{-1}$$

$$\Theta_v = \frac{hc\tilde{\omega}}{k}$$

$$= \frac{(6.626176\text{x}10^{-34}\text{Js})(2.997925\text{x}10^8\text{ms}^{-1})(4405.3\text{cm}^{-1})(10^2\text{cm m}^{-1})}{1.380\ 662 \times 10^{-23} \text{ J K}^{-1}}$$

$$= 6338.3 \text{ K}$$

$$H_v^\circ = RT\ \frac{(\Theta_v/T)}{e^{\Theta_v/T} - 1} = 7.248 \text{ kJ mol}^{-1}$$

$$(C_p^\circ)_v = R\left(\frac{\Theta_v}{T}\right)^2 \frac{e^{\Theta_v/T}}{(e^{\Theta_v/T} - 1)^2} = 5.806 \text{ J K}^{-1} \text{ mol}^{-1}$$

$$S_v^\circ = R\left[\frac{(\Theta_v/T)}{e^{\Theta_v/T} - 1} - \ell n(1 - e^{-\Theta_v/T})\right]$$

$$= 3.487 \text{ J K}^{-1} \text{ mol}^{-1}$$

$$G_v^\circ = RT\ell n(1 - e^{-\Theta_v/T})$$

$$= -3.214 \text{ kJ mol}^{-1}$$

Since $g_o = 1$, $S_e^\circ = 0$

$$H_e^\circ = G_e^\circ = -432.073 \text{ kJ mol}^{-1}$$

$$C_P^\circ = (C_P^\circ)_t + (C_P^\circ)_r + (C_P^\circ)_v + (C_P^\circ)_e$$

$$= 34.907 \text{ J K}^{-1} \text{ mol}^{-1}$$

$$H^\circ = H_t^\circ + H_r^\circ + H_v^\circ + H_e^\circ$$

$$= -337.524 \text{ kJ mol}^{-1}$$

$$S^\circ = S_t^\circ + S_r^\circ + S_v^\circ + S_e^\circ$$

$$= 201.013 \text{ J K}^{-1} \text{ mol}^{-1}$$

$$G^\circ = G_t^\circ + G_r^\circ + G_v^\circ + G_e^\circ$$

$$= -940.561 \text{ kJ mol}^{-1}$$

15.24 What are the value of ΔC_P°, ΔH°, ΔS°, and ΔG° for $H_2(g) = 2H(g)$ at 3000 K calculated in the preceeding two problems? What is the value of K_P? What is the degree of dissociation at 1 bar?

SOLUTION

$$\Delta C_P^\circ = 2C_P^\circ(H) - C_P^\circ(H_2)$$

$$= [2(20.786) - 34.907] \text{ J K}^{-1} \text{ mol}^{-1}$$

$$= 6.665 \text{ J K}^{-1} \text{ mol}^{-1}$$

$$\Delta H° = 2H°(H) - H°(H_2)$$
$$= [2(62.358) - (-337.524)] \text{ kJ mol}^{-1}$$
$$= 462.240 \text{ kJ mol}^{-1}$$

$$\Delta S° = 2S°(H) - S°(H_2)$$
$$= [2(162.707) - 201.013] \text{ J K}^{-1} \text{ mol}^{-1}$$
$$= 124.401 \text{ J K}^{-1} \text{ mol}^{-1}$$

$$\Delta G° = 2G°(H) - G°(H_2)$$
$$= [2(-425.763) - (-940.561)] \text{ kJ mol}^{-1}$$
$$= 89.035 \text{ kJ mol}^{-1}$$

$$K_P = e^{-\Delta G°/RT}$$
$$= e^{-89,035/(8.314)(3000)}$$
$$= 2.82 \times 10^{-2}$$

$$K_P = \frac{4\xi^2(P/P°)}{1 - \xi^2}$$

$$\xi = \left(\frac{K_P}{4 + K_P}\right)^{1/2}$$
$$= 0.084$$

15.25 (omitted)

15.26 Given the Lennard-Jones parameters for argon in Table 12.5, what values of the second virial coefficient do you expect at 200 K and 500 K? Do these values agree with the experimental values in Fig. 1.4?

<u>SOLUTION</u>

For argon $\varepsilon/k = 120$ K $\qquad \sigma = 341 \times 10^{-12}$ m

At 500 K $\quad T^* = \dfrac{500 \text{ K}}{120 \text{ K}} = 4.17$

$$B^* = 0$$

$$B = \frac{2\pi}{3} N_A \sigma^3 B^* = 0$$

At 200 K $\quad T^* = \frac{200\ \text{K}}{120\ \text{K}} = 1.67$

$$B^* = -1.0$$

$$B = \frac{2\pi}{3}(6.022\text{x}10^{23}\text{mol}^{-1})(341\text{x}10^{-12}\text{m})^3(-1.0)$$

$$= -50\ \text{cm}^3\ \text{mol}^{-1}$$

These values are in agreement with Fig. 1.4.

15.27 Use the law of corresponding states to estimate the Lennard-Jones parameters for methane from T_c = 191 K and V_c = 100 cm^3 mol^{-1}. The values obtained from second virial coefficient data are σ = 0.378 nm, and ε/k = 148.9 K.

SOLUTION

$$\sigma = \left(\frac{V_c}{2.7\ N_A}\right)^{1/3} = \left[\frac{(100\text{cm}^3\text{mol}^{-1})(10^{-2}\text{m cm}^{-1})^3}{(2.7)(6.022\text{x}10^{23}\text{mol}^{-1})}\right]^{1/3}$$

$$= 0.395\ \text{nm}$$

$$\varepsilon/k = T_c/1.3 = \frac{191\ \text{K}}{1.3} = 146\ \text{K}$$

15.28 Plot the probability density W(r) for random walk in three dimensions after 1000 steps with a step length of unity. Indicate the root-mean-square end-to-end distance on this plot.

SOLUTION

$$W(r) = \left(\frac{3}{2\pi n\ell^2}\right)^{3/2} \exp\left(-\frac{3r^2}{2n\ell^2}\right)$$

$$= \left(\frac{3}{2\pi 1000}\right)^{3/2} \exp\left(-\frac{3r^2}{2000}\right)$$

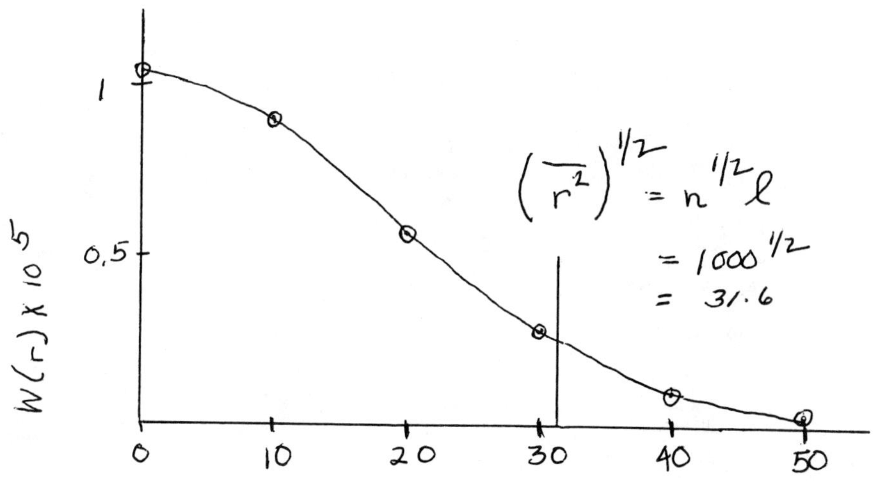

15.29 In polyethene $H(CH_2{-}CH_2)_nH$ the bond length ℓ is 0.15 nm. What is the root-mean-square end-to-end distance for a molecule with universal joints with a molar mass of 10^5 g mol^{-1}? Taking into account the fact that carbon forms tetrahedral bonds, what is $(\overline{r^2})^{1/2}$?

SOLUTION

$$N = \frac{10^5 \text{ g mol}^{-1}}{14 \text{ g mol}^{-1}}$$

$$\langle L^2\rangle^{1/2} = N^{1/2}\,b = \left(\frac{10^5}{14}\right)^{1/2} 0.15 \text{ nm}$$

$$= 12.7 \text{ nm}$$

$$\langle L^2\rangle^{1/2} = N^{1/2}\,b\left(\frac{1 + \cos\theta}{1 - \cos\theta}\right)^{1/2}$$

$$= 12.7 \text{ nm}\left(\frac{1 + \cos 71°}{1 - \cos 71°}\right)^{1/2}$$

$$= 17.8 \text{ nm}$$

15.30	(a) 8	(b) 4
15.31	(a) 3.85×10^{-173}	(b) 2.47×10^{-23}
15.32	(a) 1.253×10^{-17}	(b) 1.147×10^{-169}
	(c) 1.100×10^{-4} 2.596×10^{-40}	
15.34	1.998×10^{33} compared with 1.414×10^{30}	
15.35	(a) 1.23	(b) 2.83×10^{-11}
15.36	108.96, 141.78, 139.85 J K^{-1} mol^{-1}	
15.37	126.156 J K^{-1} mol^{-1}	
15.39	12.472, 47.822 J K^{-1} mol^{-1}	
	3.718, -10.540 kJ mol^{-1}	
15.40	2.86	
15.41	2.175×10^{-6}	
15.43	3.369 8.144	
15.44	11.526 J K^{-1} mol^{-1}	-3.437 kJ mol^{-1}
	11.634 J K^{-1} mol^{-1}	-34.902 kJ mol^{-1}
15.45	$<10^{-99}$, 1.49×10^{-23}	
15.46	251.800 J K^{-1} mol^{-1}	
15.47	222.918 J K^{-1} mol^{-1}	
15.48	29.100, 32.926 J K^{-1} mol^{-1}	
15.49	3.20×10^{-3}	
15.50	4.41×10^{-4}	
15.51	0.211	
15.52	111.770 kJ mol^{-1}	1.13×10^{-2}
15.53	3.261	
15.54	-46, 7 cm^3 mol^{-1}	

15.56

		ξ	0	25	50	75
1000 steps of 1	$W(\xi) \times 10^4$		126	92	36	7.6
500 steps of 2	$W(\xi) \times 10^4$		89.2	76.3	47.8	21.9

15.57 (a) 1100 nm (b) 13 nm (c) 18.4 nm

15.58 The maximum in the plot is at r = 10 nm.

PART THREE
CHEMICAL DYNAMICS

CHAPTER 16. Kinetic Theory of Gases

16.1 If the diameter of a gas molecule is 0.4 nm and each is imagined to be in a separate cube, what is the length of the side of the cube in molecular diameters at 0 °C and pressures of (a) 1 bar, and (b) 1 Pa.

SOLUTION

(a) $$\left[\frac{(22.4\ \text{L mol}^{-1})(1000\ \text{cm}^3\ \text{L}^{-1})(10^7\ \text{nm cm}^{-1})^3}{6.02 \times 10^{23}\ \text{mol}^{-1}}\right]^{1/3}$$

$= 3.338$ nm

$= \frac{3.338\ \text{nm}}{0.4\ \text{nm}} = 8.3$ molecular diameters

(b) $$\left[\frac{(22.4\ \text{L atm mol}^{-1})(10^{24}\text{nm}^3\text{L}^{-1})(101{,}325\ \text{Pa atm}^{-1})/(1\ \text{Pa})}{6.02 \times 10^{23}\ \text{mol}^{-1}}\right]^{\frac{1}{3}}$$

$= 155.6$ nm

$= \frac{155.6\ \text{nm}}{0.4\ \text{nm}} = 389$ molecular diameters

16.2 Calculate the root-mean-square speed of oxygen molecules having a kinetic energy of 10 kJ mol^{-1}. At what temperature would this be the root-mean-square speed?

SOLUTION

Using equation 16.17

$$\langle v^2\rangle^{1/2} = \left(\frac{2\varepsilon_t}{m}\right)^{1/2} = \left(\frac{2E}{M}\right)^{1/2}$$

$$= \left[\frac{2(10 \times 10^3 \text{ J mol}^{-1})}{32.0 \times 10^{-3} \text{ kg mol}^{-1}}\right]^{1/2}$$

$$= 7.9 \times 10^2 \text{ m s}^{-1}$$

Also using equation 16.17

$$T = \frac{2\varepsilon_t}{3k} = \frac{2E}{3R} = \frac{2(10 \times 10^3 \text{ J mol}^{-1})}{3(8.314 \text{ J K}^{-1} \text{ mol}^{-1})}$$

$$= 802 \text{ K}$$

16.3 Plot the probability density f(v) of various molecular speeds versus speed for oxygen at 25 °C.

SOLUTION

$$f(v) = 4\pi v^2 \left(\frac{M}{2\pi RT}\right)^{3/2} e^{-Mv^2/2RT}$$

$$= 4\pi v^2\left[\frac{32 \times 10^{-3} \text{ kg mol}^{-1}}{2\pi(8.314\text{JK}^{-1}\text{mol}^{-1})(298\text{K})}\right]^{3/2} e^{-\frac{(32\text{x}10^{-3}\text{kg mol}^{-1})v^2}{2(8.314\text{JK}^{-1}\text{mol}^{-1})(298\text{K})}}$$

$$= v^2(3.704\text{x}10^{-8} \text{ s m}^{-1})e^{-6.458 \times 10^{-6} v^2}$$

Where v is in ms^{-1}

$v/10^2 ms^{-1}$	$f(v)/10^{-4} s\ m^{-1}$
1	3.47
3	18.64
5	18.42
7	7.67
10	0.58

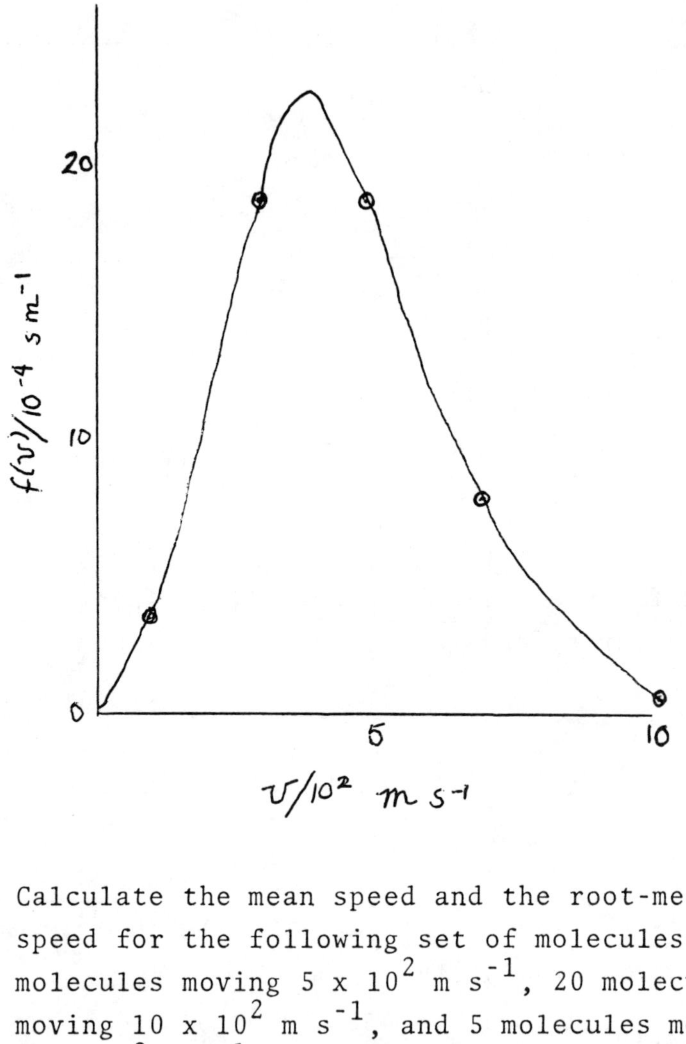

16.4 Calculate the mean speed and the root-mean-square speed for the following set of molecules: 10 molecules moving 5×10^2 m s^{-1}, 20 molecules moving 10×10^2 m s^{-1}, and 5 molecules moving 15×10^2 m s^{-1}.

SOLUTION

$$\langle v \rangle = \frac{\Sigma N_i v_i}{\Sigma N_i} = \frac{10(500) + 20(1000) + 5(1500)}{35}$$

$$= 928 \text{ m s}^{-1}$$

$$\langle v^2\rangle^{1/2} = \left(\frac{\Sigma N_i v_i^2}{\Sigma N_i}\right)^{1/2} = \left[\frac{10(500)^2+20(1000)^2+5(1500)^2}{35}\right]^{\frac{1}{2}}$$

$$= 982 \text{ m s}^{-1}$$

16.5 Calculate the ratio of the root-mean-square speed to the mean speed to the most probable speed.

SOLUTION

$$\langle v^2\rangle^{1/2} \quad : \quad \langle v\rangle \quad : \quad v_p$$

$$\left(\frac{3RT}{M}\right)^{1/2} \quad : \quad \left(\frac{8RT}{\pi M}\right)^{1/2} \quad : \quad \left(\frac{2RT}{M}\right)^{1/2}$$

$$\left(\frac{3}{2}\right)^{1/2} \quad : \quad \left(\frac{4}{\pi}\right)^{1/2} \quad : \quad 1.000$$

$$1.225 \quad : \quad 1.128 \quad : \quad 1.000$$

16.6 What is the ratio of the probability that gas molecules have five times the mean speed to the probability that they have the mean speed?

SOLUTION

$$f(\langle v\rangle) \propto \langle v\rangle^2 \; e^{-\frac{\langle v\rangle^2 m}{2kT}}$$

$$f(5\langle v\rangle) \propto 25\langle v\rangle^2 \; e^{-\frac{25\langle v\rangle^2 m}{2kT}}$$

$$\frac{f(5\langle v\rangle)}{f(\langle v\rangle)} = 25\, e^{-\frac{24\langle v\rangle^2 m}{2kT}}$$

Since $\langle v\rangle^2 = \frac{8kT}{\pi m}$

$$\frac{f(5\langle v\rangle)}{f(\langle v\rangle)} = 25\, e^{-96/\pi} = 1.24 \times 10^{-12}$$

16.7 Calculate the velocity of sound in nitrogen gas at 25 °C. (See Section 16.3.)

SOLUTION

$$v = \left(\frac{C_P RT}{C_V M}\right)^{1/2}$$

$$= \left[\frac{(29.125 \text{JK}^{-1}\text{mol}^{-1})(8.3144 \text{JK}^{-1}\text{mol}^{-1})(298.15 \text{K})}{(20.811 \text{JK}^{-1}\text{mol}^{-1})(2)(14.0067 \times 10^{-3} \text{kg mol}^{-1})}\right]^{1/2}$$

$$= 352 \text{ m s}^{-1}$$

$$= \frac{(352 \text{ m s}^{-1})(10^2 \text{ cm m}^{-1})(60 \text{ s min}^{-1})(60 \text{ min hr}^{-1})}{(2.54 \text{ cm in}^{-1})(12 \text{ in ft}^{-1})(5280 \text{ ft mile}^{-1})}$$

$$= 787 \text{ miles hr}^{-1}$$

16.8 Calculate the number of collisions per square centimeter per second of oxygen molecules with a wall at a pressure of 1 bar and 25 °C.

SOLUTION

$$\rho = \frac{PN_A}{RT} = \frac{(10^5 \text{ N m}^{-2})(6.022 \times 10^{23} \text{ mol}^{-1})}{(8.3144 \text{ J K}^{-1} \text{ mol}^{-1})(298.15 \text{ K})}$$

$$= 2.429 \times 10^{25} \text{ m}^{-3}$$

$$\langle v \rangle = \left(\frac{8RT}{\pi M}\right)^{1/2}$$

$$= \left[\frac{(8)(8.3144 \text{ J K}^{-1} \text{ mol}^{-1})(298.15 \text{ K})}{\pi(32 \times 10^{-3} \text{ kg mol}^{-1})}\right]^{1/2}$$

$$= 444.2 \text{ m s}^{-1}$$

$$\Gamma_n = \frac{\rho \langle v \rangle}{4} = \frac{1}{4}(2.429 \times 10^{25} \text{ m}^{-3})(444.2 \text{ m s}^{-1})$$

$$= 2.697 \times 10^{27} \text{ m}^{-2} \text{ s}^{-1}$$

$$= (2.697 \times 10^{27} \text{ m}^{-2} \text{ s}^{-1})(10^{-2} \text{ m cm}^{-1})^2$$

$$= 2.697 \times 10^{23} \text{ cm}^{-2} \text{ s}^{-1}$$

16.9 A fresh metal surface with 10^{15} atoms per square centimeter is prepared. This surface is exposed to oxygen at 10^{-2} Pa. If every oxygen molecule that strikes the surface reacts so that there is one oxygen atom per metal atom in the surface, how long will it take for half of the surface to become oxidized at 25 °C?

SOLUTION

$$m = \frac{32 \times 10^{-3}\ \text{kg mol}^{-1}}{6.022 \times 10^{23}\ \text{mol}^{-1}} = 5.31 \times 10^{-26}\ \text{kg}$$

$$\Gamma_n = \frac{P}{(2\pi mkT)^{1/2}}$$

$$= \frac{10^{-2}\ \text{N m}^{-2}}{[2\pi(5.31\text{x}10^{-26}\text{kg})(1.38\text{x}10^{-23}\text{JK}^{-1})(298\text{K})]^{1/2}}$$

$$= 2.70 \times 10^{20}\ \text{m}^{-2}\ \text{s}^{-1}$$

In the surface there are $(10^{15}\ \text{cm}^{-2})(100\ \text{cm m}^{-1})^2 = 10^{19}$ atoms m^{-2}. Oxidizing half of the surface will require 0.25×10^{19} oxygen molecules

$$\frac{0.25 \times 10^{19}\ \text{m}^{-2}}{2.7 \times 10^{20}\ \text{m}^{-2}\ \text{s}^{-1}} = 9.3 \times 10^{-3}\ \text{s}$$

16.10 A Knudsen cell containing crystalline benzoic acid ($M = 122\ \text{g mol}^{-1}$) is carefully weighed and placed in an evacuated chamber thermostated at 70 °C for 1 hr. The circular hole through which effusion occurs is 0.60 mm in diameter. Calculate the sublimation pressure of benzoic acid at 70 °C in Pa from the fact that the weight loss is 56.7 mg.

SOLUTION

$$P = W\left(\frac{2\pi RT}{M}\right)^{1/2}$$

$$= \frac{56.7 \times 10^{-6}\ \text{kg}}{(60\text{x}60\text{s})\pi(0.3\text{x}10^{-3}\text{m})^2}\left[\frac{2\pi(8.314\text{JK}^{-1}\text{mol}^{-1})(343\text{K})}{122 \times 10^{-3}\ \text{kg mol}^{-1}}\right]^{1/2}$$

$= 21.3\ \text{N m}^{-2}$

$= 21.3$ Pa

16.11 R. B. Holden, R. Speiser, and H. L. Johnston [J.Am.Chem.Soc.,70,3897(1948)] found the rate of loss of weight of a Knudsen effusion cell containing finely divided beryllium to be 19.8×10^{-7} g cm^{-2} s^{-1} at 1320 K and 1210×10^{-7} g cm^{-2} s^{-1} at 1537 K. Calculate ΔH_{sub} for this temperature range.

SOLUTION

According to equations 16.43 and 16.44 the two vapor pressures are proportional to $\Delta gT^{1/2}$.

$$\Delta H° = \frac{RT_1T_2}{(T_2 - T_1)}\ln\frac{P_2}{P_1}$$

$$= \frac{(8.314)(1320)(1537)}{(217)}\ln\frac{1210(1537)^{1/2}}{19.8(1320)^{1/2}}$$

$= 326\ \text{kJ mol}^{-1}$

16.12 A 5 mL container with a hole 10 μm in diameter is filled with hydrogen. This container is placed in an evacuated chamber at 0 °C. How long will it take for 90% of the hydrogen to effuse out?

SOLUTION

$\Gamma_n = \frac{\rho \langle v \rangle}{4}$ may be written $-\frac{1}{A}\frac{dN}{dt} = \frac{N}{V}\frac{\langle v \rangle}{4}$

$\langle v \rangle = 1.69 \times 10^3$ m s^{-1} from Example 16.1

$$-\frac{dN}{dt} = \frac{\pi(5 \times 10^{-6}\ \text{m})^2(1.69 \times 10^3\ \text{m s}^{-1})N}{(5 \times 10^{-6}\ \text{m}^3)(4)}$$

$$= (6.64 \times 10^{-3}\ \text{s}^{-1})N$$

$$\int_{N_o}^{N_t} \frac{dN}{N} = -(6.64 \times 10^{-3}\ \text{s}^{-1}) \int_o^t dt$$

$$\ell n\ \frac{N_t}{N_o} = -(6.64 \times 10^{-3}\ \text{s}^{-1})t$$

$$t = \frac{\ell n\ 0.1}{-6.64 \times 10^{-3}\ \text{s}^{-1}} = 347\ \text{s}$$

16.13 (a) Calculate the number of collisions per second undergone by a single nitrogen molecule in nitrogen at 1 bar pressure and 25 °C. (b) What is the number of collisions per cubic centimeter per second? What is the effect on the number of collisions (c) of doubling the absolute temperature at constant pressure, and (d) of doubling the pressure at constant temperature?

SOLUTION

(a) $Z_{1(1)} = 2^{1/2} \rho\pi d^2 \langle v \rangle$

$$\rho = \frac{PN_A}{RT} = \frac{(10^5\ \text{N m}^{-2})(6.022 \times 10^{23}\ \text{mol}^{-1})}{(8.314\ \text{J K}^{-1}\ \text{mol}^{-1})(298.15\ \text{K})}$$

$$= 2.429 \times 10^{25}\ \text{m}^{-3}$$

$$\langle v \rangle = \left(\frac{8RT}{\pi M}\right)^{1/2}$$

$$= \left[\frac{8(8.314)(298.15)}{\pi(28 \times 10^{-3})}\right]^{1/2} = 475 \text{ m s}^{-1}$$

$$Z_{1(1)} = 2^{1/2}(2.429 \times 10^{25})\pi(0.375 \times 10^{-9})^2(475)$$
$$= 7.21 \times 10^9 \text{ s}^{-1}$$

(b) $Z_{11} = \frac{1}{2^{1/2}} \rho^2 \pi d^2 \langle v \rangle$

$$= 2^{-1/2}(2.429 \times 10^{25})^2 \pi(0.375 \times 10^{-9})^2(475)$$
$$= 8.75 \times 10^{34} \text{ m}^{-3} \text{ s}^{-1}$$
$$= (8.75 \times 10^{34} \text{ m}^{-3} \text{ s}^{-1})(10^{-2}\text{m cm}^{-1})^3$$
$$= 8.75 \times 10^{28} \text{ cm}^2 \text{ s}^{-1}$$

(c) $\rho \propto T^{-1}$ $\quad \langle v \rangle \propto T^{1/2}$

$$Z_{11} \propto \rho^2 \langle v \rangle \propto T^{-2}\, T^{1/2} \propto T^{-1.5}$$

$$\frac{Z_{11}(2T)}{Z_{11}(T)} = \frac{T^{1.5}}{(2T)^{1.5}} = \frac{1}{2^{1.5}} = 0.354$$

(d) $Z_{11} \propto P^2$

$$\frac{Z_{11}(2P)}{Z_{11}(P)} = 4$$

16.14 Methyl radicals appear to combine without activation energy $2CH_3 \longrightarrow C_2H_6$
The second order rate constant at 300 K is $10^{10.5}$ L mol^{-1} s^{-1}. Assuming the collision diameter is 0.40 nm, what rate constant is expected from the rate of collisions?

SOLUTION

$$-\frac{d(CH_3)}{dt} = 2k(CH_3)^2$$

The rate constant is equal to 1/2 the rate of disappearance of CH_3 when $(CH_3) = 1$ mol L^{-1}

$$\rho = 6.02 \times 10^{23}\ \text{m}^{-3}$$

$$\langle v\rangle = \left(\frac{8RT}{\pi M}\right)^{1/2} = \left[\frac{8(8.314)(300)}{\pi(15 \times 10^{-3})}\right]^{1/2}$$

$$= 651\ \text{m s}^{-1}$$

$$Z_{11} = 2^{-1/2}\rho^2\pi d^2\langle v\rangle$$

$$= 2^{-1/2}(6.02\text{x}10^{23})^2\pi(0.4\text{x}10^{-9})^2(651)$$

$$= 8.39 \times 10^{37}\ \text{m}^{-3}\ \text{s}^{-1}$$

$$= \frac{(8.39 \times 10^{37}\ \text{m}^{-3}\ \text{s}^{-1})(10^{-3}\ \text{m}^{-3}\ \text{L}^{-1})}{(6.02 \times 10^{23}\ \text{mol}^{-1})}$$

$$= 1.39 \times 10^{11}\ \text{mol L}^{-1}\ \text{s}^{-1}$$

$$-\frac{d(CH_3)}{dt} = 2.78 \times 10^{11}\ \text{mol L}^{-1}\ \text{s}^{-1} = 2k\ (1\ \text{mol L}^{-1})^2$$

$$k = 1.39 \times 10^{11}\ \text{L mol}^{-1}\ \text{s}^{-1} = 10^{11.1}\ \text{L mol}^{-1}\ \text{s}^{-1}$$

16.15 What is the mean free path of nitrogen at 1 bar and 25 °C? What is the average time between collisions?

SOLUTION

$$\rho = \frac{PN_A}{RT} = \frac{(10^5\ \text{N m}^{-2})(6.02 \times 10^{23}\ \text{mol}^{-1})}{(8.314\ \text{J K}^{-1}\ \text{mol}^{-1})(298\ \text{K})}$$

$$= 2.43 \times 10^{25}\ \text{m}^{-3}$$

$$\ell = [2^{1/2}\rho\pi d^2]^{-1}$$

$$= [2^{1/2}(2.43 \times 10^{25}\ \text{m}^{-3})\pi(0.375 \times 10^{-9}\ \text{m})^2]^{-1}$$

$$= 65.9\ \text{nm}$$

$\langle v\rangle = 475\ \text{m s}^{-1}$ from Problem 16.13

$$t = \frac{65.9 \times 10^{-9}\ \text{m}}{475\ \text{m s}^{-1}} = 1.4 \times 10^{-10}\ \text{s}$$

16.16 (a) Calculate the mean free path for hydrogen gas (σ = 0.247 nm) at 1 bar and 0.1 Pa at 25 °C.

(b) Repeat the calculation for chlorine gas (σ = 0.496 nm).

<u>SOLUTION</u>

At 1 bar

$$\rho = \frac{N_A P}{RT} = \frac{(6.022 \times 10^{23}\ \text{mol}^{-1})(10^5\ \text{N m}^{-2})}{(8.314\ \text{J K}^{-1}\ \text{mol}^{-1})(298\ \text{K})}$$

$= 2.44 \times 10^{25}\ \text{m}^{-3}$

At 0.1 Pa

$$\rho = \frac{(6.022 \times 10^{23}\ \text{mol}^{-1})(0.1\ \text{N m}^{-2})}{(8.314\ \text{J K}^{-1}\ \text{mol}^{-1})(298\ \text{K})}$$

$= 2.43 \times 10^{19}\ \text{m}^{-3}$

<u>(a)</u>

$$\ell = \frac{1}{2^{1/2}\pi\sigma^2\rho}$$

$$= \frac{1}{2^{1/2}\pi(0.247 \times 10^{-9}\ \text{m})\ (2.44 \times 10^{25}\ \text{m}^{-3})}$$

$= 1.52 \times 10^{-7}$ m at 1 bar

$$\ell = \frac{1}{2^{1/2}\pi(0.247 \times 10^{-9}\ \text{m})^2(2.43 \times 10^{19}\ \text{m}^{-3})}$$

$= 0.152$ m at 0.1 Pa

<u>(b)</u>

$$\ell = \frac{1}{2^{1/2}\pi(0.496 \times 10^{-9}\ \text{m})^2(2.44 \times 10^{25}\ \text{m}^{-3})}$$

$= 3.77 \times 10^{-8}$ m at 1 bar

$$\ell = \frac{1}{2^{1/2}\pi(0.496 \times 10^{-9}\ \text{m})^2(2.43 \times 10^{19}\ \text{m}^{-3})}$$

$= 0.037$ m at 0.1 Pa

16.17 Large vacuum chambers have been built for testing space vehicles at 10^{-6} Pa. Calculate

(a) the mean-free path of nitrogen at this pressure, and (b) the number of molecular impacts per square meter of wall per second at 25 °C. $\sigma_{N_2} = 0.375$ nm.

SOLUTION

$$n = \frac{PN_A}{RT} = \frac{(10^{-6}\ \text{Pa})(6.022 \times 10^{23}\ \text{mol}^{-1})}{(8.314\ \text{J K}^{-1}\ \text{mol}^{-1})(298\ \text{K})}$$

$$= 2.43 \times 10^{14}\ \text{m}^{-3}$$

(a)

$$\ell = \frac{1}{\sqrt{2}\ \pi\sigma^2 n}$$

$$= \frac{1}{\sqrt{2}\ \pi(3.75 \times 10^{-10}\ \text{m})^2(2.43 \times 10^{14}\ \text{m}^{-3})}$$

$$= 6590\ \text{m}$$

(b)

$$Z = \rho\left(\frac{RT}{2\pi M}\right)^{1/2}$$

$$= (2.43\text{x}10^{14}\ \text{m}^{-3})\left[\frac{(8.314\text{JK}^{-1}\text{mol}^{-1})(298\text{K})}{2\pi\ 28 \times 10^{-3}\ \text{kg mol}^{-1}}\right]^{1/2}$$

$$= 2.88 \times 10^{16}\ \text{m}^{-2}\ \text{s}^{-1}$$

16.18 The pressure in interplanetary space is estimated to be of the order of 10^{-14} Pa. Calculate (a) the average number of molecules per cubic centimeter, (b) the number of collisions per second per molecule, and (c) the mean free path in miles. Assume that only hydrogen atoms are present and that the temperature is 1000 K. Assume $\sigma = 0.2$ nm.

SOLUTION

(a) $$\rho = \frac{N_A P}{RT} = \frac{(6.022 \times 10^{23}\ \text{mol}^{-1})(10^{-14}\text{N m}^{-2})}{(8.314\ \text{J K}^{-1}\ \text{mol}^{-1})(10^3\ \text{K})}$$

$$= 0.724 \times 10^6\ \text{m}^{-3}$$

$$= 0.724\ \text{cm}^{-3}$$

(b) $$\langle v\rangle = \left(\frac{8RT}{\pi M}\right)^{1/2}$$

$$= \left[\frac{8(8.314\ \text{J K}^{-1}\ \text{mol}^{-1})(10^3\ \text{K})}{\pi(1 \times 10^{-3}\ \text{kg mol}^{-1})}\right]^{1/2}$$

$$= 4600\ \text{m s}^{-1}$$

$$Z_1 = 2^{1/2}\rho\pi\sigma^2\langle v\rangle$$

$$= 2^{1/2}(0.724 \text{x} 10^6\ \text{m}^{-3})\ (0.2 \text{x} 10^{-9}\ \text{m})^2(4600\ \text{ms}^{-1})$$

$$= 5.92 \times 10^{-10}\ \text{s}^{-1}$$

(c) $$\ell = \frac{1}{2^{1/2}\pi\sigma^2\rho}$$

$$= \frac{1}{2^{1/2}\pi(0.2 \times 10^{-9}\ \text{m})^2(0.724 \times 10^6)}$$

$$= 7.77 \times 10^{12}\ \text{m}$$

$$= \frac{(7.77 \times 10^{12}\ \text{m})(10^2\ \text{cm m}^{-1})}{(2.54\ \text{cm m}^{-1})(12\ \text{in ft}^{-1})(5280\ \text{ft mile}^{-1})}$$

$$= 4.83 \times 10^9\ \text{miles}$$

16.19 Consider an atomic beam of potassium passing through a scattering gas of Ar contained in a cell of 1 cm length at 0 °C. Assuming a collision cross section of $6 \times 10^{-18}\ \text{m}^2$ for potassium-argon collisions, calculate the pressure of argon required to produce an attenuation of the beam of 25%.

SOLUTION

$$I_A = I_A^\circ\ e^{-Q_{AB}n_B L}$$

$$n_B = \frac{\ell n(I_A^\circ/I_A)}{Q_{AB}L}$$

$$= \frac{2.303 \log (100/75)}{(600 \times 10^{-20} \text{ m}^2)(10^{-2} \text{ m})}$$

$$= 4.80 \times 10^{18} \text{ m}^{-3} = 4.80 \times 10^{15} \text{ L}^{-1}$$

$$P = \frac{n_B RT}{N_A} = \frac{(4.80 \times 10^{15} \text{ L}^{-1})(8.314 \text{ JK}^{-1}\text{mol}^{-1})(273 \text{ K})}{(6.022 \times 10^{23} \text{ mol}^{-1})}$$

$$= 1.81 \times 10^{-5} \text{ Pa}$$

16.20 The coefficient of viscosity of helium is 1.88×10^{-5} Pa s at 0 °C. Calculate (a) the collision diameter, and (b) the diffusion coefficient at 1 bar.

SOLUTION

(a) $$d = \left[\frac{5}{16} \frac{(\pi mkT)^{1/2}}{\pi\eta}\right]^{1/2}$$

$$m = \frac{4.0026 \times 10^{-3} \text{ kg mol}^{-1}}{6.022 \times 10^{23} \text{ mol}^{-1}} = 6.647 \times 10^{-27} \text{ kg}$$

$$d = \left[\frac{5}{16} \frac{(\pi 6.647 \times 10^{-27} \times 1.38 \times 10^{-23} \times 273)^{1/2}}{\pi (1.88 \times 10^{-5})}\right]^{1/2}$$

$$= 0.217 \text{ nm}$$

(b) $$\rho = \frac{PN_A}{RT} = \frac{(10^5 \text{ Pa m}^{-2})(6.022 \times 10^{23} \text{ mol}^{-1})}{(8.314 \text{ J K}^{-1} \text{ mol}^{-1})(273 \text{ K})}$$

$$= 2.65 \times 10^{25} \text{ m}^{-3}$$

$$D = \frac{3}{8} \frac{(\pi mkT)^{1/2}}{\pi d^2 \rho m}$$

$$= \frac{3}{8} \frac{[\pi(6.647 \times 10^{-27} \text{ kg})(1.38 \times 10^{-23} \text{ JK}^{-1})(273 \text{ K})]^{1/2}}{\pi(0.217 \times 10^{-9} \text{ m})^2(2.65 \times 10^{25} \text{ m}^{-3})(6.647 \times 10^{-27} \text{ kg})}$$

$$= 1.28 \times 10^{-4} \text{ m}^2 \text{ s}^{-1}$$

16.21 What is the self-diffusion coefficient of radioactive CO_2 in ordinary CO_2 at 1 bar and 25 °C? The collision diameter is 0.40 nm.

SOLUTION

$$m = \frac{44 \times 10^{-3}\ \text{kg mol}^{-1}}{6.022 \times 10^{23}\ \text{mol}^{-1}} = 7.307 \times 10^{-26}\ \text{kg}$$

$$\rho = \frac{N}{V} = \frac{PN_A}{RT} = \frac{(1\ \text{bar})(6.022 \times 10^{23}\ \text{mol}^{-1})(10^3\ \text{L mol}^{-1})}{(0.08314\ \text{L bar K}^{-1}\ \text{mol}^{-1})(298\ \text{K})}$$

$$= 2.43 \times 10^{25}\ \text{m}^{-3}$$

$$D = \frac{3}{8}\frac{(\pi mkT)^{1/2}}{\pi d^2 \rho m}$$

$$D = \frac{3[\pi(7.307 \times 10^{-26}\ \text{kg})(1.38 \times 10^{-23}\ \text{JK}^{-1})(298\ \text{K})]^{1/2}}{\pi(0.4 \times 10^{-9}\ \text{m})^2(2.43 \times 10^{25}\ \text{m}^{-3})(7.307 \times 10^{-26}\ \text{kg})}$$

$$= 1.26 \times 10^{-5}\ \text{m}^2\ \text{s}^{-1}$$

16.22 3.89×10^{-2} eV

16.23 281 m s^{-1}

16.24 0.1991

16.25 418, 400, and 440 m s^{-1}

16.27 481, 445, and 394 m s^{-1}

16.28 (a) 1020 m s^{-1} (b) 352 m s^{-1}

16.29 3.7×10^3 s

16.30 (a) 1.136×10^{22} molecules cm^{-2} s^{-1}

(b) 0.0629 g cm^{-2} min^{-1}

16.31 0.0713 g

16.32 8.1×10^{-12}, 2×10^{-10} kg

16.33 78 g mol^{-1}

16.34 (a) $4.75 \times 10^{29}\ m^{-3}\ s^{-1}$

(b) 2.86×10^{-5} m

16.35 $5.8 \times 10^{28}\ mL^{-1}\ s^{-1}$

16.36 1.58×10^{-10} s   7489 vibrations

16.37 (a) $2.65 \times 10^{6}\ cm^{-3}$ (b) 6.52×10^{5} m

16.38 $$\ell_1 = \frac{\langle v_1 \rangle}{2^{1/2}\rho_1 \pi d_1^2 \langle v_1 \rangle + \rho_2 \pi d_{12}^2 \langle v_{12} \rangle}$$

$$\ell_2 = \frac{\langle v_2 \rangle}{2^{1/2}\rho_2 \pi d_2^2 \langle v_2 \rangle + \rho_1 \pi d_{12}^2 \langle v_{12} \rangle}$$

16.39 (a) 28 m m^2 (b) 14 nm^2

(c) The cis compound has a large dipole moment while the trans compound has no permanent dipole moment. Since CsCl is a dipolar molecule, it interacts more strongly with the polar isomer of dichloroethylene.

16.40 646 Pa

16.41 1.66×10^{-5} Pa s

16.42 1.302×10^{3} Pa s

CHAPTER 17. Kinetics: Gas Phase and Basic Relationships

17.1 The half-life of a first-order chemical reaction A ⟶ B is 10 min. What percent of A remains after 1 hr?

SOLUTION

$$A = (A_0)e^{-kT} = (A)_0\, e^{-\frac{0.693t}{t_{1/2}}}$$

$$\frac{(A)}{(A)_o} = \exp\left[\frac{-0.693\ (60\ m)}{10\ m}\right] = 0.0156$$

Thus 1.56% remains after one hour.

17.2 The following data were obtained on the rate of hydrolysis of 17% sucrose in 0.099 mol L^{-1} HCl aqueous solutions at 35 °C

t/min	9.82	59.60	93.18	142.9	294.8	589.4
Sucrose remaining, %	96.5	80.3	71.0	59.1	32.8	11.1

What is the order of the reaction with respect to sucrose and the value of the rate constant k?

SOLUTION

$$\log \frac{(A)}{(A)_0} = \frac{-kT}{2.303}$$

(please see graph, page 259)

$$\text{slope} = \frac{-1}{612\ m} = \frac{-k}{2.303}$$

$$k = \frac{2.303}{612\ m} = 3.76 \times 10^{-3}\ m^{-1}$$

$$= (3.76 \times 10^{-3}\ m^{-1})(\frac{1}{60}\ m\ s^{-1})$$
$$= 6.27 \times 10^{-5}\ s^{-1}$$

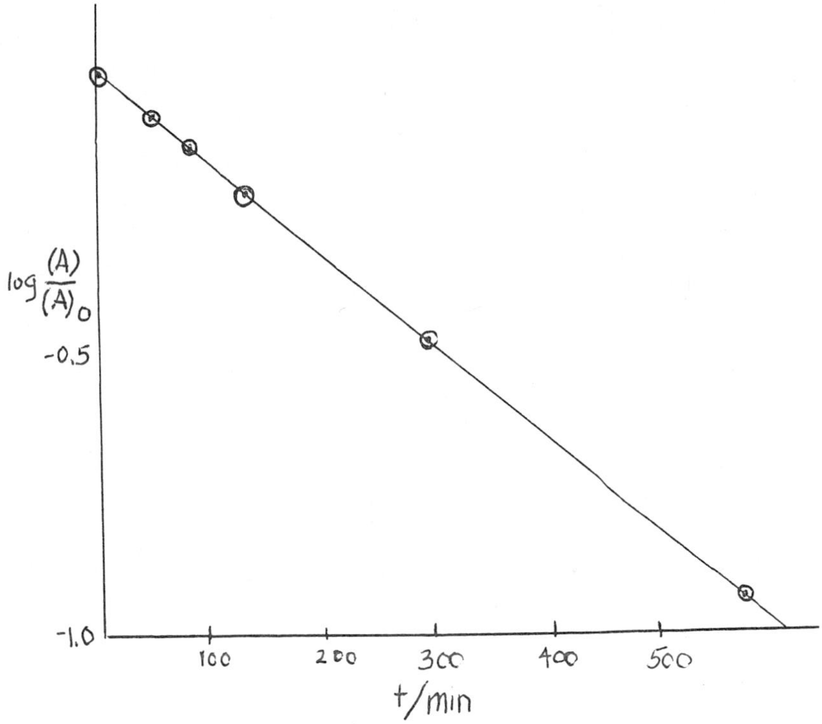

17.3 Methyl acetate is hydrolyzed in approximately 1 mol L^{-1} HCl at 25 °C. Aliquots of equal volume are removed at intervals and titrated with a solution of NaOH. Calculate the first-order rate constant from the following experimental data.

t/s	339	1242	2745	4546	∞
v/cm^3	26.34	27.80	29.70	31.81	39.81

SOLUTION

Any quantity proportional to the concentration of reactant A that remains may be used in equation

$$\log (A) = \frac{-kt}{2.303} + \log (A)_0$$

In this case the concentration of A remaining is proportional to $V - 39.81 \text{ cm}^3$. Therefore log $(V - 39.81 \text{ cm}^3)$ is plotted versus t.

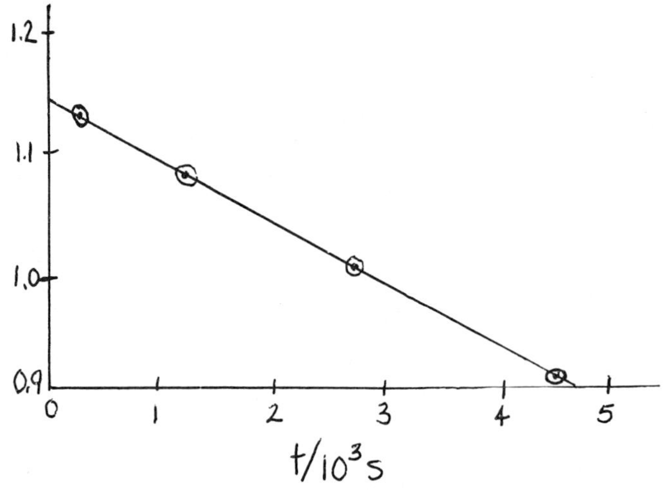

$$\text{slope} = \frac{-0.247}{4.65 \times 10^3 \text{ s}} = \frac{-k}{2.303}$$

$$k = \frac{(2.303)(0.247)}{4.65 \times 10^3 \text{ s}} = 1.22 \times 10^{-4} \text{ s}^{-1}$$

17.4 Prove that in a first-order reaction, where $dn/dt = -kn$, the average life, that is the average life expectancy of the molecules, is equal to $1/k$.

SOLUTION Average life $= \dfrac{\int_0^\infty n dt}{n_0}$

$$\text{Average life} = \frac{-\frac{1}{k}\int_0^{\infty} \frac{dn}{dt}\, dt}{n_0}$$

$$= \frac{-\frac{1}{k}(0 - n_0)}{n_0}$$

$$= \frac{1}{k}$$

17.5 Since radioactive decay is a first-order process, the decay rate for a particular nuclide is commonly given as the half life. Given that potassium contains 0.0118% ^{40}K which has a half life of 1.27×10^9 years, how many disintegrations per second are there in a gram of KCl?

SOLUTION

$$t_{1/2} = (1.27\text{x}10^9 \text{ y})(365 \text{ dy}^{-1})(24 \text{ hd}^{-1})(60 \text{ min h}^{-1})(60 \text{ s min}^{-1})$$

$$= 4.01 \times 10^{16} \text{ s}$$

$$\frac{dN}{dt} = \frac{0.693}{t_{1/2}} N$$

$$= \frac{(0.693)[(39.1/76.6)\text{g}](6.02\text{x}10^{23} \text{ mol}^{-1})(1.18\text{x}10^{-4})}{(4.01 \times 10^{16} \text{ s})(40.0 \text{ g mol}^{-1})}$$

$$= 15.7 \text{ s}^{-1}$$

17.6 It is found that the decomposition of HI to $H_2 + I_2$ at 508 °C has a half-life of 135 min when the pressure is 1 atm. (a) Show that this proves that the reaction is second order. (b) What is the value of the rate constant in L mol^{-1} s^{-1}? (c) What is the value of the rate constant in bar^{-1} s^{-1}? (d) What is the value of the rate constant in cm^3 s^{-1}?

SOLUTION

(a) For a second order reaction $t_{1/2} = 1/k(A)_0$ where $(A)_0$ is the initial concentration or pressure of the reactant. This is in agreement with the fact that the half-life is reduced by a factor of 10 when the pressure is increased by a factor of 10.

(b) $$c = P/RT = \frac{(101,325 \text{ Pa})}{(8.314 \text{ J K}^{-1} \text{ mol}^{-1})(781.15 \text{ K})}$$

$$= 1.56 \times 10^{-2} \text{ mol L}^{-1}$$

$$k = 1/(A)_0 t_{1/2} = 1/(1.56\text{x}10^{-2} \text{ mol L}^{-1})(13.5\text{x}60 \text{ s})$$

$$= 7.91 \times 10^{-2} \text{ L mol}^{-1} \text{ s}^{-1}$$

(c) $$k = \frac{1}{t_{1/2} P_0} = \frac{1}{(13.5\text{x}60 \text{ s})(1.013 \text{ bar})}$$

$$= 1.22 \times 10^{-3} \text{ bar}^{-1} \text{ s}^{-1}$$

(d) $$\frac{(7.91 \times 10^{-2} \text{ L mol}^{-1} \text{ s}^{-1})(10^3 \text{ cm}^3 \text{ L}^{-1})}{(6.022 \times 10^{23} \text{ mol}^{-1})}$$

$$= 1.31 \times 10^{-22} \text{ cm}^3 \text{ s}^{-1}$$

17.7 The reaction between propionaldehyde and hydrocyanic acid has been studied at 25 °C by W. J. Svirbely and J. F. Roth [J. Am. Chem. Soc., 75, 3106 (1953)]. In a certain aqueous solution at 25 °C the concentrations at various times were as follows.

t/min	2.78	5.33	8.17	15.23	19.80	∞
(HCN)/mol L^{-1}	0.0990	0.0906	0.0830	0.0706	0.0653	0.0424
(C_3H_7CHO)/mol L^{-1}	0.0566	0.0482	0.0406	0.0282	0.0229	0.0000

What is the order of the reaction and the value of the rate constant k?

SOLUTION

The data does not give a linear log versus t plot, and so the data are tested in the integrated equation for a second order reaction. Equation 17.22 is

$$\frac{1}{[(A)_0 - (B)_o]} \ln \frac{(A)(B)_0}{(A)_0(B)} = kt$$

In order to get $(A)_0$ and $(B)_0$ for t = 0, the clock is started at the first experimental point. This yields the following data to be plotted.

t/min	0	2.55	5.39	12.45	17.02
$\frac{1}{[(A)_0-(B)_o]} \ln \frac{(A)(B)_0}{(A)_0(B)}$	0	1.695	3.677	8.458	11.523

The slope of this plot, 0.675 L mol^{-1} min^{-1}, is the second order rate constant.

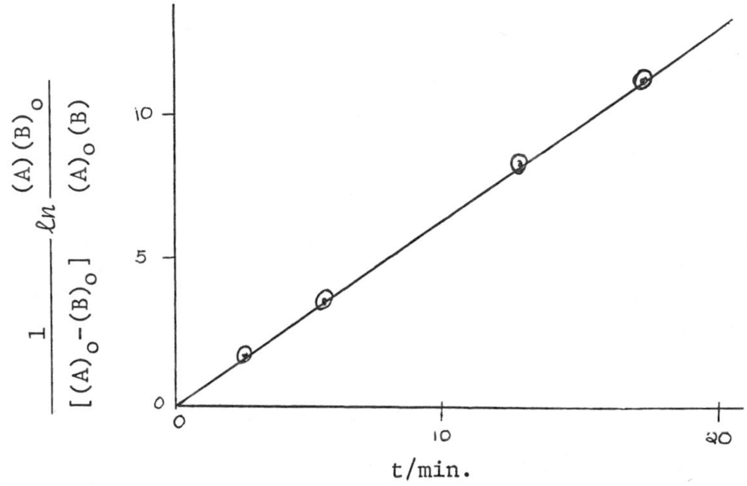

17.8 The reaction $CH_3CH_2NO_2 + OH^- \longrightarrow H_2O + CH_3CHNO_2^-$ is of second order, and k at 0 °C is 39.1 L mol^{-1} min^{-1}. An aqueous solution is made 0.004 molar in nitroethane and 0.005 molar in NaOH. How long will it take for 90% of the nitroethane to react?

SOLUTION

$$t = \frac{1}{k(a - b)} \ell n \frac{b(a - x)}{a(b - x)}$$

$$= \frac{1}{(39.1\ \text{L mol}^{-1}\ \text{min}^{-1})(0.001\ \text{mol L}^{-1})} \ell n \frac{(0.004)(0.005-0.0036)}{(0.005)(0.004 - 0.0036)}$$

$= 26.3$ min

17.9 Hydrogen peroxide reacts with thiosulfate ion in slightly acidic solution as follows.

$$H_2O_2 + 2S_2O_3^{2-} + 2H^+ \longrightarrow 2H_2O + S_4O_6^{2-}$$

This reaction rate is independent of the hydrogen-ion concentration in the pH range 4 to 6. The following data were obtained at 25 °C and pH 5.0.

Initial concentrations: $(H_2O_2) = 0.036\ 80$ mol L^{-1};

$(S_2O_3^{2-}) = 0.020\ 40$ mol L^{-1}

t/min	16	36	43	52
$(S_2O_3^{2-})/10^{-3}$ mol L^{-1}	10.30	5.18	4.16	3.13

(a) What is the order of the reaction? (b) What is the rate constant?

SOLUTION

(a) In the first 16 minutes, $(S_2O_3^{2-})$ is approximately halved. In the next 20 minutes, $(S_2O_3^{2-})$ is approximately halved. In the next 16 minutes $(S_2O_3^{2-})$ is considerably less than halved. Therefore, the order is higher than one. The next section shows that the reaction is first order in H_2O_2, first order in $S_2O_3^{2-}$, and second order overall.

(b) Let A = H_2O_2 and B = $S_2O_3^{2-}$. Equation 15.19 becomes

$$kT = \frac{1}{[2(A)_0 - (B)_0]} \ln \frac{(A)(B)_0}{(A)_0(B)}$$

$$\ln \frac{(A)}{(B)} = \ln \frac{(A)_0}{(B)_0} + [2(A)_0 - (B)_0]\ kT$$

t/min	0	16	36	43	52
$(B)/10^{-3}$ mol L^{-1}	20.40	10.30	5.18	4.16	3.13
$(A)/10^{-3}$ mol L^{-1}	36.80	31.75	29.19	28.68	28.17
$\ln \frac{(A)}{(B)}$	0.590	1.126	1.729	1.931	2.197

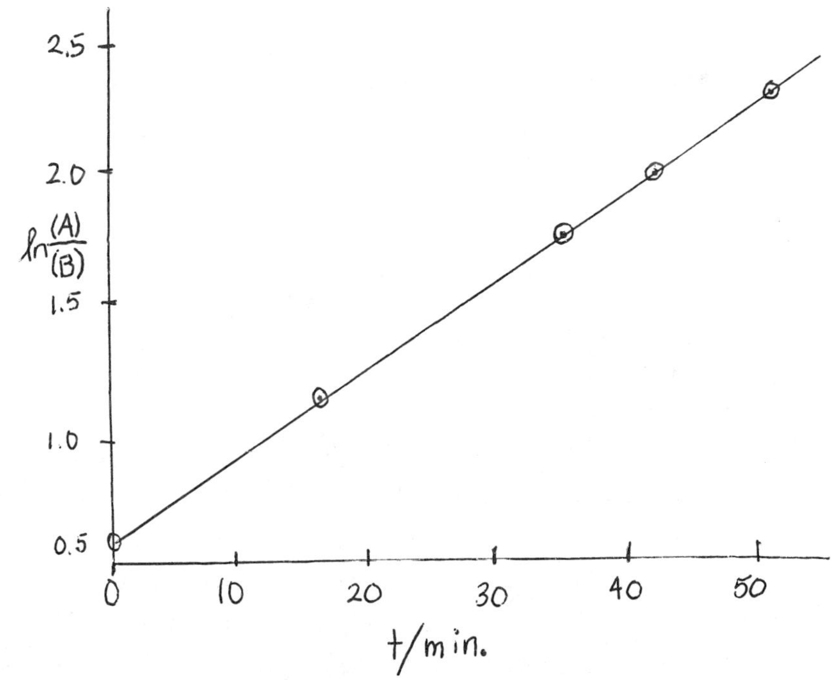

$$\text{slope} = \frac{2.46 - 0.59}{60 \text{ min}} = 0.0312 \text{ min}^{-1}$$

$$= [2(A)_0 - (B)_0]\, k$$

$$k = \frac{0.0312 \text{ min}^{-1}}{2(36.80\times10^{-3}\text{mol L}^{-1}) - 20.40\times10^{-3}\text{mol L}^{-1}}$$

$$= 0.59 \text{ L mol}^{-1} \text{ min}^{-1}$$

17.10 The reaction $2NO + O_2 \longrightarrow 2NO_2$ is third order. Assuming that a small amount of NO_3 exists in rapid reversible equilibrium with NO and O_2 and that the rate-determining step is the slow bimolecular reaction $NO_3 + NO \longrightarrow 2NO_2$, derive the rate equation for this mechanism.

SOLUTION

$$NO + O_2 \rightleftharpoons NO_3 \qquad K_{eq} = \frac{(NO_3)}{(NO)(O_2)}$$

$$NO_3 + NO \xrightarrow{k_2} 2NO_2 \quad \text{(rate determining)}$$

$$\frac{d(NO_2)}{dt} = 2k_2(NO_3)(NO)$$

$$= 2k_2\, K_{eq}\, (NO)^2\, (O_2)$$

$$= k'\, (NO)^2 (O_2)$$

17.11 The preexponential factor for the termolecular reaction $2NO + O_2 \longrightarrow 2NO_2$ is 10^9 cm^6 mol^{-2} s^{-1}. What is the value in L^2 mol^{-2} s^{-1} and cm^6 s^{-1}?

SOLUTION

$$(10^9 \text{ cm}^6 \text{ mol}^{-2} \text{ s}^{-1})(10^{-1} \text{ dm cm}^{-1})^6$$

$$= 10^3 \text{ dm}^6 \text{ mol}^{-2} \text{ s}^{-1}$$

$$= 10^3 \text{ L}^2 \text{ mol}^{-2} \text{ s}^{-1}$$

$$= \frac{(10^3\ L^2\ mol^{-2}\ s^{-1})(10^3\ cm^3\ L^{-1})^2}{(6.02 \times 10^{23}\ mol^{-1})^2}$$

$$= 2.76 \times 10^{-39}\ cm^6\ s^{-1}$$

17.12 A solution of A is mixed with an equal volume of a solution of B containing the same number of moles, and the reaction A + B = C occurs. At the end of 1 hr A is 75% reacted. How much of A will be left unreacted at the end of 2 hr if the reaction is (a) first order in A and zero order in B; (b) first order in both A and B; and (c) zero order in both A and B?

SOLUTION

(a) If the reaction is first order in A and zero order in B, the half time is 1/2 hour, and the following table may be constructed doing calculations in your head.

t/hours	0	1/2	1	1-1/2	2
% Unreacted	100	50	25	12.5	6.25
% Reacted	0	50	75	8.75	93.75

(b) If the reaction is first order in both A and B, and the initial concentrations are equal, and the stoichiometry is 1:1, the concentration of A will follow

$$k = \frac{1}{t}\left[\frac{1}{(A)} - \frac{1}{(A)_0}\right] = \frac{1}{t(A)_0}\left[\frac{(A)_0}{(A)} - 1\right]$$

$$k(A)_0 = \frac{1}{t}\left[\frac{(A)_0}{(A)} - 1\right] = \frac{1}{1\ hr}\left[\frac{100}{25} - 1\right] = 3\ hr^{-1}$$

After 2 hr

$$3\ hr^{-1} = \frac{1}{2\ hr}\left[\frac{100}{(A)} - 0\right]$$

$$(A) = \frac{100}{7} = 14.3\%$$

(c) If the reaction is zero order in both A and B, both will be completely gone in 1-1/3 hr.

17.13 Derive the integrated rate equation for a reaction of 1/2 order. Derive the expression for the half-life of such a reaction.

SOLUTION

$$\frac{d(A)}{dt} = -k(A)^{1/2}$$

$$\int_{(A)_0}^{(A)} \frac{d(A)}{(A)^{\frac{1}{2}}} = -k\int_0^t dt = -kt = [2(A)^{\frac{1}{2}}]_{(A)_0}^{(A)} = 2[(A)^{\frac{1}{2}} - (A)_0^{\frac{1}{2}}]$$

$$(A)_0^{\frac{1}{2}} - (A)^{\frac{1}{2}} = \frac{k}{2}\,t$$

At $t_{1/2}$, $(A) = (A)_0/2$

$$t_{\frac{1}{2}} = \left[(A)_0^{\frac{1}{2}} - \left(\frac{(A)_0}{2}\right)^{\frac{1}{2}}\right]\frac{2}{k} = \frac{2}{k}\left(1 - \frac{1}{\sqrt{2}}\right)(A)_0^{\frac{1}{2}}$$

$$t_{\frac{1}{2}} = \frac{\sqrt{2}}{k}(\sqrt{2} - 1)(A)_0^{\frac{1}{2}}$$

17.14 Show that for a reaction following the rate law $-d(A)/dt = k(A)^{\alpha}$, the half life is given by

$$t_{\frac{1}{2}} = \frac{2^{\alpha-1} - 1}{k(A)_0^{\alpha-1}(\alpha - 1)}$$

where α is an integer not equal to unity.

SOLUTION

$$-\frac{d(A)}{(A)^{\alpha}} = kdt = -(A)^{-\alpha}\,d(A)$$

$$\left[\frac{-(A)^{-\alpha+1}}{1-\alpha}\right]_{(A)_o}^{(A)} = kT$$

$$\frac{1}{(A)^{\alpha-1}} - \frac{1}{(A_o)^{\alpha-1}} = (\alpha - 1)kt$$

If $(A) = \frac{1}{2}(A)_o$

$$\frac{1}{(A)_o^{\alpha-1}}(2^{\alpha-1} - 1) = (\alpha - 1)kt_{\frac{1}{2}}$$

17.15 When an optically active substance is isomerized the optical rotation decreases from that of the original isomer to zero in a first-order manner. In a given case the half-time for this process is found to be 10 min. Calculate the rate constant for the conversion of one isomer to another.

SOLUTION

$$t_{\frac{1}{2}} = \frac{0.693}{k_1 + k_{-1}} = 600\ s$$

Since $K = 1$, $k_1 = k_{-1}$

$$t_{\frac{1}{2}} = \frac{0.693}{2k_1} = 600\ s$$

$$k_1 = \frac{0.693}{2(600\ s)} = 5.78 \times 10^{-4}\ s^{-1}$$

17.16 The following table gives kinetic data [Y. T. Chia and R. E. Connick, J. Phys. Chem., 63, 1518 (1959)] for the following reaction at 25 °C.

$$OCl^- + I^- = OI^- + Cl^-$$

(OCl^-)	(I^-)	(OH^-)	$\frac{d(IO^-)}{dt}/10^{-4}$
	mol L^{-1}		mol L^{-1} s^{-1}
0.0017	0.0017	1.00	1.75
0.0034	0.0017	1.00	3.50
0.0017	0.0034	1.00	3.50
0.0017	0.0017	0.5	3.50

What is the rate law for the reaction and what is the value of the rate constant?

SOLUTION

When other concentrations are held constant, doubling (OCl^-) doubles the rate, doubling (I^-) doubles the rate, and halving (OH^-) doubles the rate. Therefore the rate law is

$$\frac{d\,(OI^-)}{dt} = \frac{k(OCl^-)(I^-)}{(OH^-)}$$

Substituting the values for the first experiment

$$1.75\text{x}10^{-4}\ \text{mol L}^{-1}\ \text{s}^{-1} = \frac{k(0.0017\ \text{mol L}^{-1})(0.0017\ \text{mol L}^{-1})}{(1\ \text{mol L}^{-1})}$$

$k = 61\ s^{-1}$

17.17 For the reaction $2A = B + C$ the rate law for the forward reaction is

$$-\frac{d(A)}{dt} = k(A)$$

Give two possible rate laws for the reverse reaction.

SOLUTION

$$K = \frac{(B)(C)}{(A)^2}$$

At equilibrium $-\frac{d(A)}{dt} = 0 = k_f(A)^2 - k_r(B)(C)$

$$0 = k_f(A) - k_r(B)(C)/(A)$$

Possible rate law for the reverse reaction $k_r(B)(C)/(A)$

$$K = \frac{(B)^{\frac{1}{2}}(C)^{\frac{1}{2}}}{(A)}$$

At equilibrium $-\frac{d(A)}{dt} = 0 = k_f(A) - k_r(B)^{\frac{1}{2}}(C)^{\frac{1}{2}}$

Possible rate law for the reverse reaction $k_r(B)^{\frac{1}{2}}(C)^{\frac{1}{2}}$

$$K = \frac{(B)^2(C)^2}{(A)^4}$$

At equilibrium $\frac{-d(A)}{dt} = 0 = k_f(A)^4 - k_r(B)^2(C)^2$

$$0 = k_f(A) - k_r\frac{(B)^2(C)^2}{(A)^3}$$

Possible rate law for the reverse reaction $k_r(B)^2(C)^2/(A)^3$

17.18 Suppose the transformation of A to B occurs by both a reversible first-order reaction and a reversible second-order reaction involving hydrogen ion.

$$A \underset{k_2}{\overset{k_1}{\rightleftharpoons}} B \qquad A + H^+ \underset{k_4}{\overset{k_3}{\rightleftharpoons}} B + H^+$$

What is the relationship between these four rate constants?

SOLUTION
$$\frac{(B)_{eq}}{(A)_{eq}} = \frac{k_1}{k_2}$$

$$\frac{(B)_{eq}(H^+)_{eq}}{(A)_{eq}(H^+)_{eq}} = \frac{k_3}{k_4} \quad \text{or} \quad \frac{(B)_{eq}}{(A)_{eq}} = \frac{k_3}{k_4}$$

Therefore $\dfrac{k_1}{k_2} = \dfrac{k_3}{k_4}$ or $k_1k_4 = k_2k_3$

17.19 The first three steps in the decay of ^{238}U are

$$^{238}U \xrightarrow[4.5\times10^9\ y]{\alpha} {}^{234}Th \xrightarrow[24.1\ d]{\beta} {}^{234}Pa \xrightarrow[1.14\ m]{\beta} {}^{234}U$$

If we start with pure ^{238}U, what fraction will be ^{234}Th after 10, 20, 40, and 80 days?

SOLUTION

$$k_1 = \frac{0.693}{(4.5\times10^9\ y)(365\ dy^{-1})} = 4.2 \times 10^{-13}\ d^{-1}$$

$$k_2 = \frac{0.693}{24.1d} = 0.0288\ d^{-1}$$

$$F = \frac{(^{234}Th)}{(^{238}U)_o} = \frac{k_1}{k_2 - k_1}\left(e^{-k_1t} - e^{-k_2t}\right)$$

t/d	10	20	40	80
F	3.65×10^{-12}	6.39×10^{-12}	9.98×10^{-12}	13.14×10^{-12}

17.20 Set up the rate expressions for the following mechanism

$$A \underset{k_2}{\overset{k_1}{\rightleftharpoons}} B \qquad B + C \xrightarrow{k_3} D$$

If the concentration of B is small compared with the concentrations of A, C, and D, the steady-state approximation may be used to derive the rate law. Show that this reaction may follow the first-order equation at high pressures and the second-order equation at low pressures.

SOLUTION

Since the concentration of B is small, it is assumed to be in a steady state.

$$\frac{d(B)}{dt} = k_1(A) - [k_2 + k_3(C)](B) = 0$$

$$(B) = \frac{k_1(A)}{k_2 + k_3(C)}$$

$$\frac{d(D)}{dt} = k_3(B)(C) = \frac{k_1\ k_3\ (A)(C)}{k_2 + k_3(C)}$$

At high pressures,

$$k_3(C) \gg k_2, \quad \frac{d(D)}{dt} = k_1(A)$$

At low pressures,

$$k_2 \gg k_3(C), \quad \frac{d(D)}{dt} = \frac{k_1\ k_3}{k_2}(A)(C)$$

17.21 The reaction $NO_2Cl = NO_2 + \frac{1}{2}Cl_2$ is first order and appears to follow the mechanism

$$NO_2Cl \xrightarrow{k_1} NO_2 + Cl \qquad NO_2Cl + Cl \xrightarrow{k_2} NO_2 + Cl_2$$

(a) Assuming a steady state for the chlorine atom concentration, show that the empirical first-order rate constant can be identified with $2k_1$. (b) The following data were obtained by H. F. Cordes and H. S. Johnston [J.Am.Chem. Soc., 76, 4264 (1954)] at 180 °C. In a single experiment the reaction is first order, and the empirical rate constant is represented by k. Show that the reaction is second order at these low gas pressures and calculate the second-order rate constant.

$c/10^{-8}$ mol cm^{-3}	5	10	15	20
$k/10^{-4}$ s^{-1}	1.7	3.4	5.2	6.9

SOLUTION

(a) $$\frac{d(Cl)}{dt} = k_1(NO_2Cl) - k_2(NO_2Cl)(Cl) = 0$$

Therefore $(Cl) = \dfrac{k_1}{k_2}$

$$\frac{-d(NO_2Cl)}{dt} = k_1(NO_2Cl) + k_2(NO_2Cl)(Cl)$$

$$= k_1(NO_2Cl) + k_2(NO_2Cl)k_1/k_2$$

$$= 2k_1(NO_2Cl)$$

(b) At low pressures the rate determining step is the first step, and the reaction becomes second order.

$$NO_2Cl + NO_2Cl \xrightarrow{k_1'} NO_2Cl + NO_2 + Cl$$

$$\frac{-d(NO_2Cl)}{dt} = k_1'(NO_2Cl)^2$$

assuming (NO_2Cl) is approximately constant

$$k_{obs} = k_1'(NO_2Cl)$$

$$1.7 \times 10^{-4} = k_1'(5 \times 10^{-8})$$

$$k_1' = 3.4 \times 10^3 \text{ cm}^3 \text{ mol}^{-1} \text{ s}^{-1}$$

17.22 If a first-order reaction has an activation energy of 104,600 J mol^{-1} and, in the equation $k = Ae^{-E_a/RT}$, A has a value of 5×10^{13} s^{-1}, at what temperature will the reaction have a half-life of (a) 1 min, and (b) 30 days?

SOLUTION

$$k = \frac{0.693}{t_{1/2}} = Ae^{-E_a/RT}$$

$$T = \frac{-E_a}{R\ell n\left(\frac{0.693}{At_{\frac{1}{2}}}\right)}$$

(a) $$T = \frac{-104,600 \text{ J mol}^{-1}}{(8.314 \text{ JK}^{-1}\text{mol}^{-1})\ell n\left[\frac{0.693}{(5\text{x}10^{13} \text{ s}^{-1})(60 \text{ s})}\right]}$$

$= 349$ K or 76 °C

(b) $$T = \frac{-104,600 \text{ J mol}^{-1}}{(8.314 \text{ JK}^{-1}\text{mol}^{-1})\ell n\left[\frac{0.693}{(5\text{x}10^{13})(60)^2(24)(30)}\right]}$$

$= 270$ K or -3 °C

17.23 Isopropenyl allyl ether in the vapor state isomerizes to allyl acetone according to a first-order rate equation. The following equation gives the influence of temperature on the rate constant (in s^{-1}).

$$k = 5.4 \times 10^{11}\, e^{-123,000/RT}$$

where the activation energy is expressed in J mol^{-1}. At 150 °C, how long will it take to build up a partial pressure of 0.395 bar of allyl acetone, starting with 1 bar of isopropenyl allyl ether [L. Stein and G. W. Murphy, J.Am.Chem.Soc., 74, 1041 (1952)]?

SOLUTION

$$k = (5.4 \times 10^{11} \text{ s}^{-1})e^{-\frac{123,000}{(8.314)(423.15)}}$$
$$= 3.53 \times 10^{-4} \text{ s}^{-1}$$
$$t = \frac{1}{k}\, \ell n\, \frac{P°}{P}$$

$$= \frac{1}{3.53 \times 10^{-4}\ s^{-1}} \ln \frac{1\ bar}{0.605}$$

$$= 1420\ s$$

17.24 The hydrolysis of $(CH_2)_6C\langle^{Cl}_{CH_3}$ in 80% ethanol follows the first-order rate equation. The values of the specific reaction-rate constants, as determined by H. C. Brown and M. Borkowski [*J. Am. Chem. Soc.*, *74*, 1896 (1952)], are as follows.

t/°C	0	25	35	45
k/s^{-1}	1.06×10^{-5}	3.19×10^{-4}	9.86×10^{-4}	2.92×10^{-3}

(a) Plot log k against 1/T; (b) calculate the activation energy; (c) calculate the pre-exponential factor.

SOLUTION

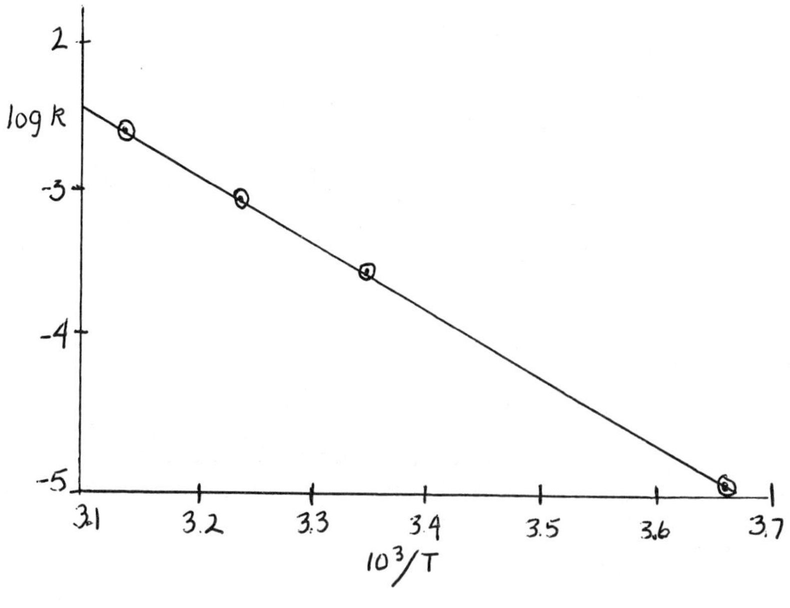

(b) $\text{slope} = \dfrac{-E_a}{2.303\ R} = -4600\ K$

$E_a = (4600\ K)(2.303)\ (8.314\ K^{-1}\ mol^{-1})$

$= 88.1\ kJ\ mol^{-1}$

(c) Taking the 35 °C point and $E_a = 88.1\ kJ\ mol^{-1}$

$$\log k = \frac{-E_a}{2.303\ RT} + \log A$$

$$-3.006 = \frac{-88{,}100\ J\ mol^{-1}}{2.303(8.314\ K^{-1}\ mol^{-1})(308.15\ K)} + \log A$$

$A = 8.4 \times 10^{11}\ s^{-1}$

17.25 (a) The viscosity of water changes about 2% per degree at room temperature. What is the activation energy for this process? (b) the activation energy for a reaction is 62.8 kJ mol^{-1}. Calculate k_{35}/k_{25}.

SOLUTION

(a) $\dfrac{1}{\eta} = Ae^{-E_a/RT}$

$$-\ln\eta = \ln A - \frac{E_a}{RT}$$

$$-\frac{d\eta}{\eta} = \frac{E_a}{RT^2}\,dT$$

$$E_a = -RT^2\,\frac{\Delta\eta/\eta}{\Delta T}$$

$$= -(8.314\ J\ K^{-1}\ mol^{-1})(300\ K)^2(-0.02)/(1/K)$$

$$= 15.1\ kJ\ mol^{-1}$$

(b) $\ln\dfrac{k_{35}}{k_{25}} = \dfrac{\Delta H(T_2 - T_1)}{RT_1T_2}$

$$= \frac{(62{,}800\ J\ mol^{-1})(10\ K)}{(8.314\ J\ K^{-1}\ mol^{-1})(298.15\ K)(308.15\ K)}$$

$$\frac{k_{35}}{k_{25}} = 2.27$$

17.26 For the two parallel reactions $A \xrightarrow{k_1} B$ and $A \xrightarrow{k_2} C$, show that the activation energy E' for the disappearance of A is given in terms of the activation energies E_1 and E_2 for the two paths by

$$E' = \frac{k_1E_1 + k_2E_2}{k_1 + k_2}$$

SOLUTION

The rate equation for A is

$$\frac{-d(A)}{dt} = k_1(A) + k_2(A) = (k_1 + k_2)(A) = k'(A)$$

where $k' = k_1 + k_2 = Ae^{-E'/RT}$

$$\frac{d\ln k'}{dT} = \frac{E'}{RT^2} = \frac{d\ln(k_1+k_2)}{dT} = \frac{d(k_1+k_2)}{(k_1+k_2)dT} = \frac{1}{(k_1+k_2)}\left(\frac{dk_1}{dT} + \frac{dk_2}{dT}\right)$$

$$= \frac{1}{(k_1 + k_2)}\left[k_1\frac{d\ln k_1}{dT} + k_2\frac{d\ln k_2}{dT}\right]$$

$$= \frac{1}{(k_1 + k_2)}\left[\frac{k_1E_1}{RT^2} + \frac{k_2E_2}{RT^2}\right]$$

$$E' = \frac{k_1E_1 + k_2E_2}{k_1 + k_2}$$

17.27 For the mechanism

$$A + B \underset{k_2}{\overset{k_1}{\rightleftharpoons}} C \qquad C \xrightarrow{k_3} D$$

(A) Derive the rate law using the steady-state approximation to eliminate the concentration of C. (b) Assuming that $k_3 << k_2$, express the preexponential factor A and E_a for the apparent second-order rate constant in terms of A_1, A_2, and A_3 and E_{a1}, E_{a2}, and E_{a3} for the three steps.

SOLUTION

(a) $$\frac{d(C)}{dt} = k_1(A)(B) - (k_2 + k_3)(C) = 0$$

$$\frac{d(D)}{dt} = k_3(C) = \frac{k_1\ k_3\ (A)(B)}{k_2 + k_3}$$

(b) For $k_2 >> k_3$,

$$k_{app} = \frac{k_1\ k_3}{k_2} = \frac{A_1 e^{-E_{a1}/RT}\ A_3 e^{-E_{a3}/RT}}{A_2 e^{-E_{a2}/RT}}$$

$$= \frac{A_1\ A_3}{A_2} e^{-(E_{a1} + E_{a3} - E_{a2})/RT}$$

$$A_{app} = \frac{A_1\ A_3}{A_2} \qquad E_{app} = E_{a1} + E_{a3} - E_{a2}$$

17.28 The thermal decomposition of gaseous acetaldehyde is a second-order reaction. The value of E_a is 190,400 J mol^{-1}, and the molecular diameter of the acetaldehyde molecule is 5×10^{-8} cm. (a) Calculate the number of molecules colliding per cm^3 per second at 800 K and 1 bar pressure. (b) Calculate k in L mol^{-1} s^{-1}.

SOLUTION

(a) $$\rho = \frac{PN_A}{RT} = \frac{(10^5\ N\ m^{-2})(6.02 \times 10^{23}\ mol^{-1})}{(8.314\ J\ K^{-1}\ mol^{-1})(800\ K)}$$

$$= 9.05 \times 10^{24}\ m^{-3}$$

$$\langle v \rangle = \left(\frac{\pi RT}{M}\right)^{1/2} = 689\ m\ s^{-1}$$

$$Z_{11} = 2^{-1/2}\rho^2 \pi d^2 \langle v \rangle$$

$$= \frac{(9.05 \times 10^{24}\ m^{-3})^2 \pi (5 \times 10^{-10}\ m)^2 (689)}{2^{1/2}}$$

$$= 3.13 \times 10^{34}\ m^{-3}\ s^{-1}$$

$$= (3.13 \times 10^{34}\ m^{-3}\ s^{-1})(10^{-2}\ m\ cm^{-1})^3$$

$$= 3.13 \times 10^{28}\ cm^{-3}\ s^{-1}$$

(b) $k = \dfrac{(10^3\ L\ m^{-3}) N_A Z_{11}}{\rho^2} e^{-E_0/RT}$

$$= \frac{(10^3\ L\ m^{-3})(6.022\times10^{23}\ mol^{-1})(3.13\times10^{34}\ m^{-3}\ s^{-1})}{(9.17 \times 10^{24}\ m^{-3})} \times e^{-\frac{190{,}400}{(8.314)(800)}}$$

$$= 0.083\ L\ mol^{-1}\ s^{-1}$$

17.29 Use the preexponential factor $A = 3 \times 10^{13}\ cm^3\ mol^{-1}\ s^{-1}$ for the reaction $Br + H_2 \longrightarrow HBr + H$ to calculate the cross section and collision diameter for this reaction at 400 K.

SOLUTION

$$N_A = \frac{1}{\dfrac{1}{79.904 \times 10^{-3}\ kg} + \dfrac{1}{2.0158 \times 10^{-3}\ kg}}$$

$$= 1.966 \times 10^{-3}\ kg\ mol^{-1}$$

$$\left(\frac{8RT}{\pi\mu N_A}\right)^{1/2} = \left[\frac{8(8.314\ J\ K^{-1}\ mol^{-1})(400\ K)}{\pi(1.966 \times 10^{-3}\ kg\ mol^{-1})}\right]^{1/2}$$

$$= 2075\ m\ s^{-1}$$

$$A = \pi d_{12}^2 \left(\frac{8RT}{\pi\mu N_A}\right)^{1/2}$$

$$d_{12} = \left[\frac{A}{\pi\left(\frac{8RT}{\pi\mu N_A}\right)^{1/2}}\right]^{1/2} = \left[\frac{(3\times10^{13}\ cm^3 mol^{-1}s^{-1})(10^{-2}\ m\ cm^{-1})^3}{\pi(6.02\times10^{23}\ mol^{-1})(2075\ m\ s^{-1})}\right]^{1/2}$$

$= 87.4$ pm

The cross section is $\pi d_{12}^2 = 2.40 \times 10^{-20}\ m^2$

17.30 The preexponential factor for the reaction $H_2 + I_2 = 2\ HI$ is 10^{11} L mol^{-1} s^{-1} and the activation energy is 165 kJ mol^{-1} in the range 300 - 500 °C. If the collision diameter is 320 pm, what value of the preexponential factor is expected from collision theory at 600 K and what is the value of the steric factor p? (At higher temperatures this reaction goes by the unbranched chain mechanism described in Section 17.22.)

SOLUTION

$$\mu N_A = \frac{1}{\frac{1}{2(1.0079\times10^{-3}\ kg)} + \frac{1}{2(126.9045\times10^{-3}\ kg}}$$

$$= 1.999 \times 10^{-3}\ kg$$

$$A = \pi d_{12}^2 \left(\frac{8RT}{\pi\mu N_A}\right)^{1/2}$$

$$A = \pi(320 \times 10^{-12}\ m)^2 \left[\frac{8(8.314\ JK^{-1}mol^{-1})(600\ K)}{\pi(1.999 \times 10^{-3}\ kg)}\right]^{1/2} (6.02\times10^{23}\ mol^{-1}) \times (10^3\ L\ m^{-3})$$

$= 4.87 \times 10^{11}$ L mol^{-1} s^{-1}

$A_{exp} = 10^{11}$ L mol^{-1} s^{-1} $= (4.87 \times 10^{11}$ L mol^{-1} s$^{-1})p$

$p = 0.21$

17.31 Show that the activated complex theory yields the simple collision theory result when it is applied to the reaction of two rigid spherical molecules.

SOLUTION

Assuming that the molecules react on their first collision (that is, the activation energy is zero), activated complex theory yields

$$k = \frac{RT}{h}\frac{q''_{AB\ddagger}}{q_A'\, q_B'} \qquad (1)$$

where q_A' and q_B' are molecular partition functions without the volume factor, and $q''_{AB\ddagger}$ is the molecular partition function for the activated complex without the volume factor and without the vibrational factor.

$$q_A' = \left(\frac{2\pi m_A kT}{h^2}\right)^{3/2} \quad (2) \qquad q_B' = \left(\frac{2\pi m_B kT}{h^2}\right)^{3/2} \quad (3)$$

$$q_{AB}'' = \left[\frac{2\pi(m_A + m_B)kT}{h^2}\right]^{3/2} \frac{8\pi^2\mu(R_A + R_B)^2 kT}{h^2} \qquad (4)$$

$$\mu = \frac{m_A\, m_B}{m_A + m_B}$$

Substituting equations 2, 3, 4 and 5 in equation 1 yields

$$k = N_A\left(\frac{8\pi kT}{\mu}\right)^{1/2} (R_A + R_B)^2$$

which may be compared with equation 17.84

$$k = \pi d_{12}^{\;2}\left(\frac{8RT}{\pi\mu N_A}\right)^{1/2}$$

$$= \left(\frac{8\pi kT}{\mu}\right)^{1/2} (R_A + R_B)^2$$

The difference between these expressions of a factor of N_A is simply a matter of units.

17.32 F. W. Schuler and G. W. Murphy [J. Am. Chem. Soc., 72, 3155 (1950)] studied the thermal rearrangement of vinyl allyl ether to allyl acetaldehyde in the range 150 to 200 °C and found that

$$k = 5 \times 10^{11}\ e^{-128,000/RT}$$

where k is in s^{-1} and the activation energy is in J mol^{-1}. Calculate (a) the enthalpy of activation, and (b) the entropy of activation, and (c) give an interpretation of the latter.

SOLUTION

The values of $\Delta H^{\ddagger}$ and $\Delta S^{\ddagger}$ are calculated for the mean temperature of the range 273 + 175 = 448 k.

(a) $\Delta H^{\ddagger} = E_a - RT$

$= 128.0\ \text{kJ mol}^{-1} - (8.314\text{x}10^{-3}\ \text{kJ K}^{-1}\ \text{mol}^{-1})(448\text{K})$

$= 124.3\ \text{kJ mol}^{-1}$

(b) $\Delta S^{\ddagger} = R\left[\ln\dfrac{N_A hA}{RT} - 1\right]$

$$= (8.314\ \text{JK}^{-1}\text{mol}^{-1})\left[\ln\frac{(6.022\text{x}10^{23}\ \text{mol}^{-1})(6.626\text{x}10^{-34}\ \text{Js})\ A}{(8.314\ \text{J K}^{-1}\ \text{mol}^{-1})(448\ \text{K})} - 1\right]$$

$= -32.6\ \text{J K}^{-1}\ \text{mol}^{-1}$ $\qquad A = 5 \times 10^{11}\ s^{-1}$

(c) The activated complex may be an improbable ring structure which is formed with a decrease in entropy.

17.33 The vapor-phase decomposition of di-t-butyl peroxide is first order in the range 110 to 280 °C °C and follows the equation

$$k = 3.2 \times 10^{16}\ e^{-163,600/RT}$$

where k is in s^{-1} and E_a is in J mol^{-1}. Calculate (a) $\Delta H^{\ddagger}$, and (b) $\Delta S^{\ddagger}$.

SOLUTION

(a) $\Delta H^{\ddagger} = E_a - RT$

$= 163{,}600 \text{ J mol}^{-1} - (8.314 \text{ J K}^{-1} \text{ mol}^{-1})(448 \text{ K})$

$= 159{,}900 \text{ J mol}^{-1}$

(b) $\Delta S^{\ddagger} = R \, \ell n \dfrac{N_A hA}{e^{RT}}$

$$= (8.314 \text{ JK}^{-1}\text{mol}^{-1}) \left[\ell n \frac{(6.02\text{x}10^{23} \text{ mol}^{-1})(6.626\text{x}10^{-34} \text{ Js}) \text{ A}}{(2.718)(8.314 \text{ J K}^{-1} \text{ mol}^{-1})(448 \text{ K})} \right]$$

$= 59.4 \text{ J K}^{-1} \text{ mol}^{-1}$ $\qquad A = 3.2 \times 10^{16} \text{ s}^{-1}$

17.34 The apparent activation energy for the recombination of iodine atoms in argon is -5.9 kJ mol^{-1}. This negative temperature coefficient may result from the following mechanism.

$$I + M = IM \qquad K = \frac{(IM)}{(I)(M)} \qquad IM + I \underset{k_{-1}}{\overset{k_1}{\rightleftharpoons}} I_2 + M$$

Assuming that the first step remains at equilibrium, derive the rate equation that includes both the forward and reverse reactions. Show that the reverse reaction is bimolecular and the equilibrium constant expression for the dissociation of iodine is independent of the concentration of the third body.

SOLUTION

$$\frac{d(I_2)}{dt} = k_1(I)(IM) - k_{-1}(I_2)(M)$$

Since $(IM) = K(I)(M)$

$$\frac{d(I_2)}{dt} = k_1K(I)^2(M) - k_1(I_2)(M)$$

Thus the rate law for the reverse reaction is $k_1(I_2)(M)$. At equilibrium $d(I_2)/dt = 0$ and

$$\frac{(I_2)}{(I)^2} = \frac{k_1}{k_{-1}} K$$

17.35 For the gas reaction $O + O_2 + M \underset{k'}{\overset{k}{\rightleftharpoons}} O_3 + M$

When $M = O_2$, Benson and Axworthy [J.Chem.Phys., 26, 1718 (1957)] obtained

$$k = (6.0 \times 10^7 \text{ L}^2 \text{ mol}^{-2} \text{ s}^{-1})\ e^{2.5/RT}$$

where the activation energy is in kJ mol^{-1}. Calculate the values of the parameters in the Arrhenius equation for the reverse reaction.

SOLUTION

$$\Delta H° = 142.7 - 249.170 = -106.5 \text{ kJ mol}^{-1}$$

$$\Delta S° = 238.82 - 160.946 - 205.029 = -127.16 \text{ J K}^{-1} \text{ mol}^{-1}$$

$$K_p = e^{-\Delta H°/RT}\ e^{\Delta S°/R}$$

$$= e^{106,500/RT}\ e^{-127.16/R}$$

$$K_c = K_p\left(\frac{P°}{c°RT}\right)^{\Sigma\nu_i} = K_p\left(\frac{1}{24.46}\right)^{-1} \text{ L mol}^{-1} = \frac{k}{k'}$$

$$k' = \frac{k}{24.46\ K_p}$$

$$k' = \frac{6 \times 10^7\ e^{2500/RT}}{24.46\ e^{106,500/RT}\ e^{-127.16/R}}$$

$$= \left[\frac{6 \times 10^7}{24.46}\ e^{127.16R}\right] e^{(2500 - 106,500)/RT}$$

$$= (1.1\times10^{13}\ \text{L mol}^{-1}\ \text{s}^{-1})\ e^{-104,000/RT}$$

17.36 Derive the steady-state rate equation for the following mechanism for a termolecular reaction.

$$A + A \underset{k_{-1}}{\overset{k_1}{\rightleftharpoons}} A_2^* \qquad A_2^* + M \xrightarrow{k_2} A_2 + M$$

SOLUTION

$$\frac{d(A_2^*)}{dt} = k_1(A)^2 - [k_{-1} + k_2(M)](A_2^*) = 0$$

$$\frac{d(A_2)}{dt} = k_2(A_2^*)(M) = \frac{k_1\ k_2(M)(A)^2}{k_{-1} + k_2(M)}$$

17.37 For the mechanism

$$H_2 + X_2 \underset{k_{-1}}{\overset{k_1}{\rightleftharpoons}} 2HX \qquad X + H_2 \underset{k_{-3}}{\overset{k_3}{\rightleftharpoons}} HX + H$$

$$X_2 \underset{k_{-2}}{\overset{k_2}{\rightleftharpoons}} 2X \qquad H + X_2 \underset{k_{-4}}{\overset{k_4}{\rightleftharpoons}} HX + X$$

show that the steady-state rate law is

$$\frac{d(HX)}{dt} = 2k_1(H_2)(X_2)\left[1 - \frac{(HX)^2}{K(H_2)(X_2)}\right]\left[1 + \frac{\dfrac{k_3}{k_1}\sqrt{\dfrac{2k_2}{k_{-2}(X_2)}}}{1 + \dfrac{k_{-3}(HX)}{k_4(X_2)}}\right]$$

SOLUTION

There are five species, but there are two conservation equations

$$(H) + (HX) + 2(H_2) = \text{const.}$$

$$(X) + (HX) + 2(X_2) = \text{const.}$$

Therefore there are only three independent rate equations

$$\frac{d(HX)}{dt} = 2k_1(H_2)(X_2) + k_3(X)(H_2) + k_4(H)(X_2) - k_{-1}(HX)^2 - k_{-3}(HX)(H) - k_{-4}(HX)(X) \quad (1)$$

$$\frac{d(X)}{dt} = 2k_2(X_2) + k_{-3}(HX)(H) + k_4(H)(X_2) - k_{-2}(X)^2 - k_3(X)(H_2) - k_{-4}(HX)(X) = 0 \quad (2)$$

$$\frac{d(H)}{dt} = k_3(X)(H_2) + k_{-4}(HX)(X) - k_{-3}(HX)(H) - k_4(H)(X_2) = 0 \quad (3)$$

Adding equations 2 and 3

$$(X) = \sqrt{\frac{2k_2}{k_{-2}}(X_2)}$$

Substituting in equation 3 yields

$$(H) = \frac{k_3(H_2) + k_{-4}(HX)}{k_{-3}(HX) + k_4(X_2)} \sqrt{\frac{2k_2(X_2)}{k_{-2}}}$$

Substituting in equation 1

$$\frac{d(HX)}{dt} = 2k_1(H_2)(X_2) - k_{-1}(HX)^2 [k_4(X_2) - k_3(HX)] \text{ x } \frac{[k_3(H_2) + k_{-4}(HX)]\sqrt{\dfrac{2k_2(X_2)}{k_{-2}}}}{k_{-3}(HX) + k_4(X_2)} + [k_3(H_2) - k_{-4}(HX)]\sqrt{\frac{2k_2(X_2)}{k_{-2}}} \quad \frac{[k_{-3}(HX) + k_4(X_2)]}{[k_3(HX) + k_4(X_2)]}$$

$$\frac{d(HX)}{dt} = 2k_1(H_2)(X_2) - 2k_{-1}(HX)^2 + \{k_3k_4(H_2)(X_2)$$

$$+ k_4k_{-4}(X_2)(HX) - k_3k_{-3}(HX)(H_2) - k_{-3}k_{-4}(HX)^2$$

$$+ k_3k_{-3}(H_2)(HX) + k_3k_4(H_2)(X_2) - k_{-3}k_{-4}(HX)^2$$

$$- k_4k_{-4}(HX)(X_2)\} \times \frac{\sqrt{\dfrac{2k_2(X_2)}{k_{-2}}}}{k_{-3}(HX) + k_4(X_2)}$$

$$= 2k_1(H_2)(X_2) - 2k_{-1}(HX)^2 + [2k_3k_4(H_2)(X_2) - 2k_{-3}k_{-4}(HX)^2]$$

$$\times \frac{\sqrt{\dfrac{2k_2(X_2)}{k_{-2}}}}{k_{-3}(HX) + k_4(X_2)}$$

$$= 2k_1(H_2)(X_2)\left[1 - \frac{k_1(HX)^2}{k_1(H_2)(X_2)}\right] + 2k_3k_4(H_2)(X_2)\left[1 - \frac{k_{-3}k_{-4}(HX)^2}{k_3k_4(H_2)(X_2)}\right]$$

$$\times \frac{\sqrt{\dfrac{2k_2(X_2)}{k_{-2}}}}{k_3(HX) + k_4(X_2)}$$

$$\frac{d(HX)}{dt} = 2k_1(H_2)(X_2)\left[1 - \frac{(HX)^2}{K(H_2)(X_2)}\right]\left[1 + \frac{k_3k_4}{k_1}\,\frac{\sqrt{\dfrac{2k_2(X_2)}{k_{-2}}}}{k_{-3}(HX) + k_4(X_2)}\right]$$

$$= 2k_1(H_2)(X_2)\left[1 - \frac{(HX)^2}{K(H_2)(X_2)}\right]\left[1 + \frac{\dfrac{k_3}{k}\sqrt{\dfrac{2k_2}{k_{-2}(X_2)}}}{1 + \dfrac{k_{-3}(HX)}{k_4(X_2)}}\right]$$

17.38 The mechanism of the pyrolysis of acetaldehyde at 520 °C and 0.2 bar is

$$CH_3CHO \xrightarrow{k_1} CH_3 + CHO$$

$$CH_3 + CH_3CHO \xrightarrow{k_2} CH_4 + CH_3CO$$

$$CH_3CO \xrightarrow{k_3} CO + CH_3$$

$$CH_3 + CH_3 \xrightarrow{k_4} C_2H_6$$

What is the rate law for the reaction of acetaldehyde, using the usual assumptions? (As a simplification further reactions of the radical CHO have been omitted and its rate equation may be ignored.)

SOLUTION

$$\frac{d(CH_3CHO)}{dt} = -[k_1 + k_2(CH_3)](CH_3CHO) = 0 \qquad (1)$$

$$\frac{d(CH_3)}{dt} = k_1(CH_3CHO) - k_2(CH_3)(CH_3CHO) + k_3(CH_3CO) - 2k_4(CH_3)^2 = 0 \qquad (2)$$

$$\frac{d(CH_3CO)}{dt} = k_2(CH_3)(CH_3CHO) - k_3(CH_3CO) = 0 \qquad (3)$$

Equation 3 yields

$$(CH_3CO) = k_2(CH_3)(CH_3CHO)/k_3$$

Substituting this in equation 2 yields

$$(CH_3) = \left[\frac{k_1(CH_3CHO)}{2k_4}\right]^{1/2}$$

Substituting this in equation 1 yields

$$\frac{d(CH_3CHO)}{dt} = -\left[k_1 + k_2\left(\frac{k_1}{2k_4}\right)^{1/2}(CH_3CHO)^{1/2}\right](CH_3CHO)$$

If k_1 is small,

$$\frac{d(CH_3CHO)}{dt} = -k_2(k_1/2k_4)^{1/2}(CH_3CHO)^{3/2}$$

17.39 (a) at 27 min $k = 0.0348\ min^{-1}$
at 60 min $k = 0.0347\ min^{-1}$
(b) 20.0 min

17.40 $0.25\ min^{-1}$ 2.77 min 4.00 min

17.41 (a) first (b) $3.59 \times 10^{-5}\ s^{-1}$ (c) 0.354

17.42 $4.03 \times 10^{-9}\ min^{-1}$

17.43 14.6%

17.44 (a) 4.97×10^{-7}% (b) 6950 g

17.45 (a) $0.107\ L\ mol^{-1}\ s^{-1}$ (b) 2850 s

17.46 (a) 221 s (b) 82.5 s

17.47 $40.4\ bar^{-1}\ s^{-1}$

17.48 (a) 80% (b) 67.2% (c) 61.7%

17.49 (a) 3.5% (b) 9%

17.50 (a) 395 s (b) 32,900 ft^3

17.51 $t_{1/2} = \dfrac{2(\sqrt{2} - 1)}{k(A_0)^{1/2}}$

17.53 $-\dfrac{d(H_2SeO_3)}{dt} = K(H_2SeO_3)(H^+)^2(I^-)^3$

17.54 $\frac{d(BrO_3^-)}{dt} = k'(SO_3^{2-})^2(SO_4^{2-})^3(H^+)(Br^-)$

17.55 0.00750 mol L^{-1}

17.57 $(B) = k\ (A)_o te^{-kt}$

$(C) = (A)_o\ [1 - e^{-kt}\ (1 + kt)]$

17.58 $\frac{d(O_2)}{dt} = 2k_1(NO)(O_3)$

17.59 $O_3 \rightleftharpoons O_2 + O$ (in equilibrium)

$O + O_3 \longrightarrow 2\ O_2$

17.60 (a) 97.1 kJ mol^{-1} (b) $9.3 \times 10^{13}\ s^{-1}$

(c) 1.7 s

17.61 (a) 447 °C (b) 388 °C

17.62 (a) 5.3×10^6 s (b) 2740 s

17.63 550 s

17.64 (a) 1091 K (b) 2180 K (c) 727 K

(d) The rate of the reaction with the higher activation energy increases more rapidly with increasing temperature than the rate of the reaction with the lower activation energy.

17.65 $A = 3 \times 10^9\ L^2\ mol^{-2}\ s^{-1}$ $E_a = 5730\ J\ mol^{-1}$

17.66 (a) 1.07 $bar^{-1}\ s^{-1}$ $2.06 \times 10^{-3}\ bar^{-1}\ s^{-1}$

(b) 0.013 s 0.62 s

17.67 $K = (AB)/(A)(B) = \exp\left(\frac{\Delta S°}{R}\right) \exp\left(-\frac{\Delta H°}{RT}\right)$

$$\frac{d(D)}{dt} = k(C)(AB) = k(C)(A)(B)\exp\left(\frac{\Delta S°}{R}\right)\exp\left(-\frac{\Delta H°}{RT}\right)$$

$$= s\ \exp\left(-\frac{E_a}{RT}\right)(C)(A)(B)\exp\left(\frac{\Delta S°}{R}\right)\exp\left(-\frac{\Delta H°}{RT}\right)$$

$$k = A\ \exp\left[\frac{-E_a + \Delta H°}{RT}\right]$$

$$\text{where } A = s\ \exp\left(\frac{\Delta S°}{R}\right)$$

17.68 (a) 2.0×10^{11} L mol^{-1} s^{-1} (b) 3.2×10^{-10} s (c) 2.4×10^{-4} s

17.69 $10^{14.45}$ cm^3 mol^{-1} s^{-1}

17.71 -598, -544 and -523 J K^{-1} mol^{-1} when the standard concentration is taken as 1 mol L^{-1}. The activated complex is more organized (lower entropy) than the reactants. This is especially true for the first reaction.

17.72 (a) 6.25×10^{14} s^{-1} (b) 9×10^{-7} s^{-1}

17.73 (a) 224 kJ mol^{-1} (b) 18.0 J K^{-1} mol^{-1}

17.74 3.3×10^{-3} s

17.75 $$-\frac{d(A)}{dt} = \left(\frac{3}{4} k_1 + \frac{1}{4}\sqrt{k_1^{\,2} + 8\, k_1 k_2 k_3/k_4}\right)(A)$$

CHAPTER 18. Kinetics: Liquid Phase

18.1 Show that if A and B can be represented by spheres of the same radius that react when they touch, the second-order rate constant is given by

$$k_a = \frac{8 \times 10^3 \ RT}{3\eta} \ \text{L mol}^{-1} \ \text{s}^{-1}$$

where R is in J K^{-1} mol^{-1}. To obtain this result the diffusion coefficient is expressed in terms of the radius of a spherical particle by use of equation 20.45. For water at 25 °C, $= 8.95 \times 10^{-4}$ kg m^{-1} s^{-1}. Calculate k at 25 °C.

SOLUTION

The diffusion coefficient for a spherical particle of radius r is given by

$$D = \frac{RT}{N_A 6\pi\eta r}$$

Substituting in equation 18.4

$$k_a = 4\pi(10^3 \ \text{L m}^{-3})N_A(D_1 + D_2)R_{12}$$

$$= 4\pi(10^3 \ \text{L m}^{-3})\left(\frac{2RT}{6\pi\eta r}\right)(2r)$$

$$= \frac{8 \times 10^3 \ RT}{3\eta} \ \text{L mol}^{-1} \ \text{s}^{-1}$$

Where RT is expressed in J mol^{-1} and η is expressed in kg m^{-1} s^{-1} to obtain the indicated units for the second-order rate constant.

For water at 25 °C

$$k_a = \frac{8(10^3 \ \text{L m}^{-3})(8.314 \ \text{J K}^{-1} \ \text{mol}^{-1})(298 \ \text{K})}{3(8.95 \times 10^{-4} \ \text{kg m}^{-1} \ \text{s}^{-1})}$$

$= 7.4 \times 10^9\ L\ mol^{-1}\ s^{-1}$

18.2 The diffusion coefficient D of an ion is related to its ionic mobility u by $D = \frac{uRT}{zF}$

where D = diffusion coefficient in $m^2\ s^{-1}$

u = ion mobility in $m^2\ V^{-1}\ s^{-1}$

R = 8.314 J $K^{-1}\ mol^{-1}$

z = number of charges on ion

F = Faraday constant = 96,500 C mol^{-1}

The ionic mobilities of H^+ and OH^- are $3.63 \times 10^{-7}\ m^2\ V^{-1}\ s^{-1}$ and $2.05 \times 10^{-7}\ m^2\ V^{-1}\ s^{-1}$ at 25 °C. What is the rate constant for the reaction $H^+ + OH^- \longrightarrow H_2O$ The reaction radius is 0.75 nm, because once the proton is this close the reaction can proceed very rapidly by quantum mechanical tunneling. The electrostatic factor f is 1.70.

SOLUTION

$$D_1 = \frac{(3.63\text{x}10^{-7}\ m^2\ V^{-1}\ s^{-1})(8.314\ J\ K^{-1}\ mol^{-1})(298\ K)}{96{,}500\ C\ mol^{-1}}$$

$$= 9.31 \times 10^{-9}\ m^2\ s^{-1}$$

$$D_2 = \frac{(2.05\text{x}10^{-7}\ m^2\ V^{-1}\ s^{-1})(8.314\ J\ K^{-1}\ mol^{-1})(298\ K)}{96{,}500\ C\ mol^{-1}}$$

$$= 5.26 \times 10^{-9}\ m^2\ s^{-1}$$

$$k_a = 4\pi(10^3\ L\ m^{-3})N_A(D_1 + D_2)R_{12}f$$

$$= 4\pi(10^3\ L\ m^{-3})(6.022\text{x}10^{23}\ mol^{-1})(14.57\text{x}10^{-9}\ m^2 s^{-1}) \times (0.75\text{x}10^{-9}\ m)1.7$$

$$= 1.4 \times 10^{11}\ L\ mol^{-1}\ s^{-1}$$

18.3 What is the reaction radius for the reaction

$$H^+ + OH^- \xrightarrow{1.4 \times 10^{11}\ L\ mol^{-1}\ s^{-1}} H_2O$$

at 25 °C given that the diffusion coefficients of H^+ and OH^- at this temperature are 9.1×10^{-9} m^2 s^{-1} and 5.2×10^{-9} m^2 s^{-1}.

SOLUTION

$$k_a = 4\pi N_A(D_1 + D_2)R_{12}f$$

$$R_{12}f = \frac{k_a}{4\pi N_A(D_1 + D_2)}$$

$$= \frac{(1.4 \times 10^{11}\ L\ mol^{-1}\ s^{-1})(10^{-3}\ m^3\ L^{-1})}{4\pi(6.02 \times 10^{23}\ mol^{-1})(14.3 \times 10^{-9}\ m^2\ s^{-1})}$$

$$= 1.29 \times 10^{-9}\ m$$

To calculate the electrostatic factor f, we have to know R_{12}. Successive approximations have to be made. Let us begin by estimating R_{12} = 0.9 nm.

$$\frac{z_1 z_2 e^2}{4\pi \varepsilon_o \kappa kTR_{12}} = \frac{-(1.602 \times 10^{-19}\ C)^2(8.988 \times 10^9\ N\ C^{-2}\ m^2)}{(78.3)(1.38 \times 10^{-23}\ J\ K^{-1})(298\ K)(0.9 \times 10^{-9}\ m)}$$

$$= -0.796 = x$$

$$f = x(e^x - 1)^{-1} = -0.796(e^{-0.796} - 1)^{-1}$$

$$= 1.45$$

$$\text{Thus } R_{12} = \frac{1.29 \times 10^{-9}\ m}{1.45} = 0.89\ nm$$

18.4 Derive the relation between the relaxation time r and the rate constants for the reaction $A \underset{k_2}{\overset{k_1}{\rightleftharpoons}} B$, which is subjected to a small displacement from equilibrium.

SOLUTION

$$\frac{d(B)}{dt} = k_1(A) - k_2(B)$$

At equilibrium

$$\frac{d(B)}{dt} = 0 = k_1(A)_{eq} - k_2(B)_{eq}$$

$$\text{Let } (A) = (A)_{eq} - \Delta(B)$$
$$(B) = (B)_{eq} + \Delta(B)$$

$$\frac{d(B)}{dt} = \frac{d\Delta(B)}{dt} = k_1[(A)_{eq} - \Delta(B)] - k_2[(B)_{eq} + \Delta(B)]$$

$$= k_1(A)_{eq} - k_2(B)_{eq} - (k_1 + k_2)\Delta(B)$$

$$= -(k_1 + k_2)\Delta(B) = -\frac{\Delta(B)}{\tau}$$

$$\tau = (k_1 + k_2)^{-1}$$

18.5 For acetic acid in dilute aqueous solution at 25 °C $K = 1.73 \times 10^{-5}$, and the relaxation time is 8.5×10^{-9} seconds for a 0.1 M solution. Calculate k_a and k_d in

$$CH_3CO_2H \underset{k_a}{\overset{k_d}{\rightleftharpoons}} CH_3CO_2^- + H^+$$

SOLUTION

$$K = \frac{(H^+)(CH_3CO_2^-)}{(CH_3CO_2H)} = \frac{x^2}{0.10} = 1.73 \times 10^{-5} = \frac{k_d}{k_a}$$

$$x = (1.73 \times 10^{-6})^{1/2} = 1.32 \times 10^{-3} = (H^+) = (CH_3CO_2^-)$$

$$\tau = \frac{1}{k_d + k_a[(H^+) + (CH_3CO_2^-)]}$$

$$8.5 \times 10^{-9} \text{ s} = \frac{1}{1.73 \times 10^{-5}\, k_a + k_a(2.64 \times 10^{-3})}$$

$k_a = 4.42 \times 10^{10}$ L mol^{-1} s^{-1}
$k_d = (1.73 \times 10^{-5})k_a = 7.65 \times 10^5$ s^{-1}

18.6 Calculate the first-order rate constants for the dissociation of the following weak acids: acetic acid, acid form of imidazole $C_3N_2H_5^+$, NH_4^+. The corresponding acid dissociation constants are 1.75×10^{-5}, 1.2×10^{-7}, and 5.71×10^{-10}, respectively. The second-order rate constants for the formation of the acid forms from a proton plus the base are 4.5×10^{10}, and 4.3×10^{10} L mol^{-1} s^{-1}, respectively.

SOLUTION

For $HA \underset{k_2}{\overset{k_1}{\rightleftharpoons}} H^+ + A^-$

$$K = \frac{(H^+)(A^-)}{(HA)} = \frac{k_1}{k_2}$$

For acetic acid $k_1 = k_2K$

$= (4.5 \times 10^{10}$ L mol^{-1} s$^{-1})(1.75 \times 10^{-5}$ mol L$^{-1})$

$= 7.9 \times 10^5$ s^{-1}

For imidazole $k_1 = k_2K$

$= (1.5 \times 10^{10}$ L mol^{-1} s$^{-1})(1.2 \times 10^{-7}$ mol L$^{-1})$

$= 1.8 \times 10^3$ s^{-1}

For NH_4^+ $k_1 = k_2K$

$= (4.3 \times 10^{10}$ L mol^{-1} s$^{-1})(5.71 \times 10^{-10}$ mol L$^{-1})$

$= 25$ s^{-1}

18.7 The mutarotation of glucose is first order in glucose concentration and is catalyzed by acids (A) and bases (B). The first-order rate constant may be expressed by an equation of the type that is encountered in reactions with parallel paths.

$$k = k_o + k_{H^+}(H^+) + k_A(A) + k_B(B)$$

where k_o is the first-order rate constant in the absence of acids and bases other than water. The following data were obtained by J. N. Brönsted and E. A. Guggenheim [J.Am.Chem.Soc., 49, 2554 (1927)] at 18 °C in a medium containing 0.02 mol L^{-1} sodium acetate and various concentrations of acetic acid.

(CH_3CO_2H)/mol L^{-1}	0.020	0.105	0.199
$k/10^{-4}$ min^{-1}	1.36	1.40	1.46

Calculate k_o and k_A. The term involving k_{H^+} is negligible under these conditions.

SOLUTION

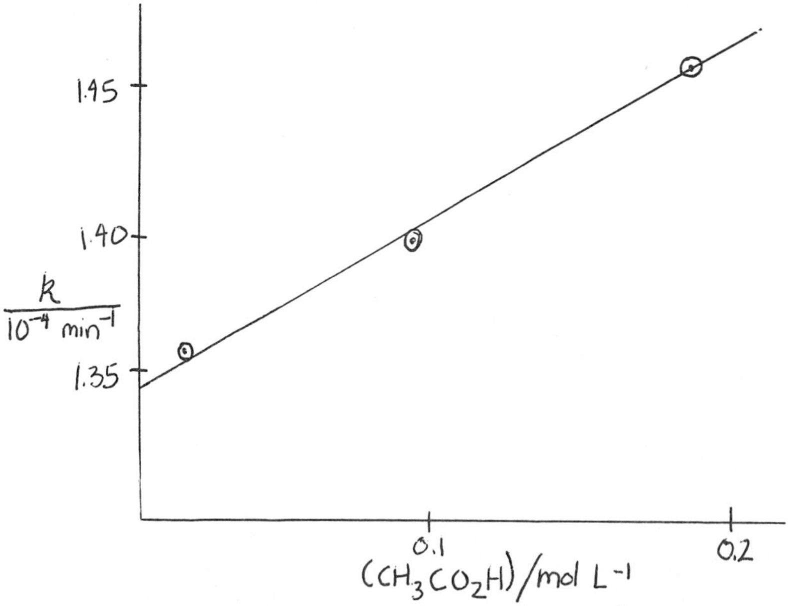

Intercept = k_o = 1.35×10^{-4} min^{-1}
Slope (see p. 299)

$$\text{slope} = \frac{(1.46 - 1.35)\text{x}10^{-4}}{0.2 \text{ mol L}^{-1}} = 5.5\text{x}10^{-5} \text{ L mol}^{-1} \text{ min}^{-1} = k_A$$

18.8 The mutarotation of glucose is catalyzed by acids and bases and is first order in the concentration of glucose. When perchloric acid is used as a catalyst, the concentration of hydrogen ions may be taken to be equal to the concentration of perchloric acid, and the catalysis by perchlorate ion may be ignored since it is such a weak base. The following first-order constants were obtained by J. N. Brönsted and E. A. Guggenheim [J. Am. Chem Soc., 49, 2554 (1927)] at 18 °C.

$(HClO_4)$/ mol L^{-1}	0.0010	0.0048	0.0099	0.0192	0.0300	0.0400
$k/10^{-4}$ min^{-1}	1.25	1.38	1.53	1.90	2.15	2.59

Calculate the values of the constants in the equation $k = k_0 + k_{H^+}(H^+)$.

SOLUTION

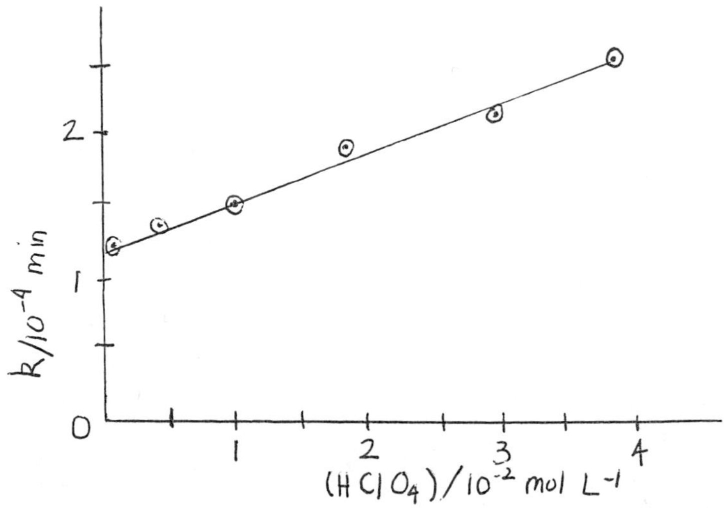

Intercept = k_0 = 1.21×10^{-4} min^{-1}
slope = k_{H+} = 3.4×10^{-3} $L\ mol^{-1}\ min^{-1}$

18.9 Adipic acid and 1,10-decamethylene glycol were mixed in equimolar amounts and polymerized at 161 °C and 191 °C (S.D. Hamann, D. H. Solomon, and J. D. Swift, J.Macromol.Sci.Chem., A2, 153 (1968)). The extents of reaction, as determined by acid titration, are as follows:

t/min	0	200	400	600	800
p(161 °C)	0.820	0.900	0.927	0.940	0.947
p(191 °C)	0.820	0.937	0.955	0.963	0.968

Time was measured from the point at which there was 82% esterification. Taking $(CO_2H)_0$ = 1.25 mol kg^{-1} at both temperatures, what are the two values of k and the activation energy?

SOLUTION

$$\frac{1}{(1-p)^2} - \frac{1}{(1-0.82)^2} = 2(CO_2H)_0^2 kt$$

At 161 °C

$$k = \frac{(1-p)^{-2} - (1-0.82)^{-2}}{2(CO_2H)_0^2 t}$$

= 0.111, 0.125, 0.132, 0.130 for the 4 successive times

= 0.13 $kg^2\ mol^{-2}\ min^{-1}$

At 191 °C

k = 0.36 $kg^2\ mol^{-2}\ min^{-1}$

$$E_a = \frac{RT_1T_2}{T_2-T_1}\ln\frac{k_2}{k_1} = \frac{(8.314)(459)(489)}{30}\ln\frac{0.36}{0.13}$$

= 63 kJ mol^{-1}

18.10 Adipic acid and diethylene glycol were polymerized at 109 °C using p-toluene sulfuric acid as a catalyst (P. J. Flory, J.Am.Chem.Soc., 61, 3334 (1939)). The extents of reaction at various times are as follows:

t/min	0	40	80	120
p	0.800	0.909	0.944	0.960

Note that the time is taken as zero at p = 0.80.

At what time will the number average molar mass reach 10,000 g mol^{-1} for this concentration of reactants and catalyst?

SOLUTION

$$\frac{1}{1 - p} = 1 + k_2(CO_2H)_o t$$

$$\frac{1}{1 - 0.800} = 1 + k_2(CO_2H)_o 0$$

$$\frac{\frac{1}{1 - p} - \frac{1}{1 - 0.800}}{t} = k_2(CO_2H)_o$$

$$k_2(CO_2H)_o = 0.16 \text{ min}^{-1}$$

$$p = 1 - \frac{M_o}{M_n} = 1 - \frac{216 \text{ g mol}^{-1}}{10^4 \text{ g mol}^{-1}} = 0.9784$$

$$\frac{1}{1 - 0.9784} - \frac{1}{1 - 0.800} = (0.16 \text{ min}^{-1})t$$

$$t = 258 \text{ min}$$

18.11 For a condensation polymerization of a hydroxyacid in which 99% of the acid groups are used up, calculate (a) the average number of monomer units in the polymer molecules, (b) the probability that a given molecule will have the number of residues given by this value, and (c) the weight fraction having this particular number of monomer units.

SOLUTION

(a) The number-average degree of polymerization is given by

$$\bar{X}_n = \frac{1}{1 - p} = \frac{1}{1 - 0.99} = 100$$

(b) The probability that a given molecule will have 100 residues is given by

$$\pi_i = p^{i-1}(1 - p)$$

$$\pi_{100} = 0.99^{100-1}(1 - 0.99) = 3.70 \times 10^{-3}$$

(c) The weight fraction having this number of residues is given by

$$W_i = ip^{i-1}(1 - p)^2$$

$$W_{100} = (100)(0.99)^{100-1}(1 - 0.99)^2 = 3.70 \times 10^{-3}$$

18.12 In the condensation polymerization of a hydroxyacid with a residue mass of 200 it is found that 99% of the acid groups are used up. Calculate (a) the number-average molar mass, and (b) the mass-average molar mass.

SOLUTION

(a) $$\bar{X}_n = \frac{1}{1 - p} = \frac{1}{1 - 0.99} = 100$$

$$M_n = (100)(200 \text{ g mol}^{-1}) = 20{,}000 \text{ g mol}^{-1}$$

(b) $$\frac{M_m}{M_n} = 1 + p = 1.99$$

$$M_m = 39{,}800 \text{ g mol}^{-1}$$

18.13 A hydroxyacid $HO-(CH_2)_5-CO_2H$ is polymerized and it is found that the product has a number average molar mass of 20,000 g mol^{-1}. (a) What is the extent of reaction p? (b) What is the degree of polymerization $\overline{X}_n$? (c) What is the mass-average molar mass?

SOLUTION

The molar mass of a monomer unit is 114 g mol^{-1} $= M_o$

(a) $M_n = M_0/(1 - p)$

$p = 1 - M_0/M_n = 1 - 114/20{,}000 = 0.9943$

(b) $$\overline{X}_n = \frac{1}{1 - p} = \frac{1}{1 - 0.9943} = 175$$

(c) $$M_m = M_0\frac{1 + p}{1 - p} = 114\frac{1.9943}{1 - 0.9943}$$

$= 39{,}900$ g mol^{-1}

18.14 Suppose an enzyme has a turnover number of 10^4 min^{-1} and a molar mass of 60,000 g mol^{-1}. How many moles of substrate can be turned over per hour per gram of enzyme if the substrate concentration is twice the Michaelis constant? It is assumed that the substrate concentration is maintained constant by a preceding enzymatic reaction and that products do not accumulate and inhibit the reaction.

SOLUTION

$$v = \frac{V}{1 + K/(S)} = \frac{(10^4\ min^{-1})(E)_0}{1 + 0.5}$$

$$= \frac{(10^4\ min^{-1})(1\ g/60{,}000\ g\ mol^{-1})(60\ min\ hr^{-1})}{1.5}$$

$= 6.7$ mol hr^{-1}

18.15 The kinetics of the fumarase reaction

fumarate + H_2O = L-malate is studied at 25 °C using an 0.01 ionic strength buffer of pH 7. The rate of the reaction is obtained using a recording ultraviolet spectrometer to measure the fumarate concentration. The following rates of the forward reaction are obtained using a fumarase concentration of 5×10^{-10} mol L^{-1}.

$(F)/10^{-6}$ mol L^{-1}	$v_F/10^{-7}$ mol L^{-1} s^{-1}
2	2.2
40	5.9

The following rates of the reverse reaction are obtained using a fumarase concentration of 5×10^{-10} mol L^{-1}.

$(M)/10^{-6}$ mol L^{-1}	$v_M/10^{-7}$ mol L^{-1} s^{-1}
5	1.3
100	3.6

(a) Calculate the Michaelis constants and turnover numbers for the two substrates. In practice many more concentrations would be studied.
(b) Calculate the four rate constants in the mechanism

$$E + F \underset{k_{-1}}{\overset{k_1}{\rightleftharpoons}} X \underset{k_{-2}}{\overset{k_2}{\rightleftharpoons}} E + M$$

where E represents the catalytic site. There are four catalytic sites per fumarase molecule.
(c) Calculate K_{eq} for the reaction catalyzed. The concentration of H_2O is omitted in the expression for the equilibrium constant because its concentration cannot be varied in dilute aqueous solutions.

SOLUTION

$$v_F = \frac{V_F}{1 + K_F/(F)}$$

$$v_F + \frac{v_F}{(F)} K_F = V_F$$

$$2.2 \times 10^{-7} + 0.110\ K_F = V_F$$

$$5.9 \times 10^{-7} + 0.0148\ K_F = V_F$$

$$-3.7 \times 10^{-7} + 0.0952\ K_F = 0$$

$$K_F = \frac{3.7 \times 10^{-7}}{0.0952} = 3.9 \times 10^{-6} \text{ mol L}^{-1}$$

$$v_F = 2.2\text{x}10^{-7} + (0.110)(3.9\text{x}10^{-6}) = 6.5\text{x}10^{-7} \text{ mol L}^{-1} \text{ s}^{-1}$$

$$v_M = \frac{V_M}{1 + K_M/(M)}$$

$$v_M + \frac{v_M}{(M)} K_M = V_M$$

$$1.3 \times 10^{-7} + 0.0260\ K_M = V_M$$

$$3.6 \times 10^{-7} + 0.0036\ K_M = V_M$$

$$-2.3 \times 10^{-7} + 0.0224\ K_M = 0$$

$$K_M = \frac{2.3 \times 10^{-7}}{0.0224} = 1.03 \times 10^{-5} \text{ mol L}^{-1}$$

$$v_M = 1.3\text{x}10^{-7} + 0.026(1.03\text{x}10^{-5}) = 4.0\text{x}10^{-7} \text{ mol L}^{-1} \text{ s}^{-1}$$

The rate constants should be expressed in terms of the concentration of enzymatic sites and so

$$k_2 = \frac{V_F}{(E)_o} = \frac{6.5 \times 10^{-7} \text{ mol L}^{-1} \text{ s}^{-1}}{4(5 \times 10^{-10} \text{ mol L}^{-1})} = 3.3 \times 10^2 \text{ s}^{-1}$$

$$k_{-1} = \frac{V_M}{(E)_o} = \frac{4.0 \times 10^{-7} \text{ mol L}^{-1} \text{ s}^{-1}}{4(5 \times 10^{-10} \text{mol L}^{-1})} = 2.0 \times 10^2 \text{ s}^{-1}$$

(b) $$K_F = \frac{k_2 + k_1}{k_1}$$

$$k_1 = \frac{k_2 + k_{-1}}{K_F} = \frac{(3.3 + 2.0) \times 10^2 \ s^{-1}}{3.9 \times 10^{-6} \ mol \ L^{-1}}$$

$$= 1.4 \times 10^8 \ L \ mol^{-1} \ s^{-1}$$

$$K_M = \frac{k_2 + k_1}{k_{-2}}$$

$$k_{-2} = \frac{k_2 + k_{-1}}{K_M} = \frac{(3.3 + 2.0) \times 10^2 \ s^{-1}}{1.03 \times 10^{-5} \ mol \ L^{-1}}$$

$$= 5.1 \times 10^7 \ L \ mol^{-1} \ s^{-1}$$

(c) $$K = \frac{(M)_{eq}}{(F)_{eq}} = \frac{V_F \ K_M}{V_M \ K_F}$$

$$= \frac{(6.5 \times 10^{-7})(1.03 \times 10^{-5})}{(4.0 \times 10^{-7})(3.9 \times 10^{-6})} = 4.3$$

$$K = \frac{k_1 \ k_2}{k_{-1} \ k_{-2}}$$

$$= \frac{(1.4 \times 10^8)(3.3 \times 10^2)}{(2.0 \times 10^2)(5.1 \times 10^7)} = 4.5$$

18.16 At pH 7 the measured Michaelis constant and maximum velocity for the enzymatic conversion of fumarate to L-malate

fumarate + H_2O = L-malate

are 4.0×10^{-6} mol L^{-1} and $(1.3 \times 10^3 \ s^{-1})(E)_o$, where $(E)_o$ is the total molar concentration of the enzyme. The Michaelis constant and maximum velocity for the reverse reaction are 10×10^{-6} mol L^{-1} and $(0.80 \times 10^3 \ s^{-1})(E)_o$. What is the equilibrium constant for this hydration reaction? (The activity of water is set equal to unity since it is in excess.)

SOLUTION

$$K = \frac{(M)_{eq}}{(F)_{eq}} = \frac{V_F\ K_M}{V_M\ K_F}$$

$$= \frac{(1.3 \times 10^3\ s^{-1})(E)_o(10 \times 10^{-6}\ mol\ L^{-1})}{(0.8 \times 10^3\ s^{-1})(E)_o(4.0 \times 10^{-6}\ mol\ L^{-1})}$$

$$= 4.1$$

18.17 Derive the steady-state rate equation for the mechanism

$$E + S \underset{k_2}{\overset{k_1}{\rightleftharpoons}} X \xrightarrow{k_3} E + P \qquad E + I \underset{k_5}{\overset{k_4}{\rightleftharpoons}} EI$$

for the case that $(S) >> (E)_o$ and $(I) >> (E)_o$.

SOLUTION

$$(E)_o = (E) + (EI) + (X)$$

$$= (E)[1 + (I)/K_I] + (X) \qquad (1)$$

since

$$K_I = \frac{(E)(I)}{(EI)} = \frac{k_5}{k_4}$$

Assuming that X is in a steady state

$$\frac{d(X)}{dt} = k_1(E)(S) - (k_2 + k_3)(X) = 0 \qquad (2)$$

Solving equation 1 for (E) and substituting this expression in equation 2 yields

$$\frac{k_1\ (S)(E)_o}{1 + (I)/K_I} = (X)\left[\frac{k_1\ (S)}{1 + (I)/K_I} + k_2 + k_3\right]$$

$$\frac{d(P)}{dt} = k_3(X) = \frac{k_3\ (E)_o}{1 + \dfrac{k_2 + k_3}{k_1\ (S)}\left[1 + \dfrac{(I)}{K_I}\right]}$$

$$= \frac{V_S}{1 + \frac{K_S}{(S)}\left[1 + \frac{(I)}{K_I}\right]}$$

18.18 The following initial velocities were determined spectrophotometrically for solutions of sodium succinate to which a constant amount of succinoxidase was added. The velocities are given as the change in absorbancy at 250 nm in 10 s. Calculate V, K_S, and K_I for malonate.

	$\frac{A \times 10^3}{10\ s}$	
(Succinate) 10^{-3} mol L^{-1}	No Inhibitor	15×10^{-6} mol L^{-1} malonate
10	16.7	14.9
2	14.2	10.0
1	11.3	7.7
0.5	8.8	4.9
0.33	7.1	—

SOLUTION

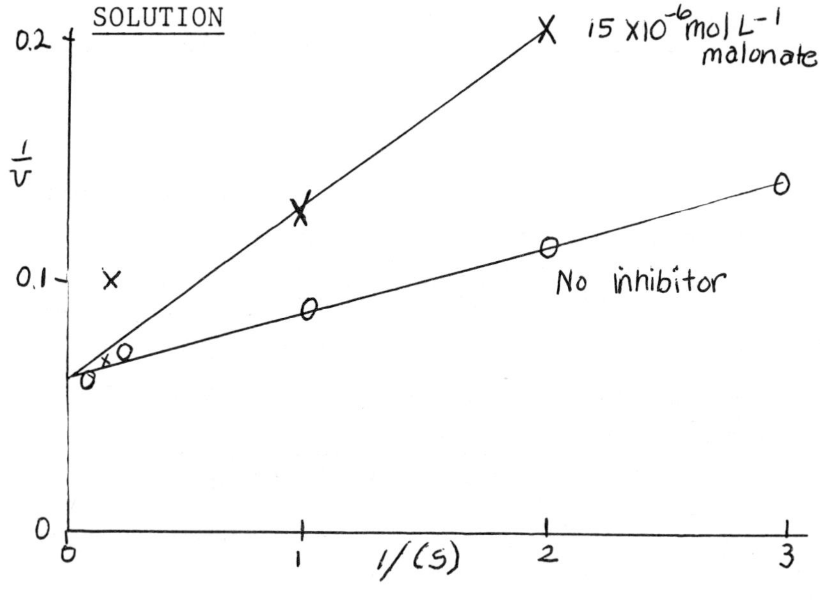

$$v = \frac{d(P)}{dt} = \frac{V_s}{1 + \dfrac{K_s}{(S)}\left[1 + \dfrac{(I)}{K_I}\right]} \tag{18.85}$$

$$\frac{1}{v} = \frac{1}{V_s} + \frac{K_s}{V_s(S)}\left[1 + \frac{(I)}{K_I}\right]$$

$$\frac{1}{V_s} = 0.0662 \qquad V_s = 15.1$$

Slope without inhibitor

$$= \frac{0.141 - 0.066}{3 \times 10^3 \text{ L mol}^{-1}} = 25 \times 10^{-6} = \frac{K_s}{V_s}$$

$$K_s = (25 \times 10^{-6})(15.1) = 0.38 \times 10^{-3} \text{ mol L}^{-1}$$

Slope with inhibitor

$$= \frac{0.204 - 0.066}{2 \times 10^3 \text{ L mol}^{-1}} = 0.069 = \frac{K_s}{V_s}\left[1 + \frac{(I)}{K_I}\right]$$

$$0.069 = \frac{0.38}{15.1}\left[1 + \frac{15 \times 10^{-6} \text{ mol L}^{-1}}{K_I}\right]$$

$$K_I = \frac{15 \times 10^{-6} \text{ mol L}^{-1}}{\dfrac{(15.1)(0.069)}{0.38} - 1} = 8.6 \times 10^{-6} \text{ mol L}^{-1}$$

18.19 The maximum initial velocities for an enzymatic reaction are determined at a series of pH values.

pH	6.0	6.5	7.0	7.5	8.0	8.5	9.0
V	11	30	74	129	147	108	53

Calculate the values of the parameters V', K_a, K_b in

$$V = \frac{V'}{1 + (H^+)/K_a + K_b/(H^+)}$$

Hint. A plot of V versus pH may be constructed

and the hydrogen ion concentration at the midpoint on the acid side referred to as $(H^+)_a$ and the hydrogen ion concentration at the midpoint on the basic side is referred to as $(H^+)_b$. Then

$$K_a = (H^+)_a + (H^+)_b - 4\sqrt{(H^+)_a(H^+)_b}$$

$$K_b = \frac{(H^+)_a(H^+)_b}{K_a}$$

<u>SOLUTION</u>

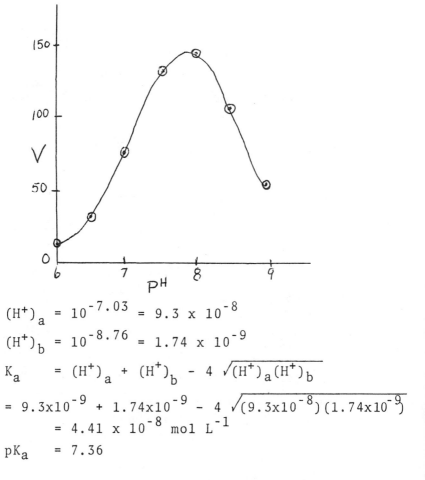

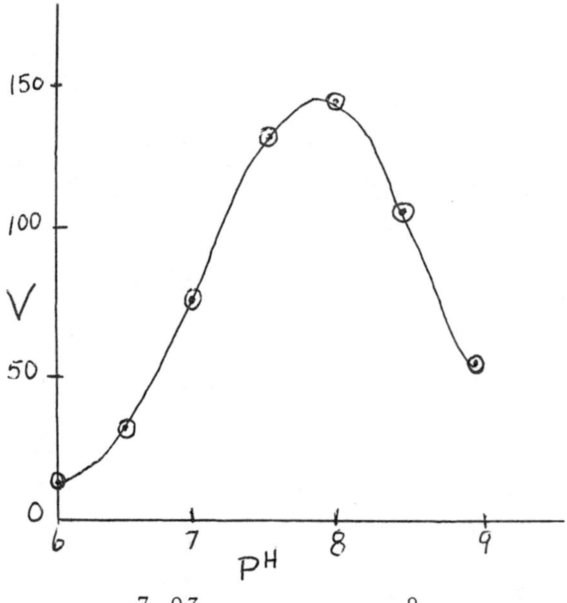

$$(H^+)_a = 10^{-7.03} = 9.3 \times 10^{-8}$$

$$(H^+)_b = 10^{-8.76} = 1.74 \times 10^{-9}$$

$$K_a = (H^+)_a + (H^+)_b - 4\sqrt{(H^+)_a(H^+)_b}$$

$$= 9.3\text{x}10^{-9} + 1.74\text{x}10^{-9} - 4\sqrt{(9.3\text{x}10^{-8})(1.74\text{x}10^{-9})}$$

$$= 4.41 \times 10^{-8} \text{ mol L}^{-1}$$

$$pK_a = 7.36$$

$$K_b = \frac{(H^+)_a(H^+)_b}{K_a} = \frac{(9.3 \times 10^{-8})(1.74 \times 10^{-9})}{4.41 \times 10^{-8}}$$

$$= 3.67 \times 10^{-9} \text{ mol L}^{-1}$$

$$pK_b = 8.45$$

$$V = \frac{V'}{1 + (H^+)/K_a + K_b/(H^+)}$$

$$147 = \frac{V'}{1 + (10^{-8})/(4.41\times10^{-8}) + (0.367\times10^{-8})/(10^{-8})}$$

$$V' = 234$$

18.20 Alcohol dehydrogenase catalyzes the reaction

$$\text{alcohol} + NAD^+ = \text{aldehyde} + NADH + H^+$$

where NAD^+ is nicotinamide adenine dinucleotide in the oxidized form and NADH in the reduced form. For enzyme isolated from yeast the maximum velocity for ethyl alcohol is given by

$$V(\text{mol L}^{-1}\text{ min}^{-1}) = 2.7 \times 10^4 (E)_o$$

where $(E)_o$ is the molar concentration of enzyme of molar mass 150,000 g mol^{-1}. The reaction is conveniently followed spectrophotometrically at 340 nm because NADH has a molar absorbancy coefficient of 620 (mol L^{-1})$^{-1}$ cm^{-1}.

$$\text{Absorbancy} = \log \frac{I_o}{I} = 6200(NADH)d$$

where d is the thickness of the spectrophotometer cuvette. How many grams of enzyme must be placed in a 3-mL reaction mixture to produce a rate of change of absorbancy of 0.001 per minute when a 1.0-cm cuvette is used?

SOLUTION

$$\frac{d(NADH)}{dt} = V = (2.7 \times 10^4\ min^{-1})(E)_o$$

$$\frac{dA}{dt} = [6200(mol\ L^{-1})^{-1}\ cm^{-1}](1\ cm)\ \frac{d(NADH)}{dt}$$

$$10^{-3}\ min^{-1} = [6200(mol\ L^{-1})^{-1}](2.7 \times 10^4\ min^{-1})(E)_o$$

$$= [6200(mol\ L^{-1})](2.7x10^4\ min^{-1})\frac{g}{(0.003\ L)(150,000\ g\ mol^{-1})}$$

$$g = 2.7 \times 10^{-9}\ g \text{ of enzyme in 3 mL}$$

18.21 What current density i will be obtained for the evolution of hydrogen gas on (a) Pt, and (b) Fe from 1 mol L^{-1} HCl at 25 °C at an overpotential of 0.1 volt?

SOLUTION

(a) $i = i_o\ e^{E_\eta/2.303a}$

$= (10^{-2.6}\ A\ cm^{-2})\ e^{0.1/(2.303)(0.028)}$

$= 0.012\ A\ cm^{-2}$

(b) $i = (10^{-6}\ A\ cm^{-2})\ e^{0.1/(2.303)(0.130)}$

$= 1.4 \times 10^{-6}\ A\ cm^{-2}$

18.22 0.6 nm

18.23 2.7×10^{-4} s

18.24 8.5 ns

18.25 $k_1 = 1.8 \times 10^{10}\ L\ mol^{-1}\ s^{-1}$

$k_{-1} = 1.1 \times 10^3\ s^{-1}$

18.26 $1/\tau = k_1[(A)_{eq} + (B)_{eq}] + k_2[(C)_{eq} + (D)_{eq}]$

18.27 $1/\tau = k_1 + k_1'\,K_A + k_{-1} + k_{-1}'\,K_B$

18.29 H^+ production: $k = 1.75\times10^5,\ 1.2\times10^2,\ 5.71\ s^{-1}$
OH^- production: $k = 5.71,\ 0.83\times10^3,\ 1.8\times10^5\ s^{-1}$
Imidazole can play both roles about equally well.

18.30 $K_1 = (HOCl)(OH^-)/(OCl^-)$

$$\frac{d(Cl^-)}{dt} = k(I^-)(HOCl) = k(I^-)K_1(OCl^-)/(OH^-)$$

18.31 0.57 kg $mol^{-1}\ min^{-1}$

18.32 (a) 20 (b) 0.0189 (c) 0.0189

18.33 20 M_o, 39.0 M_o

18.34 179 hr

18.35 $V = 1.2 \times 10^{-6}\ mol\ L^{-1}\ s^{-1}$
$K_s = 0.48 \times 10^{-3}\ mol\ L^{-1}$

18.36 (a) $k_1 = 1.3 \times 10^8\ L\ mol^{-1}\ s^{-1}$
$k_{-1} = 0.2 \times 10^3\ s^{-1}$
$k_2 = 0.33 \times 10^3\ s^{-1}$
$k_{-2} = 5.3 \times 10^7\ L\ mol^{-1}\ s^{-1}$
(b) -3.43 kJ mol^{-1}

18.37 $k_1 = 4.9 \times 10^8\ L\ mol^{-1}\ s^{-1}$
$k_{-1} = 2.6 \times 10^3\ s^{-1}$
$k_2 = 0.8 \times 10^3\ s^{-1}$
$k_{-2} = 3.4 \times 10^7\ L\ mol^{-1}\ s^{-1}$

18.38
$$\frac{d(P_2)}{dt} = \frac{[k_1\,k_2(S) - k_{-1}\,k_{-2}(P_1)(P_2)](E)_o}{k_1\,(S) + k_{-1}\,(P_1) + k_2 + k_{-2}\,(P_2)}$$

$$v_f = \frac{k_2\,(E)_o}{1 + \dfrac{k_2}{k_1\,(S)}}$$

$$v_r = \frac{k_{-1}(P_1)(E)_o}{1 + \dfrac{k_2}{k_{-2}(P_2)} + \dfrac{k_{-1}(P_1)}{k_{-2}(P_2)}}$$

Thus all four rate constants can be determined.

18.39 $V = k_2k_3(E)_0/(k_1 + k_3)$ $K_s = k_2/(k_1 + k_3)$

Since no enzyme-substrate complex is formed, this mechanism does not account for the selectivity of enzyme-catalyzed reactions.

18.40 The product completely inhibits the enzyme, putting it out of action before all the S is used up. Since the binding of product is reversible, the reaction can be restarted by adding more substrate and displacing the product.

18.41 33%

18.42 $V_s = k_2 (E)_0 / [1 + (H^+)/K_{EHS}]$

$$K_s = \frac{k_{-1} + k_{-2}}{k_1} \frac{[1 + (H^+)/K_{EH}]}{[1 + (H^+)/K_{EHS}]}$$

18.43 $$v = \frac{k_2(alc)(E)_0}{1 + \dfrac{k\ (alc)}{k_1(NAD^+)}}$$

18.44 (a) 0.045, (b) 1.31 V

18.45 (a) 78.2 kJ mol^{-1} (b) 73.6 kJ mol^{-1}

(c) 14.1 s^{-1} (d) 0.929 s^{-1}

CHAPTER 19. Photochemistry

19.1 A certain photochemical reaction requires an excitation energy of 126 kJ mol^{-1}. To what values does this correspond in the following units: (a) frequency of light, (b) wave number, (c) wavelength in nanometers, and (d) electron volts?

SOLUTION

(a) $$\nu = \frac{E}{h} = \frac{1.26 \times 10^{5}\ J\ mol^{-1}}{(6.63 x 10^{-34}\ J\ s)(6.02 x 10^{23}\ mol^{-1})}$$

$$= 3.16 \times 10^{14}\ s^{-1}$$

(b) $$\tilde{\nu} = \frac{\nu}{c} = \frac{(3.16 \times 10^{14}\ s^{-1})(10^{-2}\ m\ cm^{-1})}{2.998 \times 10^{8}\ ms^{-1}}$$

$$= 10{,}500\ cm^{-1}$$

(c) $$\lambda = \frac{1}{\tilde{\nu}} \quad \frac{1}{10{,}500\ cm^{-1}} = 9.52 \times 10^{-5}\ cm = 925\ nm$$

(d) $$\frac{126\ kJ\ mol^{-1}}{96{,}485\ C\ mol^{-1}} = 1.31\ eV$$

19.2 How many moles of photons does a 560 nm laser with an intensity of 0.1 watt at this wavelength produce in one hour?

SOLUTION

$$\frac{E\lambda}{N_A hc} = \frac{(0.1\ J\ s^{-1})(60 \times 60\ s)(560 \times 10^{-9}\ m)}{(6.022 x 10^{23}\ mol^{-1})(6.626 x 10^{-34}\ J\ s)(2.998 x 10^{8}\ ms^{-1})}$$

$$= 1.7 \times 10^{-3}\ mol$$

19.3 A sample of gaseous acetone is irradiated with monochromatic light having a wavelength of 313 nm. Light of this wavelength decomposes the acetone according to the equation

$$(CH_3)_2CO \longrightarrow C_2H_6 + CO$$

The reaction cell used has a volume of 59 cm^3. The acetone vapor absorbs 91.5% of the incident energy. During the experiment the following data are obtained.

Temperature of reaction	= 56.7 °C
Initial pressure	= 102.16 kPa
Final pressure	= 104.42 kPa
Time of radiation	= 7 hr
Incident energy	= 48.1×10^{-4} J s^{-1}

What is the quantum yield?

SOLUTION

Since PV = nRT, $V\Delta P = RT\Delta n$

$$\Delta n = \frac{V\Delta P}{RT} = \frac{(59\times10^{-6}\ m^{-3})[(104.42 - 102.16)\times10^{3}\ Pa]}{(8.314\ J\ K^{-1}\ mol^{-1})(329.85\ K)}$$

$$= 4.86 \times 10^{-5}\ mol$$

$$= (4.86 \times 10^{-5})(6.02 \times 10^{23}\ mol^{-1})$$

$$= 2.93 \times 10^{19}\ \text{molecules reacting}$$

$$E = \frac{hc}{\lambda} = \frac{(6.626 \times 10^{-34} J\ s)(2.998 \times 10^{8}\ m\ s^{-1})}{313 \times 10^{-9}\ m}$$

$$= 6.36 \times 10^{-19}\ J$$

Number of quanta absorbed

$$= \frac{(48.1 \times 10^{-4}\ J\ s^{-1})(7 \times 60 \times 60\ s)(0.915)}{6.36 \times 10^{-19}\ J}$$

$$= 1.75 \times 10^{20}$$

$$\phi = \frac{2.93 \times 10^{19}}{1.75 \times 10^{20}} = 0.167$$

19.4 A 100-cm^3 vessel containing hydrogen and chlorine was irradiated with light of 400 nm. Measurements with a thermopile showed that 11 x 10^{-7} J of light energy was absorbed by the chlorine per second. During an irradiation of 1 min the partial pressure of chlorine, as determined by the absorption of light and the application of Beer's law, decreased from 27.3 to 20.8 kPa (corrected to 0 °C). What is the quantum yield?

SOLUTION

$$E = \frac{hc}{\lambda} = \frac{(6.626 \times 10^{-34}\ \text{J s})(2.998 \times 10^{8}\ \text{m s}^{-1})}{400 \times 10^{-9}\ \text{m}}$$

$$= 4.97 \times 10^{-19}\ \text{J}$$

Number of quanta absorbed

$$= \frac{(11 \times 10^{-7}\ \text{J s}^{-1})(60\ \text{s})}{4.97 \times 10^{-19}\ \text{J}} = 1.33 \times 10^{14}$$

Since $PV = nRT$, $V\Delta P = RT\Delta n$

$$\Delta n = \frac{V\Delta P}{RT}$$

$$= \frac{(0.1 \times 10^{-3}\ \text{m}^3)[(27.3 - 20.8) \times 10^3\ \text{Pa}](6.02 \times 10^{23}\ \text{mol}^{-1})}{(8.314\ \text{J K}^{-1}\ \text{mol}^{-1})(273\ \text{K})}$$

$$= 1.72 \times 10^{20}$$

Since $H_2 + Cl_2 = 2\ HCl$

$$\phi = \frac{1.72 \times 10^{20}}{1.33 \times 10^{14}} \times 2 = 2.6 \times 10^{6}$$

19.5 Discuss the economic possibilities of using photochemical reactions to produce valuable products with electricity at 5 cents per kilowatt-hour. Assume that 5% of the electric energy consumed by a quartz-mercury-vapor lamp goes into light, and 30% of this is photochemically effective. (a) How much will it cost to produce 1 lb (453.6 g) of an organic compound having a molar mass of 100 g mol^{-1}, if the average effective wavelength is assumed to be 400 nm and the reaction has a quantum yield of 0.8 molecule per photon? (b) How much will it cost if the reaction involves a chain reaction with a quantum yield of 100?

SOLUTION

(a) Number of quanta required

$$= \frac{(453.6 \text{ g})(6.022 \times 10^{23} \text{ mol}^{-1})}{(100 \text{ g mol}^{-1})(0.8)} = 3.41 \times 10^{24}$$

$$E = \frac{hc}{\lambda} = \frac{(6.26 \times 10^{-34} \text{ J s})(2.998 \times 10^{8} \text{ m s}^{-1})}{4000 \times 10^{-10} \text{ m}}$$

$$= 4.97 \times 10^{-19} \text{ J}$$

$$\text{Energy required} = \frac{(3.41 \times 10^{24})(4.97 \times 10^{-19} \text{ J})}{(0.05)(0.3)}$$

$$= 1.13 \times 10^{8} \text{ J}$$

$$\text{Cost} = \frac{(1.13 \times 10^{8} \text{ J})(0.05\$/\text{KW hr})}{36 \times 10^{5} \text{ J/KW hr}} = \$1.55$$

$$1 \text{ KW hr} = (10^{3} \text{ J s}^{-1})(60 \times 60 \text{ s})$$

$$= 36 \times 10^{5} \text{ J/KW hr}$$

(b) $\$1.55 \frac{0.8}{100} = 1.2¢$

19.6 For 900 s, light of 436 nm was passed into a carbon tetrachloride solution containing bromine and cinnamic acid. The average power absorbed was 19.2×10^{-4} J s^{-1}. Some of the bromine reacted to give cinnamic acid dibromide, and in this experiment the total bromine content decreased by 3.83×10^{19} molecules. (a) What was the quantum yield? (b) State whether or not a chain reaction was involved. (c) If a chain mechanism was involved, suggest suitable reactions that might explain the observed quantum yield.

SOLUTION

(a) $$E = \frac{hc}{\lambda} = \frac{(6.63 \times 10^{-34}\ \text{J s})(3 \times 10^{8}\ \text{m s}^{-1})}{(436 \times 10^{-9}\ \text{m})}$$

$$= 4.56 \times 10^{-19}\ \text{J}$$

$$\frac{19.2 \times 10^{4}\ \text{J s}^{-1}}{4.56 \times 10^{-19}\ \text{J}} = 4.21 \times 10^{15}\ \text{s}^{-1}$$, rate quanta are absorbed

$$(4.21 \times 10^{15}\ \text{s}^{-1})(900\ \text{s}) = 3.79 \times 10^{18}\ \text{quanta}$$

$$\phi = \frac{3.83 \times 10^{19}}{3.79 \times 10^{18}} = 10.1$$

(b) Since $\phi > 1$, a chain reaction is involved.

(c) $\phi CH = CHCO_2H + h\nu = \phi CH = CHCO_2H^*$

$\phi CH = CHCO_2H^* + Br_2 = \phi CHBrCHCO_2H^* + Br^*$

$\phi CHBrCHCO_2H^* + Br_2 = \phi CHBrCHBrCO_2H + Br^*$

$\phi CH = CHCO_2H + Br^* = CHBrCHCO_2H^*$

19.7 The quantum yield is 2 for the photolysis of gaseous HI to $H_2 + I_2$ by light of 253.7 nm wavelength. Calculate the number of moles of HI that will be decomposed if 300 J of light of this wavelength is absorbed.

SOLUTION

$$E = \frac{hc}{\lambda} = \frac{(6.626 \times 10^{-34}\ \text{J s})(3 \times 10^{8}\ \text{m s}^{-1})}{(253.7 \times 10^{-9}\ \text{m})} = 7.84 \times 10^{-19}\ \text{J}$$

$$n = \frac{(2)(300\ J)}{(7.84 \times 10^{-19}\ J)(6.02 \times 10^{23}\ mol^{-1})} = 1.27 \times 10^{-3}\ \text{mol of HI}$$

19.8 The fluorescence quantum yield for benzene at 25 °C is 0.070. The lifetime of the excited state is 26 ns. What is the radiative lifetime τ_o?

SOLUTION

$$\tau_o = \frac{\tau_s}{\phi_F} = \frac{26 \times 10^{-9}\ s}{0.070} = 370\ ns$$

19.9 A solution of a dye is irradiated with 400 nm light to produce a steady concentration of triplet state molecules. If the triplet state yield is 0.9, and the triplet state lifetime is 20×10^{-6} s, what light intensity, expressed in watts, is required to maintain a steady triplet concentration of 5×10^{-6} mol L^{-1} in a liter of solution. Assume that all of the light is absorbed.

SOLUTION

$$\frac{d(T_1)}{dt} = 0.9\ I - \frac{1}{20 \times 10^{-6}}(T_1) = 0$$

$$I = \frac{(5 \times 10^{-6}\ mol\ L^{-1})(1\ L)}{(20 \times 10^{-6}\ s)(0.9)} = 0.278\ mol\ s^{-1}$$

$$E = \frac{N_A hc}{\lambda} = \frac{(6.022 \times 10^{23}\ mol^{-1})(6.62 \times 10^{-34}\ J\ s)(3 \times 10^{8}\ m\ s^{-1})}{400 \times 10^{-9}\ m}$$

$$= 2.99 \times 10^{5}\ J\ mol^{-1}$$

$$I = (0.278\ mol\ s^{-1})(2.99 \times 10^{5}\ J\ mol^{-1}) = 83\ kW$$

19.10 Ketone A dissolved in t-butyl alcohol is excited to a triplet state A^* by light of 320-380 nm. The triplet may then return to the ground state A or rearrange to give the isomer B, thus

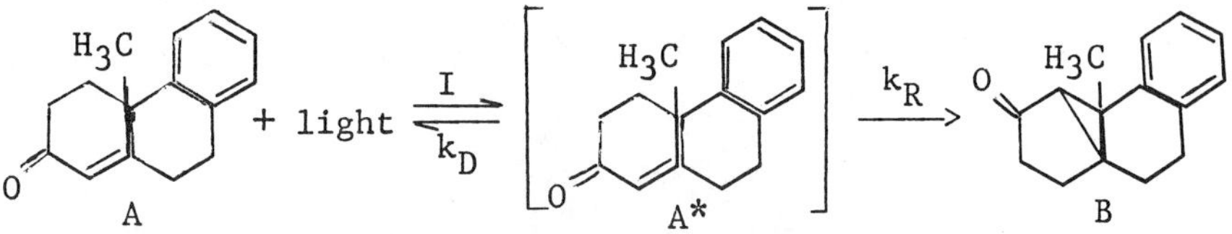

A A* B

The unimolecular rate constant k determines the rate of deactivation of excited molecules and the unimolecular rate constant k determines the rate of rearrangement. The symbol I represents the number of moles of photons per second and it is assumed that each photon absorbed produces one excited molecule of A^*. The quantum yield ϕ for the formation of B is given by

$$\phi = \frac{d(B)/dt}{I} = \frac{k_R(A^*)}{I}$$

Zimmerman, McCullough, Staley, and Padwa measured the quenching effect of dissolved napthalene and report the following data.

Moles naphthalene L^{-1}	0.0099	0.0330	0.0620	0.0680	0.0775	0.0960
ϕ	0.0049	0.0023	0.0020	0.0017	0.0014	0.0012

The quenching rate by naphthalene is controlled by the bimolecular quenching constant of A^*, which is equal to the diffusion-controlled constant $k_Q = 1.2 \times 10^9$ L mol^{-1} s^{-1}. Assuming a steady state,

$$\frac{d(A^*)}{dt} = I - k_R(A^*) - k_D(A^*) - k_Q(A^*)(N) = 0$$

where (N) is the concentration of naphthalene. Calculate k_R and k_D by plotting $1/\phi$ versus (N) and determining the slope and intercept of the line.

SOLUTION

$$\frac{I}{(A^*)} = k_R + k_D + k_Q(N)$$

$$\frac{1}{\phi} = \frac{I}{k_R(A^*)} = 1 + \frac{k_D}{k_R} + \frac{k_Q}{k_R}(N)$$

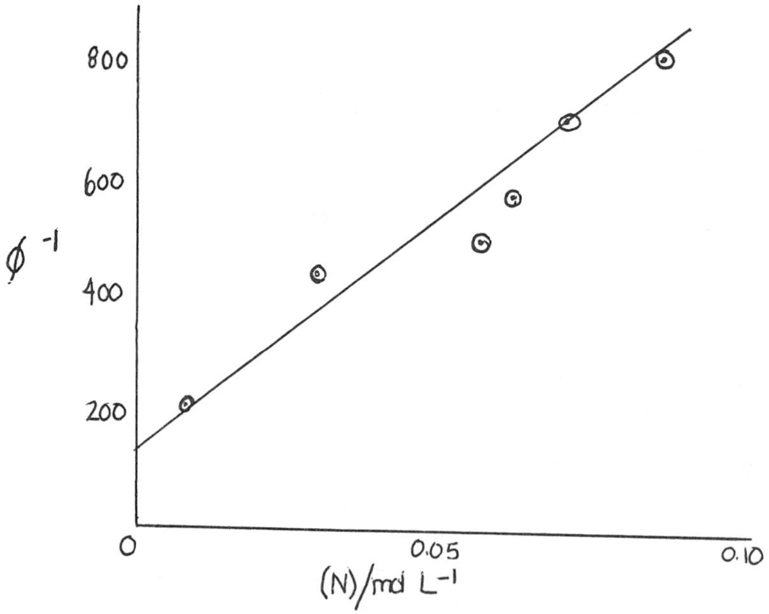

$$\text{slope} = \frac{875 - 120}{0.1} = 7.55 \times 10^3 \text{ L mol}^{-1}$$

$$= \frac{1.20 \times 10^9 \text{ L mol}^{-1} \text{ s}^{-1}}{k_R} \qquad k_R = 1.6 \times 10^5 \text{ s}^{-1}$$

$$\text{intercept} = 120 = 1 + \frac{k_D}{k_R} = 1 + \frac{k_D}{1.6 \times 10^5 \text{ s}^{-1}}$$

$k_D = 1.9 \times 10^7\ s^{-1}$

19.11 The phosphorescence of butyrophenone in acetonitrile is quenched by 1,3-pentadiene (P). The following quantum yields were measured at 25 °C (see N. J. Turro, Modern Molecular Photochemistry, The Benjamin/Cummings Publ. Co., Menlo Park, CA (1978), p. 248).

$(P)/10^{-3}$ mol L^{-1}	0	1.0	2.0
ϕ/ϕ_0	1	0.61	0.43

Assuming that the quenching reaction is diffusion controlled and the rate constant has a value of 10^{10} L mol^{-1} s^{-1}, what is the lifetime of the triplet state?

SOLUTION

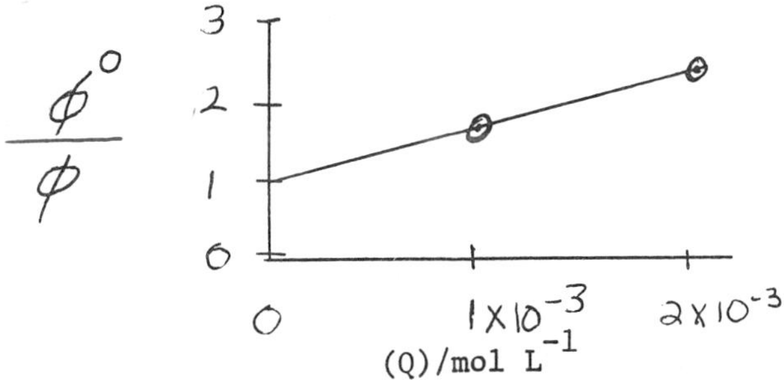

$$\text{slope} = k_Q \tau_{T_1} = \frac{1.33}{2 \times 10^{-3}\ \text{mol L}^{-1}} = 640\ \text{L mol}^{-1}$$

$$\tau_{T_1} = \frac{640\ \text{L mol}^{-1}}{10^{10}\ \text{L mol}^{-1}\ \text{s}^{-1}} = 6.4 \times 10^{-8}\ \text{s}$$

19.12 The following calculations are made on a uranyl oxalate actinometer, on the assumption that the energy of all wavelengths between 254 and 435 nm is completely absorbed. The actinometer contains 20 cm^3 of 0.05 mol L^{-1} oxalic acid, which also is 0.01 mol L^{-1} with respect to uranyl sulfate. After 2 hr of exposure to ultraviolet light, the solution required 34 cm^3 of potassium permanganate, $KMnO_4$, solution to titrate the undecomposed oxalic acid. The same volume, 20 cm^3, of unilluminated solution required 40 cm^3 of the $KMnO_4$ solution. If the average energy of the quanta in this range may be taken as corresponding to a wavelength of 350 nm, how many joules were absorbed per second in this experiment? (ϕ = 0.57)

SOLUTION

Moles of oxalic acid decomposed

$$= (0.02\ L)(0.05\ mol\ L^{-1})\left(\frac{40 - 34}{40}\right) = 1.5 \times 10^{-4}\ mol$$

Moles of photons required

$$= \frac{1.5 \times 10^{-4}\ mol}{(2 \times 60 \times 60\ s)(0.57)} = 3.65 \times 10^{-8}\ mol\ s^{-1}$$

$$E = \frac{hc}{\lambda} = \frac{(6.626 \times 10^{-34}\ J\ s)(2.998 \times 10^{8}\ m\ s^{-1})}{3.5 \times 10^{-7}\ m}$$

$$= 5.68 \times 10^{-19}\ J$$

$$\text{Energy flux} = (3.65 \times 10^{-8}\ mol^{-1}\ s^{-1})(6.02 \times 10^{23}\ mol^{-1}) \times (5.68 \times 10^{-19}\ J)$$

$$= 1.25 \times 10^{-2}\ J\ s^{-1}$$

19.13 The quantum yield for the photolysis of acetone

$$(CH_3)_2CO = C_2H_6 + CO$$

at 300 nm is 0.2. How many moles per second of CO are formed if the intensity of the 300-nm radiation absorbed is 10^{-2} J s^{-1}?

SOLUTION

$$E = \frac{hc}{\lambda} = \frac{(6.626 \times 10^{-34}\ \text{J s})(3 \times 10^{8}\ \text{m s}^{-1})}{300 \times 10^{-9}\ \text{m}}$$

$$= 6.63 \times 10^{-19}\ \text{J}$$

$$\frac{(10^{-2}\ \text{J s}^{-1})(0.2)}{(6.02\text{x}10^{23}\ \text{mol}^{-1})(6.63\text{x}10^{-19}\ \text{J})} = 5 \times 10^{-9}\ \text{mol s}^{-1}$$

19.14 The photochemical oxidation of phosgene, sensitized by chlorine, has been studied by G. K. Rollefson and C. W. Montgomery [J.Am.Chem.Soc., 55, 142, 4025 (1932)]. The overall reaction is

$$2COCl_2 + O_2 = 2CO_2 + 2Cl_2$$

and the rate expression that gives the effect of the several variables is

$$\frac{d(CO_2)}{dt} = \frac{kI_0(COCl_2)}{1 + k'(Cl_2)/(O_2)}$$

where I_0 is the intensity of the light. The quantum yield is about two molecules per quantum. Devise a series of chemical equations involving the existence of the free radicals ClO and COCl that will give a mechanism consistent with the rate expression.

SOLUTION

$$COCl_2 + h\nu \xrightarrow{k_1} COCl + Cl \qquad COCl + Cl_2 \xrightarrow{k_4} COCl_2 + Cl$$

$$COCl + O_2 \xrightarrow{k_2} CO_2 + ClO \qquad Cl + Cl \xrightarrow{k_5} Cl_2$$

$$COCl_2 + ClO \xrightarrow{k_3} CO_2 + Cl_2 + Cl$$

$$2\ COCl_2 + O_2 = 2\ CO_2 + 2\ Cl_2$$

In the steady state

$$\frac{d(COCl)}{dt} = I_o(COCl_2) - k_2(COCl)(O_2) - k_4(COCl)(Cl_2) = 0$$

$$\text{Therefore } (COCl) = \frac{I_o(COCl_2)}{k_2(O_2) + k_4(Cl_2)}$$

$$\frac{d(ClO)}{dt} = k_2(COCl)(O_2) - k_3(COCl)(ClO) = 0$$

$$\text{Therefore } k_2(COCl)(O_2) = k_3(COCl)(ClO)$$

$$\frac{d(CO_2)}{dt} = k_2(COCl)(O_2) - k_3(COCl_2)(ClO)$$

$$= 2\ k_2(COCl)(O_2) = \frac{2\ I_o(COCl_2)}{1 + \frac{k_4(Cl_2)}{k_2(O_2)}}$$

19.15 Given that solar radiation at noon at a certain place on the earth's surface is 4.2 J cm^{-2} min^{-1}, what is the maximum power output in W m^{-2}?

SOLUTION

$$\frac{(4.2\ J\ cm^{-2}\ min^{-1})(100\ cm\ m^{-1})^2}{60\ s\ min^{-1}} = 700\ W\ m^{-2}$$

19.16 If a good agricultural crop yields about 2 tons $acre^{-1}$ of dry organic material per year with a heat of combustion of about 16.7 kJ g^{-1}, what fraction of a year's solar energy is stored in an agricultural crop if the solar energy is about 4184 J min^{-1} ft^{-2} and the sun shines about 500 min day^{-1} on the average? 1 acre = 43,560 ft^2 and 1 ton = 907,000 g.

SOLUTION

$$\text{solar energy} = (43{,}560\ ft^2\ acre^{-1})(500\ min\ day^{-1})(365\ day\ year^{-1}) \times (4184\ J\ ft^{-2}\ min^{-1})$$

$$= 3.33 \times 10^{13}\ J\ acre^{-1}\ year^{-1}$$

$$\text{energy stored} = (2\ ton\ acre^{-1}\ year^{-1})(907{,}000\ g\ ton^{-1})(16{,}700\ J\ g^{-1})$$

$$= 303 \times 10^{10}\ J\ acre^{-1}\ year^{-1}$$

$$\text{fraction stored} = \frac{3.03 \times 10^{10}\ J}{3.33 \times 10^{13}\ J} \approx 10^{-3}$$

19.17 Calculate the longest wavelength of light that can theoretically decompose water at 25 °C in a one-photon electrochemical process to give $H_2(g)$ and $\frac{1}{2}$ $O_2(g)$ in their standard states. Given: $\Delta G° = 237.2$ kJ mol^{-1} for $H_2O(\ell) = H_2(g) + \frac{1}{2}O_2(g)$.

SOLUTION

$$\Delta G° = N_A hc/\lambda$$

$$\lambda = \frac{N_A hc}{\Delta G°} = \frac{(6.022\text{x}10^{23}\ mol^{-1})(6.626\text{x}10^{-34}\ J\ s)(2.998\text{x}10^{8}\ m\ s^{-1})}{237{,}200\ J\ mol^{-1}}$$

$$= 504\ nm$$

19.18 2.51×10^{-6} mol s^{-1}

19.19 (a) 0.1709 J s^{-1} (b) 0.3988 J s^{-1}
(c) 0.1709 watt (d) 0.3988 watt

19.20 2.18×10^{-19} J

19.21 3.94 tons

19.22 7.84×10^{4} s

19.24 10^{-7} mol L^{-1}

19.25 9×10^{-9} mol L^{-1}

19.26 47.1 s

19.27 0.050 J s^{-1}

19.28 $NO_2 + h\nu \longrightarrow NO_2^*$
$NO_2^* + NO_2 \longrightarrow 2\ NO + O_2$

Decreased concentration of NO_2 after long illumination makes decomposing collisions less likely. Also the reverse reaction becomes more important as the concentrations of the products build up.

19.30 (a) 10^{14} (b) 4.53×10^{-9} g day^{-1}

19.31 617 g

19.32 1.36×10^{17} cm^{-2}

19.33 24.4 tons

19.34 504 nm

CHAPTER 20. Irreversible Processes in Solution

20.1 Ten cubic centimeters of water at 25 °C is forced through 20 cm of 2-mm diameter capillary in 4 s. Calculate the pressure required and the Reynolds number.

SOLUTION

$$P = \frac{8V\ell\eta}{\pi r^4 t}$$

$$= \frac{(8)(10 \times 10^{-6}\ m^3)(0.2\ m)(8.95 \times 10^{-4}\ kg\ m^{-1}\ s^{-1})}{\pi(10^{-3}\ m)^4\ (4\ s)}$$

$$= 1.14 \times 10^3\ N\ m^{-2}$$

$$\text{Reynolds number} = \frac{d\bar{v}\rho}{\eta}$$

$$\bar{v} = \frac{\text{volume of liquid per unit time}}{\text{area of tube}}$$

$$= \frac{(10 \times 10^{-6}\ m^3)/(4\ s)}{\pi(10^{-3}\ m)^2} = 0.796\ m\ s^{-1}$$

Reynolds number

$$= \frac{(2 \times 10^{-3}\ m)(0.796\ m\ s^{-1})(10^3\ kg\ m^{-3})}{(8.95 \times 10^{-4}\ kg\ m^{-1}\ s^{-1})} = 1780$$

20.2 A steel ball (ρ = 7.86 g cm^{-3}) 0.2 cm in diameter falls 10 cm through a viscous liquid (ρ_o = 1.50 g cm^{-3}) in 25 s. What is the viscosity at this temperature?

SOLUTION

$$\eta = \frac{2r^2(\rho - \rho_o)g}{9\ \frac{dx}{dt}}$$

$$= \frac{2(1 \times 10^{-3}\ m)^2[(7.86 - 1.50) \times 10^3\ kg\ m^{-3}](9.8\ m\ s^{-2})}{9\left(\frac{0.10\ m}{25\ s}\right)}$$

$$= 3.46\ Pa\ s$$

20.3 Estimate the rate of sedimentation of water droplets of 1-μm diameter in air at 20 °C. The viscosity of air at this temperature is 1.808×10^{-5} Pa s.

SOLUTION

$$\frac{dx}{dt} = \frac{2r^2(\rho - \rho_0)g}{9\eta}$$

$$= \frac{2(0.5 \times 10^{-6}\ m)^2(0.998 \times 10^3\ kg\ m^{-3})(9.8\ m\ s^{-2})}{9(1.808 \times 10^{-5}\ Pa\ s)}$$

$$= 3.01 \times 10^{-5}\ m\ s^{-1}$$

20.4 Using data in Table 20.1 and equation 20.6, estimate the activation energy for water molecules to move into a vacancy at 25 °C.

SOLUTION

$$E_a = \frac{RT_1T_2}{T_2 - T_1}\ \ell n\ \frac{\eta_1}{\eta_2}$$

$$= \frac{(8.314\ J\ K^{-1}\ mol^{-1})(273.15\ K)(323.15\ K)}{(50\ K)(10^3\ J\ kJ^{-1})}\ \ell n\ \frac{1793}{549}$$

$$= 17.4\ kJ\ mol^{-1}$$

20.5 The relative viscosities of a series of solutions of a sample of polystyrene in toluene were determined with an Ostwald viscometer at 25 °C.

$c/10^{-2}$ g cm^{-3}	0.249	0.499	0.999	1.998
η/η_o	1.355	1.782	2.879	6.090

The ratio η_{sp}/c is plotted against c and extrapolated to zero concentration to obtain the intrinsic viscosity. If the constants in equation 20.13 are K = 3.7 x 10^{-2} and a = 0.62 for this polymer, when concentrations are expressed in g/cm^3, calculate the molar mass.

SOLUTION

$c/10^{-2}$ g cm^{-3}	0.249	0.499	0.999	1.998
$\dfrac{\eta/\eta_o - 1}{c}$	142.6	156.7	188.1	254.8

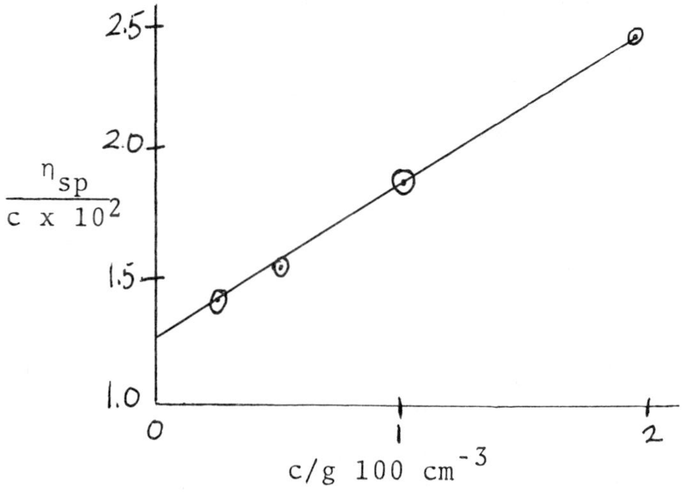

$$\ln \frac{[\eta]}{3.7 \times 10^{-2}} = 0.62 \ln M$$

$$M = \exp\left[\frac{1}{0.62} \ln \frac{126}{3.7 \times 10^{-2}}\right] = 500{,}000 \text{ g mol}^{-1}$$

20.6 At 34 °C the intrinsic viscosity of a sample of polystyrene in toluene is 84 $cm^3\ g^{-1}$. The empirical relation between the intrinsic viscosity of polystyrene in toluene and molar mass is $[\eta] = 1.15 \times 10^{-2}\ M^{0.72}$

What is the molar mass of this sample?

SOLUTION

$$\ln \frac{[\eta]}{1.15 \times 10^{-2}} = 0.72 \ln M$$

$$M = \exp\left[\frac{1}{0.72} \ln \frac{84}{1.15 \times 10^{-2}}\right] = 230{,}000\ g\ mol^{-1}$$

20.7 Given that the intrinsic viscosity of myosin is 217 $cm^3\ g^{-1}$, approximately what concentration of myosin in water would have a relative viscosity of 1.5?

SOLUTION

$$\frac{\frac{\eta}{\eta_o} - 1}{c} = 217\ cm^3\ g^{-1}$$

$$\frac{0.5}{c} = 217\ cm^3\ g^{-1}$$

$$c = \frac{0.5}{217\ cm^3\ g^{-1}} = 2.30 \times 10^{-3}\ g\ cm^{-3}$$

20.8 A conductance cell was calibrated by filling it with a 0.02 mol L^{-1} solution of potassium chloride ($\kappa = 0.2768\ \Omega^{-1}\ m^{-1}$) and measuring the resistance at 25 °C, which was found to be 457.3 Ω. The cell was then filled with a calcium chloride solution containing 0.555 g of $CaCl_2$ per liter. The measured resistance was

1050 Ω. Calculate (a) the cell constant for the cell, and (b) the conductivity of the $CaCl_2$ solution.

SOLUTION

(a) From equation 20.15

$$K_{cell} = \kappa R$$
$$= (0.2768\ \Omega^{-1}\ m^{-1})(457.3\ \Omega)$$
$$= 126.6\ m^{-1}$$

(b)

$$\kappa = \frac{K_{cell}}{R}$$
$$= \frac{126.6\ m^{-1}}{1050\ \Omega} = 0.1206\ \Omega^{-1}\ m^{-1}$$

20.9 It is desired to use a conductance apparatus to measure the concentration of dilute solutions of sodium chloride. If the electrodes in the cell are each 1 cm^2 in area and are 0.2 cm apart, calculate the resistance that will be obtained for 1, 10, and 100 ppm NaCl at 25 °C.

SOLUTION

$$1\ ppm = 1\ g\ NaCl\ in\ 10^6\ g\ H_2O$$
$$= \frac{(1\ g)/(58.45\ g\ mol^{-1})}{1\ m^3}$$
$$= 1.71 \times 10^{-2}\ mol\ m^{-3}$$

Using electric mobilities at infinite dilution,

$$\kappa = Fc(u_{Na} + u_{Cl^-})$$
$$= (96{,}485\ C\ mol^{-1})(1.71 \times 10^{-2}\ mol\ m^{-3})[(5.192 + 7.913) \times 10^{-8}\ m^2\ V^{-1}\ s^{-1}]$$
$$= 2.16 \times 10^{-4}\ \Omega^{-1}\ m^{-1}$$

$$R = \frac{\ell}{\kappa A} = \frac{0.2 \times 10^{-2}\ m}{(2.16 \times 10^{-4}\ \Omega^{-1}\ m^{-1})(0.01\ m)^2}$$

$= 9.24 \times 10^4\ \Omega$

For 10 ppm $R = 9.25 \times 10^3\ \Omega$

For 100 ppm $R = 925\ \Omega$

20.10 A moving boundary experiment is carried out with a 0.1 mol L^{-1} solution of hydrochloric acid at 25 °C ($\kappa = 4.24\ \Omega^{-1}\ m^{-1}$). Sodium ions are caused to follow the hydrogen ions. Three milliamperes is passed through the tube of 0.3 cm^2 cross-sectional area, and it is observed that the boundary moves 3.08 cm in 1 hr. Calculate (a) the hydrogen ion mobility, (b) the chloride ion mobility, and (c) the electric field strength.

SOLUTION

(a) $$E = \frac{I}{A\kappa} = \frac{3 \times 10^{-3}\ A}{(0.3 \times 10^{-4}\ m^2)(4.24\ \Omega^{-1}\ m^{-1})}$$

$$= 23.58\ V\ m^{-1}$$

$$u = \frac{\Delta x/\Delta t}{E} = \frac{3.08 \times 10^{-2}\ m}{(60 \times 60\ s)(23.58\ V\ m^{-1})}$$

$$= 3.63 \times 10^{-7}\ m^2\ V^{-1}\ s^{-1}$$

(b) $$\kappa = F_c(u_{H^+} + u_{Cl^-})$$

$$u_{Cl^-} = \frac{\kappa}{F_c} - u_{H^+} = \frac{(4.24\ \Omega^{-1}\ m^{-1})}{(96{,}485\ C\ mol^{-1})(0.1 \times 10^3\ mol^{-3})} - 3.63 \times 10^{-7}\ m^2\ V^{-1}\ s^{-1}$$

$$= 7.64 \times 10^{-8}\ m^2\ V^{-1}\ s^{-1}$$

(c) See (a)

20.11 Calculate the conductivity of 0.001 mol L^{-1} HCl at 25 °C. The limiting ion mobilities may be used for this problem.

SOLUTION

$$\kappa = F_c(u_{H^+} + u_{Cl^-})$$

$$= (96,485\ \text{C mol}^{-1})(1\ \text{mol m}^{-3})[(36.25 + 7.91) \times 10^{-8}\ \text{m}^2\ \text{V}^{-1}\ \text{s}^{-1}]$$

$$= 0.042\ 61\ \Omega^{-1}\ \text{m}^{-1}$$

20.12 One hundred grams of sodium chloride is dissolved in 10,000 L of water at 25 °C, giving a solution that may be regarded in these calculations as infinitely dilute. (a) What is the conductivity of the solution? (b) This dilute solution is placed in a glass tube of 4-cm diameter provided with electrodes filling the tube and placed 20 cm apart. How much current will flow if the potential drop between the electrodes is 80 V?

SOLUTION

(a) $$c = \frac{(100\ \text{g})/(58.5\ \text{g mol}^{-1})}{(10^7\ \text{cm}^3)(10^{-2}\ \text{m cm}^{-1})^3}$$

$$= 1.71 \times 10^{-1}\ \text{mol m}^{-3}$$

$$\kappa = (96,485\ \text{C mol}^{-1})(1.71\text{x}10^{-1}\ \text{mol m}^{-3})(13.105\text{x}10^{-8}\ \text{m}^2\ \text{V}^{-1}\ \text{s}^{-1})$$

$$= 2.16 \times 10^{-3}\ \Omega^{-1}\ \text{m}^{-1}$$

(b) $$R = \frac{\ell}{\kappa A} = \frac{0.2\ \text{m}}{(2.16 \times 10^{-3}\ \Omega^{-1}\ \text{m}^{-1})(2 \times 10^{-2}\ \text{m})^2 \pi}$$

$$= 7.36 \times 10^4\ \Omega$$

$$I = \frac{E}{R} = \frac{80\ \text{V}}{7.36 \times 10^4\ \Omega} = 1.09 \times 10^{-3}\ \text{A}$$

20.13 Using Stokes' law, calculate the effective radius of a nitrate ion from its mobility (74.0×10^{-9} m^2 V^{-1} s^{-1} at 25 °C).

SOLUTION

$$f_i = \frac{|Z_i|e}{u_i} = 6\pi\eta r$$

$$r = \frac{|Z_i|e}{u_i\ 6\pi\eta}$$

$$= \frac{1.602 \times 10^{-19}\ \text{C}}{(74\text{x}10^{-9}\ \text{m}^2\ \text{V}^{-1}\ \text{s}^{-1})(6\pi)(8.95\text{x}10^{-4}\ \text{kg m}^{-1}\ \text{s}^{-1})}$$

$$= 0.128\ \text{nm}$$

20.14 It may be shown that the diffusion coefficient at infinite dilution of an electrolyte with two univalent ions is given by

$$D = \frac{2u_1u_2RT}{(u_1 + u_2)F}$$

where u_1 and u_2 are the limiting values of the mobilities of the two ions. What is the diffusion coefficient of potassium chloride in water at 25 °C?

SOLUTION

$$D = \frac{2(7.617\text{x}10^{-8}\ \text{m}^2\text{V}^{-1}\text{s}^{-1})(7.913\text{x}10^{-8}\ \text{m}^2\text{V}^{-1}\text{s}^{-1})(\text{A})(298.15\ \text{K})}{(15.53 \times 10^{-8}\ \text{m}^2\ \text{V}^{-1}\ \text{s}^{-1})(96{,}485\ \text{C mol}^{-1})}$$

$$\text{A} = 8.314\ \text{J K}^{-1}\ \text{mol}^{-1}$$

$$= 1.99 \times 10^{-9}\ \text{m}^2\ \text{s}^{-1}$$

20.15 What is the self-diffusion coefficient of Na^+ in water at 25 °C?

SOLUTION

$$D = \frac{uRT}{|z|F}$$

$$= \frac{(5.192\times10^{-8}\ m^2\ V^{-1}\ s^{-1})(8.314\ J\ K^{-1}\ mol^{-1})(298.15\ K)}{9.6485\times 10^4\ C\ mol^{-1}}$$

$$= 1.334\times 10^{-9}\ m^2\ s^{-1}$$

20.16 Using a table of probability integral, calculate enough points on a plot of c versus x (like Fig. 20.4c) to draw in the smooth curve for diffusion of 0.1 mol L^{-1} sucrose into water at 25 °C after 4 hr and 29.83 min ($D = 5.23\times 10^{-10}$ $m^2\ s^{-1}$).

SOLUTION

Normal probability function $= \frac{1}{\sqrt{2\pi}}\ e^{-y^2/2} =$ N.P.F.

Equation 20.38

$$c = \frac{c_o}{2}\left[1 + \frac{2}{\sqrt{\pi}}\int_0^{x/2\sqrt{Dt}} e^{-\beta^2}\ d\beta\right]$$

In order to get this equation in the form of the normal probability function let $\beta^2 = t^2/2$. Thus $\beta = t/\sqrt{2}$ and $d\beta = dt/\sqrt{2}$.

$$c = \frac{c_o}{2}\left[1 + \frac{2}{\sqrt{2\pi}}\int_0^{x/\sqrt{2Dt}} e^{-t^2/2}\ dt\right]$$

$$= \frac{c_o}{2}\left[1 + \underbrace{\frac{1}{\sqrt{2\pi}}\int_{-x/\sqrt{2Dt}}^{x/\sqrt{2Dt}} e^{-t^2/2}\ dt}_{\text{This area is obtainable from}}\right]$$

tables of the normal probability integral (N.P.I.)

$$\frac{c_o}{2} = [1 \pm \text{Area of N.P.I.}]$$

The + sign is used when x > 0, and the - sign is used when x < 0.

At x = 1 cm the Gaussian parameter is

$$\frac{x}{\sqrt{2Dt}} = \frac{10^{-2}\ \text{m}}{\sqrt{2(5.23 \times 10^{-10}\ \text{m}^2\ \text{s}^{-1})(269.83 \times 60\ \text{s})}}$$

$$= 2.43$$

$\lvert x \rvert$ cm	$\frac{x}{\sqrt{2Dt}}$	Area N.P.I.	c(x < 0)	c(x > 0)
0	0	0	0.05 mol L^{-1}	0.05 mol L^{-1}
0.1	0.243	0.1922	0.040	0.060
0.2	0.486	0.3732	0.031	0.069
0.3	0.729	0.5340	0.023	0.077
0.4	0.972	0.6690	0.016	0.084
0.6	1.458	0.8556	0.007	0.093
0.8	1.944	0.9480	0.003	0.097
1.0	2.43	0.9850	0.002	0.098
1.2	2.916	0.9964	0.001	0.099

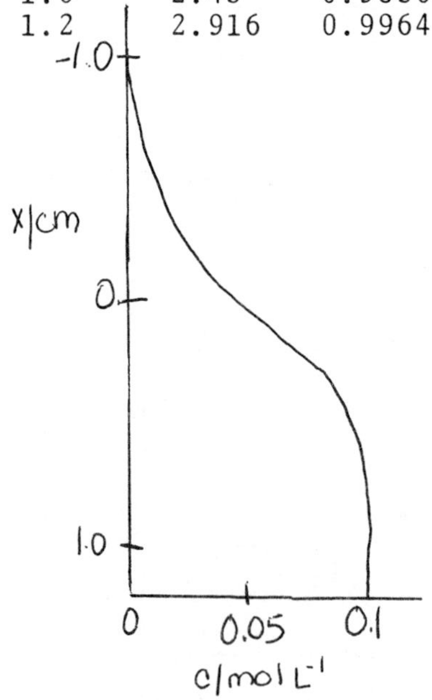

20.17 A sharp boundary is formed between a dilute aqueous solution of sucrose and water at 25 °C. After 5 hr the standard deviation of the concentration gradient is 0.434 cm. (a) What is the diffusion coefficient for sucrose under these conditions? (b) What will be the standard deviation after 10 hr?

SOLUTION

(a) $D = \frac{\sigma^2}{2t} = \frac{(4.34 \times 10^{-3}\ m)^2}{2(5 \times 60 \times 60\ s)} = 5.23 \times 10^{-10}\ m^2\ s^{-1}$

(b) $\sigma = \sqrt{2Dt}$

$= \sqrt{2(5.23 \times 10^{-10}\ m^2\ s^{-1})(10 \times 60 \times 60\ s)}$

$= 6.14 \times 10^{-3}\ m$

20.18 A sharp boundary is formed between a dilute buffered solution of hemoglobin ($D = 6.9 \times 10^{-11}\ m^2\ s^{-1}$) and the buffer at 20 °C. What is the half width of the boundary after 1 hr and 4 hr?

SOLUTION

(a) $\sigma = \sqrt{2Dt} = \sqrt{2(6.9 \times 10^{-11}\ m^2\ s^{-1})(60 \times 60\ s)}$

$= 7.05 \times 10^{-4}\ m = 0.0705\ cm$

(b) $\sigma = \sqrt{2(6.9 \times 10^{-11}\ m^2\ s^{-1})(4 \times 60 \times 60\ s)}$

$= 1.41 \times 10^{-3}\ m = 0.141\ cm$

20.19 What is the root-mean-square displacement in the x direction of a molecule of tobacco mosaic virus due to Brownian motion during one minute in water at 20 °C? (See Table 20.5)

SOLUTION

$$(\langle(\Delta x)^2\rangle)^{1/2} = \sqrt{2Dt}$$

$$= [(2)(0.53 \times 10^{-11}\ m^2\ s^{-1})(60\ s)]^{1/2}$$

$$= 0.0252\ nm$$

20.20 The diffusion coefficient for serum globulin at 20 °C in a dilute aqueous salt solution is 4.0 x 10^{-11} m^2 s^{-1}. If the molecules are assumed to be spherical, calculate their molar mass. Given: η_{H_2O} = 0.001 005 Pa s at 20 °C and $\bar{v}$ = 0.75 cm^3 g^{-1} for the protein.

SOLUTION

$$D = \frac{RT}{N_A 6\pi\eta}\left(\frac{4\pi N_A}{3M\bar{v}}\right)^{1/3}$$

$$M = \frac{4\pi N_A}{3\bar{v}}\left(\frac{RT}{N_A 6\pi\eta D}\right)^3$$

$$= \frac{4\pi(6.02\times10^{23}\ mol^{-1})}{3(0.75\times10^{-3}\ m^3\ kg^{-1})}\left[\frac{\dfrac{(8.31\ J\ K^{-1}\ mol^{-1})(293\ K)}{(6.02\times10^{23}\ mol^{-1})6\pi(1.005\times10^{-3}\ J\ m^{-3}\ s)}}{(4.0\times10^{-11}\ m^2\ s^{-1})}\right]^3$$

$$= 511\ kg\ mol^{-1}$$

$$= 511{,}000\ g\ mol^{-1}$$

20.21 The diffusion coefficient of hemoglobin at 20 °C is 6.9 x 10^{-11} m^2 s^{-1}. Assuming its molecules are spherical, what is the molar mass? Given: $\bar{v}$ = 0.749 x 10^{-3} m^3 kg^{-1} and η = 0.001 005 J m^{-3} s.

SOLUTION

$$D = \frac{RT}{N_A 6\pi\eta}\left(\frac{4\pi N_A}{3M\overline{v}}\right)^{1/3}$$

$$M = \frac{4\pi N_A}{3\overline{v}}\left(\frac{RT}{DN_A 6\pi\eta}\right)^3$$

$$M = \frac{4\pi(6.022\times10^{23}\ \mathrm{mol}^{-1})}{3(0.749\times10^{-3}\ \mathrm{m}^3\ \mathrm{kg}^{-1})}\left[\frac{(8.314\ \mathrm{J\ K^{-1}\ mol^{-1}})(293\ \mathrm{K})}{(6.9\times10^{-11}\ \mathrm{m}^2\ \mathrm{s}^{-1})(6.022\times10^{23}\ \mathrm{mol}^{-1})\,6\pi(0.001\ 005\ \mathrm{J\ m^{-3}\ s})}\right]^3$$

$$= 100\ \mathrm{kg\ mol^{-1}} = 100{,}000\ \mathrm{g\ mol^{-1}}$$

20.22 Calculate the sedimentation coefficient of tobacco mosaic virus from the fact that the boundary moves with a velocity of 0.454 cm hr^{-1} in an ultracentrifuge at a speed of 10,000 rpm at a distance of 6.5 cm from the axis of the centrifuge rotor.

SOLUTION

$$S = \frac{\dfrac{dr}{dt}}{\omega^2 r} = \frac{\dfrac{0.454\ \mathrm{cm\ hr^{-1}}}{3600\ \mathrm{s\ hr^{-1}}}}{\left(\dfrac{10{,}000\times 2\pi}{60\ \mathrm{s}}\right)^2 6.5\ \mathrm{cm}}$$

$$= 177\times10^{-13}\ \mathrm{s}$$

20.23 The sedimentation coefficient of myoglobin at 20 °C is 2.06×10^{-13} s. What molar mass would it have if the molecules were spherical? Given: $\overline{v} = 0.749\times10^{-3}\ \mathrm{m^3\ kg^{-1}}$, $\rho = 0.9982\times10^3\ \mathrm{kg\ m^{-3}}$, and $\eta = 0.001\ 005$ Pa s.

SOLUTION

$$S = \frac{M(1 - \bar{v}\rho)}{N_A f}$$

$$= \frac{M(1 - \bar{v}\rho)}{N_A 6\pi\eta}\left(\frac{4\pi N_A}{3M\bar{v}}\right)^{1/3}$$

$$M = \left(\frac{N_A 6\pi\eta S}{1 - \bar{v}\rho}\right)^{3/2}\left(\frac{3\bar{v}}{4\pi N_A}\right)^{1/2}$$

$$M = \left[\frac{(6.022\text{x}10^{23}\ \text{mol}^{-1})6\pi(0.001\ 005)(2.06\text{x}10^{-13})}{1 - (0.749\text{x}10^{-3})(0.9982\text{x}10^{3})}\right]^{3/2} \text{x} \left(\frac{3 \text{ x } 0.749 \text{ x } 10^{-3}}{4\pi 6.022 \text{ x } 10^{23}}\right)^{1/2}$$

$= 15.5\ \text{kg mol}^{-1}$

$= 15{,}500\ \text{g mol}^{-1}$

20.24 The sedimentation and diffusion coefficients for hemoglobin corrected to 20 °C in water are $4.41 \text{ x } 10^{-13}$ s and $6.3 \text{ x } 10^{-11}\ \text{m}^2\ \text{s}^{-1}$, respectively. If $\bar{v} = 0.749\ \text{cm}^3\ \text{g}^{-1}$ and $\rho_{H_2O} = 0.998$ g cm^{-3} at this temperature, calculate the molar mass of the protein. If there is 1 mol of iron per 17,000 g of protein, how many atoms of iron are there per hemoglobin molecule?

SOLUTION

$$M = \frac{RTS}{D(1 - \bar{v}\rho)}$$

$$= \frac{(8.31\ \text{J K}^{-1}\ \text{mol}^{-1})(203\ \text{K})(4.41 \text{ x } 10^{-13}\ \text{s})}{(6.3\text{x}10^{-11}\ \text{m}^2\ \text{s}^{-1})[1 - (0.749\text{x}10^{-3}\ \text{m}^3\ \text{kg}^{-1})(0.998\text{x}10^{3}\ \text{kg m}^{-3})]}$$

$= 67.6\ \text{kg mol}^{-1}$

$= 67{,}600\ \text{g mol}^{-1}$

$$\frac{67{,}600\ \text{g mol}^{-1}}{17{,}000\ \text{g g-atom Fe}} = 4\ \text{g- atom Fe mol}^{-1}$$

$= 4$ Fe per molecule

20.25 Given the diffusion coefficient for sucrose at 20 °C in water ($D = 45.5 \times 10^{-11}$ m^2 s^{-1}), calculate its sedimentation coefficient. The partial specific volume $\bar{v}$ is 0.630 cm^3 g^{-1}.

SOLUTION

$$s = \frac{MD(1 - \bar{v}\rho)}{RT}$$

$$= \frac{(342\times10^{-3}\ \text{kg mol}^{-1})(45.5\times10^{-11}\ \text{m}^2\text{s}^{-1})[1 - (0.630\ \text{cm}^3\text{g}^{-1})(A)]}{(8.31\ \text{J K}^{-1}\ \text{mol}^{-1})(293\ \text{K})}$$

$$A = 1\ \text{g cm}^{-3}$$

$$= 0.236 \times 10^{-13}\ \text{s}$$

20.26 A sedimentation equilibrium experiment is to be carried out with myoglobin ($M = 16{,}000$ g mol^{-1}) in an ultracentrifuge operating at 15,000 rpm. The bottom of the cell is 6.93 cm from the axis of rotation and the meniscus is 6.67 cm from the axis of rotation. What ratio of concentrations is expected at 20 °C if $\bar{v} = 0.75 \times 10^{-3}$ m^3 kg^{-1} and $\rho = 1.00 \times 10^3$ kg m^{-3}?

SOLUTION

$$\ln \frac{c_2}{c_1} = \frac{M(1 - v\rho)\,\omega^2\,(r_2^{\,2} - r_1^{\,2})}{2RT}$$

$$= \frac{(16\ \text{kg mol}^{-1})(0.25)(2\pi 250\ \text{s}^{-1})^2[(0.0693\ \text{m})^2 - (0.0667\ \text{m})^2]}{2(8.314\ \text{J K}^{-1}\ \text{mol}^{-1})(293.15\ \text{K})}$$

$$= 0.716$$

$$\frac{c_2}{c_1} = 2.05$$

20.27 1.14×10^3 Pa m^{-2} The Reynolds number is 447, and so the flow will be laminar.

20.28 95.7 min

20.29 0.66 s

20.30 0.00141 Pa s

20.31 8.39×10^6 g mol^{-1}

20.32 638,000 g mol^{-1}

20.33 520 nm

20.34 (a) 4.28 Ω^{-1} m^{-1} (b) 0.0768 m

(c) 58.4 V m^{-1}

20.35 0.1058 Ω^{-1} m^{-1}

20.36 1.6×10^{-3} Ω^{-1} m^{-1}

20.37 3.16×10^{-8} m^2 V^{-1} s^{-1}

20.38

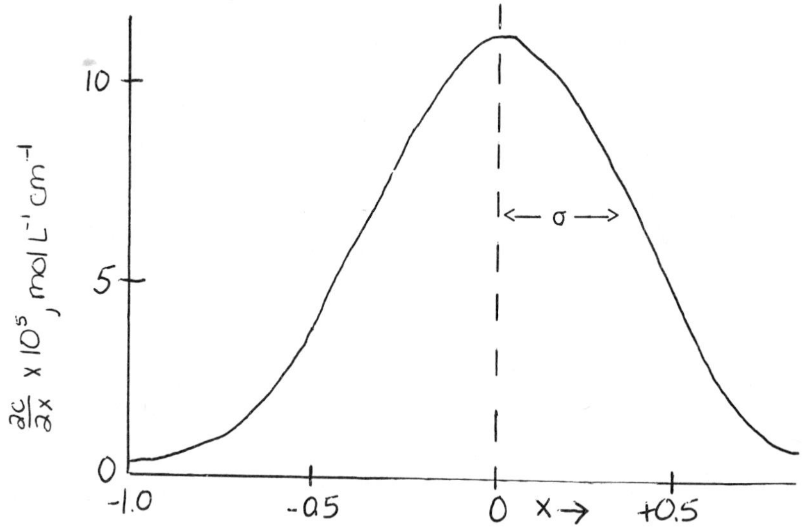

20.39 9.34×10^{-9}, 2.03×10^{-9}, 3.34×10^{-9} m^2 s^{-1}

20.40 (a) 1.96 hr. (b) 56.4 hr.

20.41 6.43×10^{-8} m^2 V^{-1} s^{-1}

20.44 (b) $\frac{\partial c}{\partial t} = D\nabla^2 c$

20.45 (a) 104 min (b) 56.5 hr.

20.46 7.09×10^{-11} $m^2\ s^{-1}$

20.48 5.41×10^{-4} cm

20.49 27.6×10^{6} g mol^{-1}

20.50 0.33 cm

20.51 64,000 g mol^{-1}

PART FOUR
STRUCTURES

CHAPTER 21. Solid-State Chemistry

21.1 Calculate the angles at which the first-, second-, and third-order reflections are obtained from planes 500 pm apart, using X rays with a wavelength of 100 pm.

SOLUTION

$$\sin\theta = \frac{\lambda}{2d_{hk\ell}} = \frac{\lambda}{2(d/n)}$$

$$\sin\theta_1 = \frac{100\text{ pm}}{2(500\text{ pm}/1)} \qquad \theta_1 = 5.74°$$

$$\sin\theta_2 = \frac{100\text{ pm}}{2(500\text{ pm}/2)} \qquad \theta_2 = 11.54°$$

$$\sin\theta_3 = \frac{100\text{ pm}}{2(500\text{ pm}/3)} \qquad \theta_3 = 17.46°$$

21.2 The crystal unit cell of magnesium oxide is a cube 420 pm on an edge. The structure is interpenetrating face centered. What is the density of crystalline MgO?

SOLUTION

$$\frac{4(40.32\times10^{-3}\text{ kg mol}^{-1})}{(6.022\times10^{23}\text{ mol}^{-1})(4.20\times10^{-10}\text{ m})^3} = 3.615\times10^3\text{ kg m}^{-3}$$

$$= 3.615\text{ g cm}^{-3}$$

21.3 Tungsten forms body-centered cubic crystals. From the fact that the density of tungsten is

$19.3\ g\ cm^{-3}$, calculate (a) the length of the side of this unit cell, and (b) d_{200}, d_{110}, and d_{222}.

SOLUTION

(a) $$19.3 \times 10^{3}\ kg\ m^{-3} = \frac{2(183.85 \times 10^{-3}\ kg\ mol^{-1})}{(6.022\ 045 \times 10^{23}\ mol^{-1})a^3}$$

$a = 316 \times 10^{-12}\ m = 316\ pm$

(b) $$d_{200} = \frac{a}{\sqrt{2^2 + 0 + 0}} = \frac{316\ pm}{2} = 158\ pm$$

$$d_{110} = \frac{a}{\sqrt{1 + 1 + 0}} = \frac{316\ pm}{\sqrt{2}} = 223\ pm$$

$$d_{222} = \frac{a}{\sqrt{4 + 4 + 4}} = \frac{316\ pm}{\sqrt{12}} = 91.2\ pm$$

21.4 Copper forms cubic crystals. When an X-ray powder pattern of crystalline copper is taken using X rays from a copper target (the wavelength of the $K\alpha$ line is 154.05 pm), reflections are found at $\theta = 21.65°, 25.21°, 44.96°, 47.58°$, and other larger angles. (a) What type of lattice is formed by copper? (b) What is the length of a side of the unit cell at this temperature? (c) What is the density of copper?

SOLUTION

$$d_{hk\ell} = \frac{\lambda}{2 \sin \theta} = \frac{154.05\ pm}{2 \sin \theta} = \frac{a}{\sqrt{h^2 + k^2 + \ell^2}}$$

(a)

θ	$d_{hk\ell}$/pm	Ratio $d_{hk\ell}$ to largest spacing
21.65°	208.77	1.0000
25.21°	180.84	0.8662
37.06°	127.81	0.6122

(continued page 348)

44.96°	109.00	0.5221
47.58°	104.34	0.4998

Ratios expected for cubic crystals

$$d_{hk\ell} = \frac{a}{\sqrt{h^2 + k^2 + \ell^2}}$$

	Primitive	Body-centered	Face-centered
100	1.0000	—	—
110	0.7071	1.0000	—
111	0.5774	—	1.0000
200	0.5000	0.7071	0.8660
210	0.4472	—	—
211	X	0.5774	—
220	X	0.5000	0.6124
310	X	0.4472	—
311	X	—	0.5222
222	X	X	0.5000

— reflection absent

X not needed for this problem

Thus copper forms face-centered cubic crystals.

(b) $$d_{111} = \frac{a}{\sqrt{1 + 1 + 1}} = 208.77 \text{ pm}$$

$$a = 361.6 \text{ pm}$$

(c) $$d = \frac{4(63.546 \times 10^{-3} \text{ kg mol}^{-1})}{(6.022\ 05 \times 10^{23} \text{ mol}^{-1})(361.6 \times 10^{-12} \text{ m}^3)^3}$$

$$= 0.8927 \times 10^3 \text{ kg m}^{-3}$$

21.5 (a) Metallic iron at 20 °C is studied by the Bragg method, in which the crystal is oriented so that a reflection is obtained from the planes parallel to the sides of the cubic crystal, then from planes cutting diagonally through opposite edges, and finally from planes cutting diagonally through opposite corners. Reflections are first obtained at θ = 11° 36', 8° 3', and 20° 26', respectively. What type of cubic lattice does iron have at 20 °C? (b) Metallic iron also forms cubic crystals at 1100 °C, but the reflections determined as described in (a) occur at θ = 9° 8', 12° 57', and 7° 55', respectively. What type of cubic lattice does iron have at 1100 °C? (c) The density of iron at 20 °C is 7.86 g cm^{-3}. What is the length of a side of the unit cell at 20 °C? (d) What is the wavelength of the X rays used? (e) What is the density of iron at 1100 °C?

SOLUTION

(a)(b) The three orientations will give reflections from planes a00, bb0, and ccc, respectively where a, b and c are small integers. The smallest suitable integers are: primitive lattice: a=b=c=1; body-centered lattice: b=1, a=c=2 (100 and 111 reflections are missing); face-centered lattice: c=1, a=b=2.

$$\sin\theta = \frac{\lambda}{2d_{hk\ell}} = \frac{\lambda\sqrt{h^2 + k^2 + \ell^2}}{2a}$$

Since $\lambda/2a$ is constant for a particular experiment, the three lattice types can be distinguished simply on the basis of the relative magnitudes of the three θ's. Thus the crystal at 20 °C is body-centered since the second angle is smaller than the first and the third is the largest of all (ratio of $\sin\theta$ is $2:\sqrt{2}:\sqrt{12}$). At 1100 °C the face-centered form is found (ratio of $\sin\theta$ is $2:\sqrt{8}:\sqrt{3}$). (Note that a primitive lattice would give ratios of $\sin\theta$ of $1:\sqrt{2}:\sqrt{3}$).

(c) $$d = 7.86\text{x}10^3 \text{ kg m}^{-3} = \frac{2(55.847 \text{ x } 10^{-3} \text{ kg mol}^{-1})}{(6.022\ 045 \text{ x } 10^{23} \text{ mol}^{-1})a^3}$$

$$a = \left[\frac{(2)(55.847 \times 10^{-3}\ \text{kg mol}^{-1})}{(6.022\ 045 \times 10^{23}\ \text{mol}^{-1})(7.86 \times 10^{3}\ \text{kg m}^{-3})}\right]^{1/3}$$

$$= 286.8 \times 10^{-12}\ \text{m} = 286.8\ \text{pm}$$

(d) $\lambda = 2d_{hk\ell} \sin\theta = \dfrac{2a \sin\theta}{\sqrt{h^2 + k^2 + \ell^2}}$

$$\frac{573.6\ \text{pm}}{2} \sin 11°\ 36' = 57.7\ \text{pm}$$

$$\frac{573.6\ \text{pm}}{\sqrt{2}} \sin 8°\ 3' = 56.8\ \text{pm}$$

$$\frac{573.6\ \text{pm}}{\sqrt{12}} \sin 20°\ 26' = 57.8\ \text{pm}$$

Average = 57.4 pm

(e) $d_{hk\ell} = \dfrac{\lambda}{2 \sin\theta}$

$$d_{200} = \frac{57.4\ \text{pm}}{2 \sin 9°\ 8'} = 180.8\ \text{pm};\ a = d_{200}\sqrt{4} = 361.6\ \text{pm}$$

$$d_{220} = \frac{57.4\ \text{pm}}{2 \sin 12°\ 57'} = 128.1\ \text{pm};\ a = d_{220}\sqrt{8} = 362.3\ \text{pm}$$

$$d_{111} = \frac{57.4\ \text{pm}}{2 \sin 7°\ 55'} = 208.4\ \text{pm};\ a = d_{111}\sqrt{3} = 360.9\ \text{pm}$$

Average = 361.6 pm

$$d = \frac{4(55.847 \times 10^{-3}\ \text{kg mol}^{-1})}{(6.022\ 045 \times 10^{23}\ \text{mol}^{-1})(361.6 \times 10^{-12}\ \text{m})^3}$$

$$= 7.846 \times 10^{3}\ \text{kg m}^{-3}$$

21.6 The only metal that crystallizes in a primitive cubic lattice is polonium, which has unit cell side of 334.5 pm. What are the perpendicular distances between planes with indices (110), (111), (210) and (211)?

SOLUTION

$$d_{hk\ell} = \frac{a}{(h^2 + k^2 + \ell^2)^{1/2}}$$

$$d_{110} = 334.5\ pm/2^{1/2} = 236.5\ pm$$

$$d_{111} = 334.5\ pm/3^{1/2} = 193.1\ pm$$

$$d_{210} = 334.5\ pm/5^{1/2} = 149.6\ pm$$

$$d_{211} = 334.5\ pm/6^{1/2} = 136.6\ pm$$

21.7 Cesium chloride, bromide, and iodide form interpenetrating simple cubic crystals instead of interpenetrating face-centered cubic crystals like the other alkali halides. The length of the side of the unit cell of CsCl is 412.1 pm. (a) What is the density? (b) Calculate the ion radius of Cs^+, assuming that the ions touch along a diagonal through the unit cell and that the ion radius of Cl^- is 181 pm.

SOLUTION

(a) $$d = \frac{(168.36 \times 10^{-3}\ kg\ mol^{-1})}{(6.022\ 045 \times 10^{23}\ mol^{-1})(412.1 \times 10^{-12}\ m)^3}$$

$$= 3.995 \times 10^3\ kg\ m^{-3}$$

(b) The diagonal plane through the cubic unit cell is as follows:

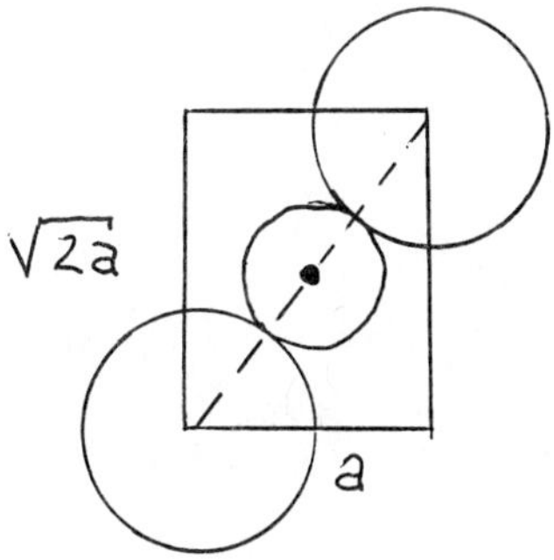

$$a^2 + (\sqrt{2}\,a)^2 = (2r_{Cs} + 2r_{C\ell})^2$$

$$r_{Cs} = \frac{1}{2}\left[a^2 - (\sqrt{2}\,a)^2\right]^{1/2} - 2(181\ pm)$$

$$= 176\ pm$$

21.8 The density of potassium chloride at 18 °C is 1.9893 g cm^{-3}, and the length of a side of the unit cell is 629.082 pm, as determined by X-ray diffraction. Calculate the Avogadro constant using the values of the relative atomic masses given in the front cover.

SOLUTION

$$1.9893 \times 10^{3}\ \mathrm{kg\ m^{-3}} = \frac{4(74.551 \times 10^{-3}\ \mathrm{kg\ mol^{-1}})}{N_A(629.082 \times 10^{-12}\ \mathrm{m})^3}$$

$$N_A = 6.0213 \times 10^{23}\ \mathrm{mol^{-1}}$$

21.9 Insulin forms crystals of the orthohombic type with unit-cell dimensions of 13.0 x 7.48 x 3.09 nm. If the density of the crystal is 1.315 g cm^{-3} and there are six insulin molecules per unit cell, what is the molar mass of the protein insulin?

SOLUTION

$$d = 1.315 \times 10^{3}\ \mathrm{kg\ m^{-3}} = \frac{6M}{(6.022045\times10^{23}\ \mathrm{mol^{-1}})(13.0\times7.48\times3.09 \times 10^{-27}\ \mathrm{m^3})}$$

$$M = \frac{1}{6}(1.315\times10^{3}\ \mathrm{kg\ m^{-3}})(6.022045\times10^{23}\ \mathrm{mol^{-1}})(13.0\times7.48\times3.09 \times 10^{-27}\ \mathrm{m^3})$$

$$= 39.7\ \mathrm{kg\ mol^{-1}} = 39{,}700\ \mathrm{g\ mol^{-1}}$$

21.10 Tungsten has a body-centered cubic structure at room temperature. Since the density is 19.35 g cm^{-3} at 20 °C, what is the atomic radius?

SOLUTION

For a body-centered cubic crystal there are two atoms per unit cell.

$$d = 19.35 \times 10^3 \text{ kg m}^{-3} = \frac{2(183.85 \times 10^{-3} \text{ kg mol}^{-1})}{(6.022\,045 \times 10^{23} \text{ mol}^{-1})a^3}$$

$$a = 316.0 \text{ pm}$$

$$R = \frac{\sqrt{3}}{4}\,a = \frac{\sqrt{3}}{4}\,(316.0 \text{ pm}) = 136.8 \text{ pm}$$

21.11 Molybdenum forms body-centered cubic crystals and, at 20 °C, the density is 10.3 g cm^{-3}. Calculate the distance between the centers of the nearest molybdenum atoms.

SOLUTION

$$10.3 \times 10^3 \text{ kg m}^3 = \frac{2(95.94 \times 10^{-3} \text{ kg mol}^{-1})}{(6.022\,045 \times 10^{23} \text{ mol}^{-1})a^3}$$

$$a = 314 \times 10^{-12} \text{ m} = 314 \text{ pm}$$

Consider a plane bisecting the cube and cutting diagonally through opposite faces.

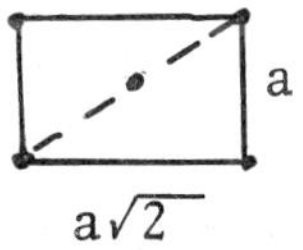

$$\text{length of diagonal} = \sqrt{a^2 + 2a^2} = a\sqrt{3}$$

$$\text{distance between atoms} = \frac{a\sqrt{3}}{2} = 273 \text{ pm}$$

21.12 Aluminum forms face-centered cubic crystals, and the length of the side of the unit cell is 405 pm at 25 °C. Calculate (a) the density of aluminum at this temperature, and (b) the distances between (200), (220), and (111) planes.

SOLUTION

$$\text{(a)}\quad d = \frac{4(26.98154 \times 10^{-3} \text{ kg mol}^{-1})}{(6.022\,045 \times 10^{23} \text{ mol}^{-1})(405.0 \times 10^{-12} \text{ m})^3}$$

$$= 2.698 \times 10^3 \text{ kg m}^{-3}$$

(b) $$d_{200} = \frac{a}{\sqrt{h^2 + k^2 + \ell^2}} = \frac{405.0 \text{ pm}}{2} = 202.5 \text{ pm}$$

$$d_{220} = \frac{a}{\sqrt{8}} = 143.2 \text{ pm}$$

$$d_{111} = \frac{a}{\sqrt{3}} = 233.8 \text{ pm}$$

21.13 Rutile (TiO_2) forms a primitive tetragonal lattice with a = 459.4 pm and c = 296.2 pm. There are two Ti atoms per unit cell, one at (000) and the other at (½½½). The four oxygen atoms are located at ± (uu0) and ± (½ + u, ½ - u, ½) with u = 0.305. What is the density of the crystal?

SOLUTION

$$V = (459.4 \text{ pm})^2 (296.2 \text{ pm})$$
$$= 62.51 \times 10^{-30} \text{ m}^3$$

$$d = \frac{2(47.90 + 2 \times 15.99) \times 10^{-3} \text{ kg mol}^{-1}}{(6.022\,045 \times 10^{23} \text{ mol}^{-1})(62.51 \times 10^{-30} \text{ m}^3)}$$

$$= 4.245 \times 10^3 \text{ kg m}^{-3}$$

21.14 A close-packed structure of uniform spheres has a cubic unit cell with a side of 800 pm. What is the radius of the spherical molecule?

SOLUTION

Since the cubic close-packed structure is face-centered cubic, a face diagonal is equal to four radii.

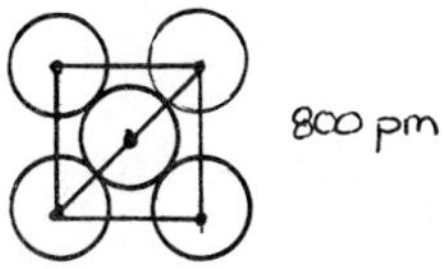

Length of face diagonal = $\sqrt{128 \times 10^4}$

Radius of sphere:

$$\frac{\sqrt{128 \times 10^4}}{4} = 282.8 \text{ pm}$$

21.15 If spherical molecules of 500 pm radius are packed in cubic close packing, and body-centered cubic, what are the lengths of the sides of the cubic unit cells in the two cases?

SOLUTION

Cubic close packing (face-centered cubic)

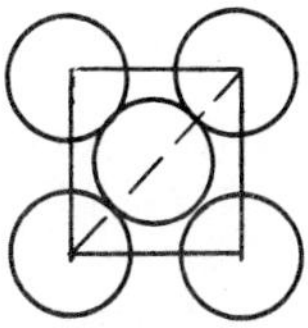

Length of diagonal = 2,000 pm

Length of side of unit cell =

$$\frac{2000}{\sqrt{2}} = 1414.2 \text{ pm}$$

Body-centered cubic

Consider a plane bisecting the cube and cutting diagonally through opposite faces.

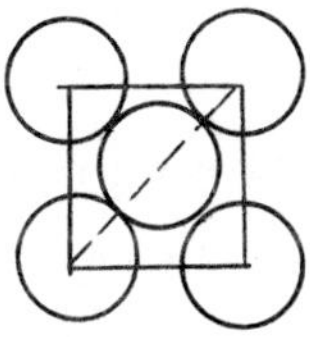

Length of diagonal = 2,000 pm

$$2000^2 = a^2 + (\sqrt{2}\ a)^2 = 3a^2$$

$$a = \frac{2000^2}{\sqrt{3}} = 1154.7 \text{ pm}$$

21.16 Titanium forms hexagonal close-packed crystals. Given the atomic radius of 146 pm, what are the unit cell dimensions and what is the density of the crystal?

SOLUTION

$a = b = 2(156 \text{ pm}) = 292 \text{ pm}$

$c = 2\sqrt{2}\, a/\sqrt{3} = 477 \text{ pm}$

$V = a^2 c(1 - \cos^2 \gamma)^{1/2} = a^2 c \sin \gamma$

$= (292 \text{ pm})^2 (477 \text{ pm}) \sin 120°$

$= 35.22 \times 10^{-30} \text{ m}^3$

$$d = \frac{2(47.90 \times 10^{-3} \text{ kg mol}^{-1})}{(6.022\ 045 \times 10^{23} \text{ mol}^{-1})(35.22 \times 10^{-30} \text{ m}^3)}$$

$= 4.517 \times 10^3 \text{ kg m}^{-3}$

21.17 Metallic sodium forms a body-centered cubic unit cell with a = 424 pm. What is the sodium atom radius?

SOLUTION

Three atoms are in contact along the diagonal of the unit cell

$(4R)^2 = (424 \text{ pm})^2 + (424 \text{ pm})^2 + (424 \text{ pm})^2$

$R = \frac{\sqrt{3}}{4} 424 \text{ pm} = 184 \text{ pm}$

Alternatively,

$\ell = a[(x_2 - x_1)^2 + (y_2 - y_1)^2 + (z_2 - z_1)^2]^{1/2}$

$= 424 \text{ pm} \left[\frac{1}{4} + \frac{1}{4} + \frac{1}{4}\right]^{1/2}$

$= \frac{\sqrt{3}}{4} = 424 \text{ pm} = 367 \text{ pm}$

The sodium atom radius is half the distance.

$\frac{367 \text{ pm}}{2} = 184 \text{ pm}$

21.18 The diamond has a face-centered cubic crystal lattice, and there are eight atoms in a unit cell. Its density is 3.51 g cm^{-3}. Calculate the first six angles at which reflections would be obtained using an X-ray beam of wavelength 71.2 pm.

SOLUTION

$$d = 3.51 \times 10^3 \text{ kg m}^{-3} = \frac{8(12.011 \times 10^{-3} \text{ kg mol}^{-1})}{(6.022\ 045 \times 10^{23} \text{ mol}^{-1})a^3}$$

$$a = 357 \times 10^{-12} \text{ m} = 357 \text{ pm}$$

$$\theta = \sin^{-1} \frac{\lambda}{2a}\sqrt{h^2 + k^2 + \ell^2} = \sin^{-1} \frac{71.2}{2(357)}\sqrt{h^2 + k^2 + \ell^2}$$

For face-centered cubic the first six reflections are:

$hk\ell$	θ
111	9.95°
200	11.50°
220	16.38°
311	19.31°
222	20.21°
400	23.51°

21.19 Calculate the density of diamond from the fact that it has a face-centered cubic structure with two atoms per lattice point and a unit cell edge of 356.7 pm.

SOLUTION

$$d = \frac{(4)(12)(12.011 \times 10^{-3} \text{ kg mol}^{-1})}{(6.022\ 045 \times 10^{23} \text{ mol}^{-1})(356.7 \times 10^{-12} \text{ m})^3}$$

$$= 3.516 \times 10^3 \text{ kg m}^{-3}$$

21.20 What neutron energy in electronvolts is required for a wavelength of 100 pm?

SOLUTION

Kinetic energy $= \frac{p^2}{2m_n} = \frac{h^2}{\lambda^2 2m_n}$ since $\lambda = h/p$

$$Ee = \frac{h^2}{2\lambda^2 m_n}$$

$$E = \frac{h^2}{2\lambda^2 m_n e}$$

$$= \frac{(6.626 \times 10^{-34}\ \mathrm{J\ s})^2}{2(10^{-10}\ \mathrm{m})^2(1.675\times10^{-27}\ \mathrm{kg})(1.602\times10^{-19}\ \mathrm{C})}$$

$= 0.0818$ V or 0.0818 eV

21.21 At 550 °C the conductivity of solid NaCl is $2 \times 10^{-4}\ \Omega^{-1}\ \mathrm{m}^{-1}$. Since the sodium ions are smaller than the chloride ions (see Table 21.1), they are responsible for most of the electric conductivity. What is the ionic mobility of Na^+ under these conditions?

SOLUTION

Since there are 4 Na^+ per unit cell of $a = 564$ pm

$$N = \frac{4}{(564 \times 10^{-12}\ \mathrm{m})^3} = 2.23 \times 10^{28}\ \mathrm{m}^{-3}$$

According to equation 21.38

$$u = \kappa/Nq$$

$$= \frac{2 \times 10^{-4}\ \Omega^{-1}\ \mathrm{m}^{-1}}{(2.23 \times 10^{28}\ \mathrm{m}^{-3})(1.60 \times 10^{-19}\ \mathrm{C})}$$

$$= 5.61 \times 10^{-14}\ \mathrm{m}^2\ \mathrm{V}^{-1}\ \mathrm{s}^{-1}$$

21.22 What fraction of the lattice sites of a crystal are vacant at 300 K if the energy required to move an atom from a lattice site in the crystal to a lattice site on the surface is 1 eV? At 100 K?

SOLUTION

At 300 K

$$\frac{n}{N} = e^{-E_V/kT}$$

$$= e^{-\frac{(1V)(1.602 \times 10^{-19}\ C)}{(1.38 \times 10^{-23}\ J\ K^{-1})(300\ K)}}$$

$$= 1.56 \times 10^{-17}$$

At 1000 K

$$\frac{n}{N} = e^{-\frac{(1V)(1.602 \times 10^{-19}\ C)}{(1.38 \times 10^{-23}\ J\ K^{-1}\ mol^{-1})(1000\ K)}}$$

$$= 9.08 \times 10^{-6}$$

21.23 A solution of carbon in face-centered cubic iron has a density of 8.105 g cm^{-3} and a unit cell edge of 358.3 pm. Are the carbon atoms interstitial, or do they substitute for iron atoms in the lattice? What is the weight percent carbon?

SOLUTION

$$d = \frac{4(55.847 \times 10^{-3}\ kg\ mol^{-1})}{(6.022\ 045 \times 10^{23}\ mol^{-1})(358.3 \times 10^{-12}\ m)^3}$$

$$= 8.064 \times 10^3\ kg\ m^{-3}$$

Since the experimental density is greater, the carbon atoms must be interstitial.

$$\begin{array}{r} 8.105 \times 10^3\ kg\ m^{-3} \\ -8.064 \times 10^3 \qquad\quad \\ \hline 0.041 \times 10^3 \qquad\quad \end{array}$$

$$\frac{0.041 \times 10^3}{8.105 \times 10^3}\ 100 = 0.51\%\ \text{by weight}$$

21.24 n = 7

21.25 (a) 6 (b) 8 (c) 12

21.27 (a) body-centered (b) 314.8 pm (c) 10.21 g cm^{-3}

21.28 (a) 628 pm (b) 2.00 g cm^{-3}

21.29 392.4 pm

21.30 74.69 g mol^{-1}

21.31 4

21.32 2826 kg m^{-3}

21.33 (a) 2 (b) 328 (c) 164 (d) 232 (e) 94.7 nm

21.34 287,124 pm

21.35 152 pm

21.36 346 pm

21.37 2.703 g cm^{-3}

21.38 3.13 g cm^{-3}

21.39 (a) $\sqrt{8/3}\ a$ (b) $2\sqrt{2}\ r$ (c) 2r

21.40 0.524

21.41 8.836×10^3 kg m^{-3}

21.42 0.414

21.43 2.33×10^3 kg m^{-3}

21.44 12

21.45 145.8 pm

21.46 490 $cm^2\ V^{-1}\ s^{-1}$

21.47 3.6×10^{-8} mole fraction